Assessing the environmental impact of agriculture

It is widely recognised that agriculture is a significant contributor to global warming and climate change. Agriculture needs to reduce its environmental impact and adapt to current climate change whilst still feeding a growing population, i.e. become more 'climate-smart'. Burleigh Dodds Science Publishing is playing its part in achieving this by bringing together key research on making the production of the world's most important crops and livestock products more sustainable.

Based on extensive research, our publications specifically target the challenge of climate-smart agriculture. In this way we are using 'smart publishing' to help achieve climate-smart agriculture.

Burleigh Dodds Science Publishing is an independent and innovative publisher delivering high quality customer-focused agricultural science content in both print and online formats for the academic and research communities. Our aim is to build a foundation of knowledge on which researchers can build to meet the challenge of climate-smart agriculture.

For more information about Burleigh Dodds Science Publishing simply call us on +44 (0) 1223 839365, email info@bdspublishing.com or alternatively please visit our website at www.bdspublishing.com.

Related titles:

Improving organic animal farming
Print (ISBN 978-1-78676-180-4) Online (ISBN 978-1-78676-182-8; 978-1-78676-183-5)

Improving organic crop cultivation
Print (ISBN 978-1-78676-184-2); Online (ISBN 978-1-78676-186-6, 978-1-78676-187-3)

Managing soil health for sustainable agriculture Volume 1: Fundamentals
Print (ISBN 978-1-78676-188-0) Online (ISBN 978-1-78676-191-0; 978-1-78676-190-3)

Managing soil health for sustainable agriculture Volume 2: Monitoring and management
Print (ISBN 978-1-78676-192-7) Online (ISBN 978-1-78676-195-8; 978-1-78676-194-1)

Precision agriculture for sustainability
Print (978-1-78676-204-7); Online (978-1-78676-206-1, 978-1-78676-207-8)

Water management for sustainable agriculture
Print (ISBN 978-1-78676-176-7); Online (ISBN 978-1-78676-178-1, 978-1-78676-179-8)

Chapters are available individually from our online bookshop:
https://shop.bdspublishing.com

climate•SMART•publishing
IN AGRICULTURAL SCIENCE

BURLEIGH DODDS SERIES IN AGRICULTURAL SCIENCE
NUMBER 57

Assessing the environmental impact of agriculture

Edited by Professor Bo P. Weidema,
Aalborg University, Denmark

burleigh dodds
SCIENCE PUBLISHING

Published by Burleigh Dodds Science Publishing Limited
82 High Street, Sawston, Cambridge CB22 3HJ, UK
www.bdspublishing.com

Burleigh Dodds Science Publishing, 1518 Walnut Street, Suite 900, Philadelphia, PA 19102-3406, USA

First published 2019 by Burleigh Dodds Science Publishing Limited

Library of Congress Control Number: 2019941170

British Library Cataloguing in Publication Data
A catalogue record for this book is available from the British Library

ISBN 978-1-78676-228-3 (Print)
ISBN 978-1-78676-231-3 (PDF)
ISBN 978-1-78676-230-6 (ePub)
ISSN 2059-6936 (print)
ISSN 2059-6944 (online)

DOI 10.19103/AS.2018.0044

Typeset by Deanta Global Publishing Services, Chennai, India

Printed and bound by CPI Group (UK) Ltd, Croydon, CR0 4YY

Contents

Part 2 Modelling particular impacts

Series list

Title	Series number
Achieving sustainable cultivation of maize - Vol 1 From improved varieties to local applications Edited by: Dr Dave Watson, CGIAR Maize Research Program Manager, CIMMYT, Mexico	001
Achieving sustainable cultivation of maize - Vol 2 Cultivation techniques, pest and disease control Edited by: Dr Dave Watson, CGIAR Maize Research Program Manager, CIMMYT, Mexico	002
Achieving sustainable cultivation of rice - Vol 1 Breeding for higher yield and quality Edited by: Prof. Takuji Sasaki, Tokyo University of Agriculture, Japan	003
Achieving sustainable cultivation of rice - Vol 2 Cultivation, pest and disease management Edited by: Prof. Takuji Sasaki, Tokyo University of Agriculture, Japan	004
Achieving sustainable cultivation of wheat - Vol 1 Breeding, quality traits, pests and diseases Edited by: Prof. Peter Langridge, The University of Adelaide, Australia	005
Achieving sustainable cultivation of wheat - Vol 2 Cultivation techniques Edited by: Prof. Peter Langridge, The University of Adelaide, Australia	006
Achieving sustainable cultivation of tomatoes Edited by: Dr Autar Mattoo, USDA-ARS, USA & Prof. Avtar Handa, Purdue University, USA	007
Achieving sustainable production of milk - Vol 1 Milk composition, genetics and breeding Edited by: Dr Nico van Belzen, International Dairy Federation (IDF), Belgium	008
Achieving sustainable production of milk - Vol 2 Safety, quality and sustainability Edited by: Dr Nico van Belzen, International Dairy Federation (IDF), Belgium	009
Achieving sustainable production of milk - Vol 3 Dairy herd management and welfare Edited by: Prof. John Webster, University of Bristol, UK	010
Ensuring safety and quality in the production of beef - Vol 1 Safety Edited by: Prof. Gary Acuff, Texas A&M University, USA & Prof. James Dickson, Iowa State University, USA	011
Ensuring safety and quality in the production of beef - Vol 2 Quality Edited by: Prof. Michael Dikeman, Kansas State University, USA	012
Achieving sustainable production of poultry meat - Vol 1 Safety, quality and sustainability Edited by: Prof. Steven C. Ricke, University of Arkansas, USA	013
Achieving sustainable production of poultry meat - Vol 2 Breeding and nutrition Edited by: Prof. Todd Applegate, University of Georgia, USA	014

Introduction

In spite of its decreasing share of global economic output, agriculture is still responsible for a large part of overall physical output, both to the rest of the economy and to the environment. For example, it has been estimated that agriculture and land use are responsible globally for around 24% of the greenhouse gas emissions fuelling climate change. A prerequisite to improving the sustainability of agriculture are reliable methods to identify and quantify types of environmental impact. These can then be used to identify priorities, set targets for improvement and monitor progress. This book provides a review of current research on the use of life cycle assessment (LCA) and other modelling techniques to measure and model the environmental impact and improve the sustainability of agriculture. The book is divided in three parts. Chapters in Part 1 review key issues in using LCA and modelling applied to farming systems. Part 2 provides more detail on the modelling of three particular impacts, namely freshwater, pesticides and social impacts. Part 3 looks in more detail on options for reducing the environmental impact and improving the performance of both crop and livestock farming.

Part 1 Life cycle assessment

Life cycle assessment (LCA) is increasingly applied to agricultural systems. However, in spite of the standardisation of this assessment technique through the ISO 14040 series, there are still many different interpretations of the requirements laid down by the standards, resulting in a plethora of ways in which LCA is applied in practice. In turn, this results in the generation of sometimes conflicting results. Chapter 1 provides a review of the most important issues in agricultural LCA and demonstrates how to deal with challenges such as system boundary delimitations, defining the functional unit, handling co-production and choice of impact assessment methods.

Chapter 2 provides a broader view on modelling of agricultural activities and their environmental impacts. The focus of the chapter is on model quality as dependent on the quality of data and the accuracy of modelling decisions to support mitigation strategies or management changes towards environmental friendlier agricultural products. Applying uncertainty and sensitivity analysis can reduce uncertainty in model outputs and direct data collection efforts towards the most important variables, thereby contributing to overall model quality. The chapter includes two detailed case studies that show the benefits of including uncertainty propagation and sensitivity analysis in model assessment. The chapter ends by describing future trends in modelling, especially in facilitating the reuse of models.

Chapter 3 describes a range of farm-level greenhouse gas modelling tools, and critically assesses their ability to function as a tool for farm-level benchmarking and mitigation assessment. The authors conclude that the majority of farm-level tools do not effectively account for their methodological choices and that the disparity in interpretation and adaptation leads to considerable differences in output. The chapter considers the challenges, limitations and opportunities offered by further development of different modelling approaches.

Chapter 4 provides a review of recent LCAs on ruminant production systems and considers the particular challenges in modelling, especially in pasture-based ruminant production systems with their high degree of variation in physical, chemical and biological processes. A case study demonstrates how uncertainty surrounding climate impacts of ruminant systems can potentially be reduced through on-farm measurements of greenhouse gas fluxes, but that not all measurements carry the same degree of value.

Chapter 5 examines the use of LCA for comparisons of intensive and extensive agricultural systems. Most often LCAs of agricultural systems compare conventional agriculture as an intensive farming system with certified organic agriculture as an extensive farming system, even though production intensity in terms of yields and inputs may vary over a wide range in both production systems. Nevertheless, comparative LCA studies on organic and conventional farming are well suited to discuss advantages and limitations of using LCA to assess the environmental sustainability of farming systems of different production intensities. The chapter assesses the validity of the modelling of reactive nitrogen and heavy metal emissions in different farming systems.

Part 2 Modelling particular impacts

Chapter 6 looks at modelling the impacts of agriculture on freshwater consumption, water quality and salinization. 80-90% of human freshwater consumption is accounted for by agriculture, mainly due to the high volumes of irrigation water for crop production. Additionally, agricultural activities affect hydrological cycles through land use changes and soil modifications, and through pollution of water bodies with fertilizers and pesticides. The chapter addresses data availability and databases, the issues of high geographical and temporal variability, and current gaps in modelling. Notable gaps are found in hydrological data and models for groundwater as well as in the assessment of impacts on ecosystems.

Chapter 7 reviews how pesticides are currently addressed in impact assessment. A real-life scenario study of four selected pesticides applied to potatoes illustrates and provides guidance on the quantification of emissions, exposure and toxicity. The chapter discusses the relevance of spatiotemporal variability in modelling emissions and the toxicity and ecotoxicity impacts of pesticides, and how substitution scenarios can be used to identify more sustainable pesticides.

Chapter 8 looks at the wider economic and social impacts of agriculture. Economic assessments remain the dominating form of socio-economic assessment in agriculture, but its practical implementation often suffer from severe limitations. The chapter describes how much of the early criticism has slowly been adopted and integrated in the form of significant improvements in the consistency and completeness of economic assessment techniques. Nevertheless, a better integration with the qualitative understandings developed in the social impact assessment community is still largely missing, notably with respect to power, politics and institutions, and viewing human and social capabilities as direct contributors to wellbeing, i.e. beyond being just economic factors of production. The chapter suggests that these limitations can be overcome by combining existing models. The unique importance of agriculture in sustainable development is highlighted through a discussion of how the benefits of increases in agricultural productivity are distributed, as illustrated by the role of inequality, land tenure, and food pricing. Most of these issues can only be effectively addressed by well-functioning public institutions. The final section of the chapter discusses to what extent certification and fair-trade schemes might address these issues in the absence of public governance.

Part 3 Improvement options

Part 3 looks assesses the available options for mitigating environmental impacts within agricultural crop production (Chapter 9), horticulture (Chapter 10), and pork and poultry production (Chapter 11). A final chapter looks at the options for valorising by-products. In general, the four chapters point out the large variation between local contexts, crops, and types of impacts. Nevertheless, some general conclusions can be drawn. The authors of Chapter 9 highlight the promising results from research on agro-ecology and other quantitative evidence on designing highly efficient production systems. Chapter 10 examines key findings from LCA of horticultural crops and production systems and finds a relatively low environmental impact compared to other food types. For fruit crops, the largest impacts occur at the farm level, most often related to machinery use, pesticides and fertilizers.

Currently, pork and chicken account for nearly three quarters of global meat consumption. With increasing concern for the environmental impacts associated with human activity, and due to the magnitude of the pork and poultry sectors, there is mounting pressure to meet the growing demand for these animal-based proteins with fewer resources and lower emissions. Chapter 11 points out that the emission intensity of pork and poultry operations is largely determined by the feed conversion ratio and manure handling. The authors outline improvement options and provide case studies for making pig and poultry production more sustainable. In their outlook for the future, they highlight the role of data science, data-driven decision-making and automation in order to continue improving efficiency in agriculture.

Chapter 12 provides an overview of opportunities to reduce the environmental impact of valorising agricultural by-products, i.e. crop residues, industrial by-products, and animal manure. The authors estimate the available by-products and the maximum and realistic greenhouse gas reduction potentials in North Western Europe, assessing the relative importance of availability and collectability, as well as broader issues such as land use changes, soil carbon sequestration and pollution swapping.

To the reader

The chapters of this book can be read individually. As editor, I have sought to ensure consistency in the use of terms and methodological recommendations. It is my hope that this book can serve as a standard reference for the agricultural science and LCA research community and that it will support further research to improve measuring and modelling the environmental impact of agriculture.

Part 1

Life cycle assessment

Chapter 1

Life cycle assessment methodology for agriculture: some considerations for best practices

Seyyed Hassan Pishgar-Komleh, Institute of Soil Science and Plant Cultivation, Poland and Czech University of Life Sciences Prague, Czech Republic; Paria Sefeedpari, Institute of Soil Science and Plant Cultivation, Poland; Nathan Pelletier, University of British Columbia, Canada; and Miguel Brandão, KTH Royal Institute of Technology, Sweden

1 Introduction

Along with the increased demand by a growing and wealthier human population for healthy and nutritious food, growing awareness and concerns in today's society over the environmental impacts associated with agricultural systems have shifted focus to sustainable consumption and production (Mazur-Wierzbicka, 2015). As a result, agriculture has been the subject of considerable public debate and scrutiny (Jansen and Vellema, 2004).

Life cycle assessment (LCA) is an approach formalized in the 14040 and 14044 standards of the International Organisation for Standardisation (ISO) (ISO, 2006a,b). These standards provide a methodological framework for quantitatively evaluating the environmental impacts and as a further step, the trade-offs among the corresponding LCA results, economic and social impacts can be considered for sustainability assessment (Lee and Inaba, 2004). While focus on socio-economic objectives, such as employment effects and contribution to gross domestic product (GDP), historically took predominance over environmental concerns (FAO, 2006), the latter are swiftly catching

http://dx.doi.org/10.19103/AS.2018.0044.01

up by drawing the interest of policy makers and expressed preferences of consumers. One of the limitations of typical LCA studies is the exclusion of social and economic aspects of product systems, which may favour a decision that otherwise would not be favoured. A sustainability assessment study that excludes these impacts is incomplete. However, incorporating life cycle costing (LCC) and social impacts has proven challenging, as pointed out by Dreyer et al. (2006), who noted that 'recommendations based on LCA fail to address possible trade-offs between environmental protection and both social and economic concerns in the product life cycle'. In an attempt to overcome this issue, a wide range of studies have been conducted on life cycle social assessment (LCSA) and LCC (Klöpffer and Renner, 2008; Finkbeiner et al., 2010). Although including a social assessment into agricultural LCA studies is still in its infancy, literature exists on some case studies and applications. In this regard, agriculture and agro-food systems are the first sectors in which social LCA has been applied – for example, tomatoes and citrus fruits from Italy, milk and eggs from Canada, bananas from Cameroon, roses from Ecuador and the Netherlands, and oysters from Denmark (Petti et al., 2018; Pelletier, 2018; Revéret et al., 2015; De Luca et al., 2015; Franze and Ciroth, 2011; Wangel, 2014; Feschet et al., 2013).

In order to manage the social performance of a production system, it is required to ensure that a wide variety of sustainability concerns are met, including labour issues, human health, resource availability, animal welfare, biodiversity, and food security, in addition to the quantification of the impacts of emissions to air, soil, and water. In this context, LCA can be an important tool for measuring, benchmarking and, in turn, improving the social performance of the agri-food sector. This is relevant, in particular, when seeking to understand and make comparisons between traditional subsistence and industry and farming, organic and conventional production systems, or animal-based and animal-free diets.

The concept of social responsibility has been defined as 'the commitment of business to contribute to sustainable economic development, working with employees, their families, the local community and society at large to improve their quality of life' (Holme and Watts, 2000). In ISO 26000 (ISO, 2010), this concept has been defined more comprehensively as: 'responsibility of an organization for the impacts of its decisions and activities on society and the environment, through transparent and ethical behaviour that (1) contributes to sustainable development, including health and the welfare of society; (2) takes into account the expectations of stakeholders; (3) is in compliance with applicable law and consistent with international norms of behaviour; and (4) is integrated throughout the organisation and practised in its relationships'. Responsibility allocation is an issue on which there is no general agreement and the literature shows different viewpoints (Mankiw and Taylor, 2006; Ferng,

2003; Lenzen et al., 2007; Rodrigues et al., 2006). Based on the literature, allocation of responsibility can be done according to criteria representing either *producer responsibility*, *consumer responsibility*, or *income responsibility* (Weidema et al., 2018). Although these three types of responsibility are discussed in the context of global regulation of greenhouse gas (GHG) emissions, they can be generally applied to all environmental impacts of the product life cycle. According to the first alternative, the producer bears the main responsibility for the environmental impact of a goods item or a service (we use the term *product* to denote both goods and services) throughout the entire life cycle. This allocation method ignores the important role of the final consumers as the main drivers for environmental impacts and that outsources the environmental impacts to countries with fewer commitments (Pedersen and de Haan, 2006; Weidema et al., 2018). The second alternative ascribes all environmental impacts related to production and consumption of a product to the consumers. This encourages the consumers to purchase products with the least environmental impacts. The last alternative - income responsibility - allocates all of the environmental impacts in the value chain to the activities that receive income from the value chain (Weidema et al., 2018). Since LCA studies are performed in the context of life cycle management and social responsibility, the relationship between LCA models and social responsibility needs to be investigated. Accordingly, Weidema et al. (2018) identified three social responsibility paradigms, namely *value chain responsibility*, *supply chain responsibility*, and *consequential responsibility* for social assessments.

In summary, the current commitments to continuous improvements over time in social responsibility and environmental performance are major challenges for policy makers, producers, and consumers, who rely on scientifically robust tools with which to quantify sustainability impacts and meet self-imposed targets. The LCA can play a key role in supporting the transition to socially and environmentally responsible systems to obtain the systematic and scientifically robust sustainability results that it can produce.

2 Identifying a functional unit

The functional unit (FU) is a key element in LCA and must be defined clearly. ISO (2006a) states that 'LCA is a relative approach, which is structured around a FU. This FU defines what is being studied. All subsequent analyses are then relative to that FU, as all inputs and outputs in the life cycle inventory (LCI) and consequently the life cycle impact assessment (LCIA) profile are related to the FU'. In other words, an FU for the product of an agricultural system ought to represent a quantifiable value associated with the function of the agricultural production system under study (Caffrey and Veal, 2013a). Defining the appropriate FU is dependent on the system being analysed, the goal of the study, and the system boundary of the LCA and is generally chosen to reflect the way the product is traded (Harris and Narayanaswamy, 2009); in case of comparing different alternatives, an equivalent basis of the alternatives can be considered (e.g. energy content in biofuels and gasoline or diesel as a transport fuel). Weidema et al. (2004) provided some useful recommendations on key issues in LCA in the form of a guideline. Based on this guideline, the FU needs to be, as far as possible, related to the function of the product rather than to the quantity of the product. For example, for agricultural tillage machinery 'the tillage area (with specific depth) per hour' can be selected rather than choosing 'the mass of tillage machine'. Failing to define an appropriate FU makes it impossible to compare different LCA results (Caffrey and Veal, 2013b) (e.g. comparing organic and conventional systems). Depending on the type of study and product, different product attributes can form the basis for defining the proper FU in the modelling of agricultural systems.

2.1 Mass

The majority of agricultural studies use a mass-based FU. In this category, the environmental impacts of a product are related to 1 kg or tonne of a product, regardless of the quality and commercial value of the product (Cerutti et al., 2013; Pishgar-Komleh et al., 2017a). For simplicity, the mass-based FU is widely used in LCA of agricultural products. Studies on animal production tend to use mass-based FUs for milk, butter, yogurt, cheese, ice cream and whey products (e.g. 1 kg of raw milk) (Basset-Mens et al., 2009). Examples of agricultural LCAs that adopt mass-based FU are (Basset-Mens et al., 2009; Haas et al., 2001; Mollenhorst et al., 2006; Fantin et al., 2017; Anton et al., 2005; Hayashi, 2005; Nienhuis and de Vreede, 1996; Pishgar-Komleh et al., 2013).

2.2 Energy

In studies that deal with bioenergy and biofuel production, an energy-based (or a proportional function, such as distance driven) FU can be used, for example, 1 MJ, since the function of the system is to deliver energy. Studies that have employed an energy-based FU in the LCA of agricultural crops include Hanegraaf and Biewinga (1998), Heller et al. (2003), Monti et al. (2009) and Fazio and Monti (2011).

2.3 Other functional units

In some studies, alternative FUs are relevant (Cerutti et al., 2013). Since land use *per se* is not seen directly as a production function, land-based FUs are not commonly used in LCA. However, this FU can be correctly applied when the goal of the study is, for example, related to supporting the minimization of environmental impacts associated with cultivating a certain area (Mouron et al., 2006a,b). When comparing different land-management options (e.g. food, fuel, feed, fibre timber, or carbon sequestration), given that the functions of the different land uses are different, it is necessary to balance each alternative so that they can be compared. The FU (e.g. 1 ha) would thus reflect no net production (Brandão, 2012). The land-based FU has been used to compare low input–low output systems with high input–high output systems (Cerutti et al., 2013), although the outputs across alternatives must be equal to enable comparability. In addition, this FU has been used to compare crop-rotation systems including fallow land (Karlsson and Sund, 2016). Defining the appropriate FU for comparing and contrasting farming systems, such as organic and conventional systems, is very important. Applying different FUs may reverse the ranking of the options being compared (e.g. organic and conventional product systems seen from a mass- or land-area-based FU). Intuitively, it is widely believed that organic farming is an environmentally superior alternative to conventional farming; however, this viewpoint is often challenged by the insights gained from LCA studies, where results are normalised by the quantity of product. Tuomisto et al. (2012) found that organic systems may decrease nitrate leaching, while increasing land requirements and soil erosion. Nemecek et al. (2011) showed organic farming has better environmental outcomes at the farm scale (per ha), but not per unit of output (per kg product). Several LCAs have used land-based FUs in the modelling of agricultural and animal products (Haas et al., 2001; Nemecek et al., 2001; Huguenin-Elie and Nemecek, 2004; Basset-Mens and van der Werf, 2005; Hayashi, 2005; Mouron et al., 2006a; Xiao et al., 2019; Pirlo and Lolli, 2019); however, results must be interpreted with caution because the comparability between alternative systems, which requires that systems to produce an equivalent amount of all products, is not always ensured.

In other studies, value-based or economic-based FUs have been adopted, which can be useful because they integrate product quantity and quality in a single index (Mouron et al., 2006b; Cerutti et al., 2013). The disadvantage of this FU is that it does not enable comparison across countries and years. It is strongly influenced by the economic context in which the farm is located and can change significantly from one year to another (Cerutti et al., 2013).

Despite the need to adapt FUs to the product properties and purpose of the study, the majority of LCA studies on agricultural products use the above FUs. For example, in studies on animal products, FUs have been defined as:

- 1 kg or tonne of energy corrected milk (ECM) leaving the farm-gate over a specific period of time (Cederberg and Stadig, 2003; Cederberg and Mattsson, 2000; Cederberg and Flysjö, 2004; Casey and Holden, 2005);

- 1 kg or tonne of fat and protein corrected milk (FPCM) (Pirlo and Lolli, 2019; Gerber et al., 2011; Thomassen et al., 2008b, 2009; Basset-Mens et al., 2009).
- 1 litre or tonne of milk produced and sold per year per herd size (Hospido et al., 2003; Williams et al., 2006; Eide, 2002);
- number of products (such as 1,000 eggs produced) (Williams et al., 2006);
- 1 kg of bone-free meat leaving the farm-gate (Basset-Mens et al., 2009; Cederberg and Darelius, 2002; Cederberg and Stadig, 2003).
- 1 kg or tonne of live or dead animal carcass, such as pig, chicken, and beef leaving the farm-gate (Basset-Mens et al., 2009; Haas et al., 2000; Eriksson et al., 2005; Basset-Mens and van der Werf, 2005; Williams et al., 2006; Casey and Holden, 2006; Katajajuuri, 2007; Pishgar-Komleh et al., 2017b).
- 1 kg or tonne of animal protein, such as pork, chicken, or beef protein (Zhu and van Ierland, 2004).

3 Unit processes as the building blocks of LCI

3.1 Collecting data on inputs and outputs: completeness vs. precision

A 'unit process' is one or more grouped operations in a production system that can be defined and separated from others. According to ISO (2006a,b), unit processes are the smallest elements in the LCI of product systems for which input and output data are quantified. With high data resolution (i.e. geographical, technological, and temporal), it can refer to an activity such as transportation, planting, harvesting, tillage, cleaning, and separation; if less resolved data are available, it can refer to a collection of activities that aggregate into a higher-level process, like crop production, post-harvest and so on. The outputs of a unit process are main products, by-products, emissions, and wastes, while inputs refer to resources, such as raw materials and energy, intermediate products, or other products (e.g. fertilizers or pesticides).

Like any assessment tool, data are a driving force behind LCA (Curran, 2015). Data acquisition is one of the most complex issues facing LCA practitioners (Klöpffer and Grahl, 2014). Large numbers of interlinked processes need to be considered to complete the LCI. Raw material inputs, energy use, ratio of main product to co-products, production rates, and environmental releases must all be quantified for each process in the production system (Curran, 2015). LCI data can be classified into two categories: foreground and background. Foreground data are typically those unique to the product system being modelled and are collected directly from sources including on-site measurements, interviews, questionnaires or surveys, and online or offline data collection tools. Background data represent common industrial processes that may be shared across multiple product systems and typically come from databases, open literature, and statistics. Once raw data are collected, data modelling approaches are applied to transform raw data into unit process datasets. Unit process datasets are the basis of every LCI database.

Depending on the LCI data collection method, LCA can be classified into three types: process-based, input–output (IO) based, and hybrid LCA (Heijungs and Suh, 2002). To select the most appropriate method for a given study, important factors such as boundary completeness and precision should be considered carefully with respect to the goals of the study (Lenzen and Crawford, 2009).

In process-based LCA, all the resources and emissions over the entire life cycle (or specific life cycle stages) of a product are considered. This allows calculating the complete environmental impact of a product. However, finding physical data for all inputs and outputs of a background process can be challenging, which can lead to the underestimation of impacts (Curran, 2015; Mattila, 2018; Crawford et al., 2018). It is challenging, if not impossible, to have a complete LCI with process data.

To overcome this problem, IO-based LCA can be applied. The IO-based LCA relies on sectoral-exchange information in a given year for a country's entire economy. This approach uses an IO table containing the annual transactions between all the sectors of the economy, along with data for sector-specific resource inputs and emissions (Curran, 2015; Mattila, 2018). Some of the limitations of this approach include price variability, sector aggregation, and the lack of input-specific data (Teh et al., 2018). The IO-based LCA requires detailed sector-level data regarding resource inputs and emissions. These are often available for certain kinds of emissions that have particular policy relevance – for example, GHG emissions – but will often be unavailable for many standard impact categories in LCA. Since the IO-based models use aggregated sectoral data, product-level LCIs will be less accurate (Huijbregts et al., 2001a). For instance, different agricultural systems (notorious for high variability among products and regions), such as organic and conventional farming systems, belong to the same sector, and thus the results of environmental impact assessments for products from these sectors will be similar, which is not valid (Curran, 2015). This concern applies to all agricultural commodities, which are considered equal in terms of resource use and emissions, when IO LCA results are calculated. Numerous LCA studies of agricultural products used process-based models (Cellura et al., 2012; Pelletier et al., 2010; Stoessel et al., 2012; Lardon et al., 2009) while few studies used IO-based models (Yan et al., 2013; Joshi, 1999).

In order to have both the completeness of the IO-based LCA and the precision of process-based LCA, a hybrid LCA approach can be applied (Suh and Huppes, 2005; Curran, 2015; Yang et al., 2017; Pomponi and Lenzen,

2018). In a hybrid-based LCA, the most important aspects of the process are modelled using either process data or IO data (Curran, 2015). For example, in a crop production LCA study, it may be assumed that agricultural machinery is not an environmentally important input, while chemical fertilizer is important. In this situation, to estimate the LCI of agricultural machinery and chemical fertilizer inputs, IO data and process data could be used, respectively (Finkbeiner, 2016). Recently, hybrid LCA has become a major tool of LCI and many practitioners have applied hybrid LCA methods to model agricultural products (e.g. food and energy crops and animal products) (Bright et al., 2010; Peters et al., 2010a,b; Acquaye et al., 2011; Virtanen et al., 2011; Palma-Rojas et al., 2017; Watanabe et al., 2016). According to Crawford et al. (2018), the complexity of hybrid LCA is likely to be a key factor in preventing this method from becoming commonplace. Although use of hybrid LCA requires less data for unit processes, the results have less precision and provide less detail for the product supply chain (e.g. use of crop and food packaging materials) (Cellura et al., 2012). Packaging of food products (e.g. red wine and milk) may make a significant contribution to environmental impacts, but it would not be possible to disaggregate the impacts of the food and packaging materials using an IO approach (Cellura et al., 2012; Ardente et al., 2006). The higher accuracy claim of hybrid LCA is often under scrutiny (Gibon and Schaubroeck, 2017; Yang, 2017a,b; Schaubroeck and Gibon, 2017). For example, Yang et al. (2017) demonstrated using a counter-example that hybrid LCA does not necessarily provide more accurate results than process-based LCA. However, Pomponi and Lenzen (2018) rejected Yang et al. (2017), demonstrating that this argument may not hold in a real economy with technological and sectorial interdependencies.

3.2 The consequential and attributional modelling perspectives and their application areas

Attributional life cycle assessment (ALCA) and consequential life cycle assessment (CLCA) are two distinct life cycle inventory modelling methodologies that may be applied in agricultural LCAs. Estimating the environmental consequences of an increased demand or supply of the FU, or of changes in the management of the agricultural system, such as intensifying production, use of co-products that substitute marginal animal-feed feedstocks, or expanding biofuel production requires a CLCA modelling approach that needs to be distinguished from ALCA in precision, accuracy, uncertainty, and the ability to support policy decisions. The consequential life cycle inventory modelling approach is used to assess the potential environmental consequences of a change in product output compared to a reference situation, including any related, market-mediated product substitution effects (Heijungs, 1997; Frischknecht, 1998; Tillman, 2000; Weidema, 2003; Wiedemann et al., 2019;

Finnveden et al., 2009). The CLCA approach relies on substitution to resolve multi-functionality problems (Thomassen et al., 2008b; Weidema, 2003; Ekvall and Weidema, 2004; Rebitzer et al., 2004). In contrast, ALCA describes the material or energy flows to and the outputs from an otherwise static model of a product system and is defined with respect to a specific product property (Rebitzer et al., 2004; Weidema, 2018). ALCA multi-functionality problems are usually solved via allocation, except for situations of closed-loop recycling.

In LCA studies, the goal and scope definition phase specifies the type of LCA (attributional or consequential). As above mentioned, ALCA and CLCA assess the environmental burden of a product in status quo and as a consequence of a change in demand, respectively (Finnveden et al., 2009). Since ALCA does not model consequences of decisions, it is less relevant for supporting policy makers. However, ALCA results are used to identify the hotspots of environmental impacts through the life cycle of agricultural products. Moreover, ALCA results provide the possibility to compare different production systems of a product and the contribution of certain stressors to an impact category. CLCA shows the changes of relevant environmental flows due to various decisions (Finnveden et al., 2009). Examples of the application of CLCA to agricultural case studies are Sánchez et al. (2012), Cederberg and Stadig (2003), Wiedemann et al. (2019), Dalgaard et al. (2008) and Schmidt and Weidema (2008). A number of studies have applied CLCA in livestock farming for the purpose of assessing the impacts of intensification of production systems (Chobtang et al., 2017), alternative production systems (organic and conventional dairy farming) (Thomassen et al., 2008b), and substituting diet ingredients by using co-products (e.g. wheat middling, beet tails, rapeseed meal, and soybean meal) (van Zanten et al., 2014, 2018).

3.3 Ensuring balanced systems when dealing with co-production

After data acquisition, it is necessary to normalize the inputs and outputs (emissions, energy, materials, etc.) to the determining product of every unit process. Often, however, agricultural unit processes will have more than one product output. For example, manure is a by-product of dairy farming in addition to milk and meat. In order to have an appropriate interpretation of the environmental impacts of the system, it is essential to distinguish the determining product (i.e. milk) from the by-products (i.e. meat and manure). In such cases, it is essential to isolate the determining product on a justifiable basis. Allocating the environmental impacts to co-products is a key concept and one of the most controversial issues in LCA (Suh et al., 2010). Agricultural LCAs are very sensitive to the algorithm adopted for handling co-production. Due to existence of several co-production activities in agricultural LCAs, allocation is an important source of uncertainty in LCA results. For example, wheat production:

at harvest, grain and straw are co-produced. At milling, flour, bran and middling are co-produced from the grain. For each step, different approaches have inconsistently been applied, which highlight large uncertainties in LCA results. Chiaramonti and Recchia (2010) reported up to approximately 300% variations in LCA results due to application of different allocation approaches. In a case study of soybean biodiesel LCA carried out by Cavalett and Ortega (2010), it was concluded that the LCA results are strongly affected by the different allocation choices. Although often ignored in published LCA studies, ISO 14044 provides a mandatory hierarchy for dealing with co-production as follows (ISO, 2006b):

> '*Step 1*: wherever possible, allocation should be avoided by (1) dividing the unit process to be allocated into two or more sub-processes and collecting the input and output data related to these sub-processes or (2) expanding the product system to include the additional functions related to the co-products.
>
> *Step 2*: Where allocation cannot be avoided, the inputs and outputs of the system should be partitioned between its different products or functions in a way that reflects the underlying physical relationships between them; i.e. they should reflect the way in which the inputs and outputs are changed by quantitative changes in the products or functions delivered by the system.
>
> *Step 3*: Where physical relationship alone cannot be established or used as the basis for allocation, the inputs should be allocated between the products and functions in a way that reflects other relationships between them. For example, input and output data might be allocated between co-products in proportion to the economic value of the products.'

Before describing the methods for allocating the environmental impacts to the co-products, it is essential to define two modes of co-production: combined and joint production. In combined production, the share of co-products is independently variable, while in joint production, the share of co-product outputs is fixed (Frischknecht, 2000). In combined production, sub-division according to physical causalities (modelling the consequences of increasing one output while keeping the others constant) can be observed. ISO Step 1, where allocation is avoided by using system expansion, is applicable for all cases of joint production in CLCA and for closed loop recycling in ALCA. Step 2 is applicable for combined production in the same way in both CLCA and ALCA, while step 3 is applicable exclusively in ALCA.

3.4 Avoiding allocation via system expansion

ISO 14044 recommends, first and foremost, the avoidance of allocation by sub-division or system expansion (also known as substitution) where possible (ISO, 2006b; Klöpffer and Grahl, 2014). The idea of avoiding allocation by expanding the system has been put forward by Tillman et al. (1991) and Vigon et al. (1993). System expansion is primarily applicable to joint production.

Due to the nature of the CLCA approach in considering indirect (i.e. market-mediated) effects on product system, system expansion may be the only effective solution to multi-functionality problems in CLCA (Pelletier et al., 2015). In this method, allocation is avoided and all co-products, together with what they displace, remain within the system boundary. To do this, the inventory includes the flows for all products (determining and dependent). Subsequently, the dependent products are considered to displace marginal production of an equivalent quantity of the same function, where the displaced product is the determining product of that alternative system. By subtracting the inventory of the displaced product from the multi-functional system, the net inputs and outputs of the determining product of the system under study can be calculated (Weidema, 2000). For example, in a milk production system, the determining product is milk and the dependent product or by-product is meat (that arises from the calves that the dairy cows must inevitably produce in order to give milk). In this system, allocation is avoided by expanding the system to involve the alternative way of producing an equivalent amount of meat, which is dedicated beef production. By subtracting the results of the dedicated meat production system from the joint milk and meat production system, the inputs and outputs can be estimated for milk (Harris and Narayanaswamy, 2009). A major difference between avoiding and performing allocation is that, as demonstrated by Weidema and Schmidt (2010), the substitution approach preserves all mass and energy balances, while any form of allocation does not, which is a major drawback of the latter approach. Examples of LCA studies of agricultural products where allocation is avoided by system expansion include Cederberg and Stadig (2003), Hospido and Sonesson (2005), Botha and Von Blottnitz (2006), Eriksson et al. (2007), Kim and Dale (2008), Searcy and Flynn (2008), Spatari et al. (2010), Uihlein et al. (2008), Reckmann (2013) and Cherubini et al. (2015).

3.5 Allocation

Allocation (also known as partitioning) is the attribution of inputs and outputs between co-products (Klöpffer and Grahl, 2014). According to the ISO (2006b) standard for LCA, allocation according to the causal relationship between inputs and outputs is preferred when allocation cannot be avoided. Since causal physical relationships between the individual inputs and outputs can only be established in the case of combined production, situations of combined production call for allocation based on economic value. A more broad (lax) interpretation of the ISO standard has appeared, where 'relationships' are interpreted also as non-causal, thus allowing for a more general use of physical properties such as mass, energy, exergy, and a particular chemical element as allocation keys. A special case of this is the use of 'biophysical' allocation factor for livestock, which applies metabolic relationships to establish allocation factors. The descriptions of these methodologies are as follows.

3.5.1 Causal physical allocation

When the co-products of a system are physically independent of each other (combined production), allocation according to the causal relationships should be conducted (Ekvall and Finnveden, 2001). Allocation according to causal physical relationship between co-products has rarely been applied in agricultural LCAs. Audsley et al. (1997) applied causal physical allocation to divide the environmental burdens of wheat production system between wheat grain and straw. Ayer et al. (2007) studied the possibility of using allocation according to a causal physical relationship in aquaculture LCAs and found that since co-products of these systems are physically dependent, this type of allocation is not applicable to the LCA of aquaculture systems.

3.5.2 Allocation according to economic, mass, and energy content

Allocation or partitioning in joint production is a very important issue in ALCAs of agricultural products due to multiple production processes (Cavalett and Ortega, 2010). Various allocation approaches have been developed for joint production namely economic allocation, mass allocation, particular chemical element in the products, and energy or exergy allocation. In this approach, environmental impacts of the production system are allocated among the co-products using a specific index according to the allocation criterion (such as economic and mass). For example, according to economic allocation, the specific index is the relative economic value of co-products to the entire economic value of products.

Economic allocation is the most commonly reported allocation method in published agricultural LCA studies (de Vries and de Boer, 2010; Ardente and Cellura, 2012; Thomassen et al., 2008a, 2009; Williams et al., 2006). Energy content is another allocation criterion which is widely applied in bioenergy-related LCAs (Lam et al., 2009; Ndong et al., 2009; Angarita et al., 2009). Some agricultural LCAs applied mass allocation (Hu et al., 2008; Harding et al., 2008).

Allocations according to allocation key (such as economic and mass) have some problems that need to be considered. For example, manure and straw in animal and crop production as co-products do not have any economic value in some regions and countries. By economic allocation, all environmental burdens of these production systems are allocated to the main product. However, farmers may return straw to soil or apply manure as fertilizer. Since application of straw and manure decreases the amount of chemical fertilizer required for further crop rotation, these co-products possess hidden economic values. In this case, it is necessary to apply a value-corrected allocation as was suggested by Werner and Richter (2000). Another problem of applying economic allocation is that the economic value of the products can be highly variable over time and location. However, in reality, the material and energy flows remain constant over time. Despite this limitation, many researchers argue that since the economic value of co-products reflects the incentive for production, this method is one of the best choices for partitioning the environmental burdens over the co-products.

3.5.3 Metabolic 'biophysical' allocation

As stated above, economic allocation is the most common approach in agricultural LCAs. However, a variety of studies on biophysical allocation have been published in recent years. These are generally based on quantifiable relationships with respect to the metabolic allocation of resource inputs between co-products. For example, in a dairy farm, there is a causal relationship between the feed inputs and their respective allocation to the production of milk, meat, and calves. In one example, the ratio for this method is based on official recommendations that shows the proportion of inputs (such as feed in dairy cows) needed for production (e.g. of milk) (Harris and Narayanaswamy, 2009; Pelletier and Tyedmers, 2011). Similar methods have been developed and applied in a variety of animal production systems (such as milk and egg production systems). Cederberg and Stadig (2003) applied an early version of the metabolic biophysical allocation approach to divide the environmental burdens of milk production system between the co-products. The basis for their choice was the fact that there is causal relationship between feed mix and the production of milk, meat, and calves. Calculations were conducted according to the Swedish fodder tables and the algorithm resulted in an allocation of 85% to milk and 15% to meat and the surplus calves. Based on the International Dairy Federation (IDF) methodology, the allocation factor between meat and milk is recommended as Eq. (1) (IDF, 2010). Researchers have subsequently refined this equation on a regional basis (e.g. see Thoma et al., 2013b):

$$AF = 1 - 5.77R \quad (1)$$

where AF is the allocation factor for milk, R is the fraction of meat and milk (kg beef/kg milk), and kg milk is corrected to 4% fat and 3.3% protein.

The rationale supporting this equation is that 'feed energy available for growth, for a given feed, is more readily available than that available for milk production' (IDF, 2010). Many published LCA studies of agricultural products have applied biophysical allocation (Thoma et al., 2013a; Cederberg and Mattsson, 2000; Dollé and Gac, 2012).

An important issue in the biophysical allocation approach is that, to establish physical causality between co-products and environmental burdens, the co-products of a system should be physically independent of each other (Ekvall and Finnveden, 2001). However, in this allocation approach, the co-products cannot be considered independently (we consider them as joint products) and therefore the question arises as to whether 'causality' is or can be demonstrated in any of the biophysical allocation approaches (Mackenzie et al., 2017). One of the limitations of the biophysical allocation methodology is the sensitivity of this method to the economic value of co-products (Mackenzie et al., 2017). As it was observed in the FAO (2014) guidelines on the environmental impact of livestock production system, when manure (as co-product in egg production system) does not have any economic value, 0% of environmental burden associated with egg production would be allocated to manure. The other limitation is due to the complex nature of agricultural systems and the interconnectivity that exists between their outputs. In this situation, modellers are not able to adequately address the physical relationships by choosing the proper parameters in each case to correctly show the causal relationship mechanisms. Therefore, modellers should always be aware of the arbitrary nature of selecting the allocation methodology where systems are too complex to be modelled (Mackenzie et al., 2017). Moreover, biophysical models depend on various parameters such as characteristics of the production system (breed of animals, variety of crops, etc.) and time and place of assessment, which influence the accuracy of the results.

According to the ISO 14044 recommendation, when several alternative allocation methods are applicable, a sensitivity analysis should be conducted to evaluate the impact of different allocation methods on the LCA results (ISO, 2006b).

4 Categories of environmental pressures and impacts

The LCI reports the sum of flows, by elementary flow (e.g. kg CO_2 to the atmosphere), that cross the technosphere/ecosphere boundary. These flows may contribute to a variety of environmental impacts. According to ISO 14040/14044 (ISO, 2006a,b), LCIA is the third phase of the LCA, where the results of the LCI are transformed (by modelling, calculation, and valuation procedures) into their potential environmental impacts. The results of LCIA support the interpretation phase to find the answer to the study question. According to ISO 14040/14044 (ISO, 2006a,b), five steps including selection, classification, characterization, normalization, and weighting are important in LCIA. It should be noted that according to ISO 14040 (ISO, 2006a), the first three steps (selection, classification, and characterization) are essential in an LCA, while the latter two steps (normalization and weighting) are optional. First of all, the appropriate impact categories are selected in accordance with the goal of the study. During the classification step, inventory flows such as emissions and resource usage are assigned to the selected impact categories, according to their potential environmental impacts. In the characterization step, all elementary flows of the inventory are multiplied by corresponding characterization factors, which indicate their relative contribution to the overall environmental impact. During the normalization step, the characterized results of each impact category are divided by a reference value. For example, a reference value could be defined as the annual impact per person within a target group. Uncertainties related to the selection of the appropriate normalization references is the main criticism regarding the normalization step (Pizzol et al., 2017) since it affects the LCA results and final conclusions. The final, and optional, stage in LCIA is the weighting of impacts. There is legitimate criticism for the weighting step, as ISO 14044 considers it to be 'not scientifically based' (ISO, 2006a,b). Although normalization and weighting are optional steps in the LCIA phase, they have been applied in LCA studies. Normalization and weighting steps help to identify the important impact categories, better understand the meaning of LCA results, and assist practitioners in the trade-off between different impact categories (Pizzol et al., 2017). For example, biofuel production may mitigate climate change but increase impacts on biodiversity, acidification, and eutrophication. Thus, in this case, weighting is vital for aggregating the results of different impact categories (Hauschild and Huijbregts, 2015).

According to ISO 14044 (ISO, 2006b), depending on the location of the impact category indicator along the impact pathway, midpoint and endpoint impact categories are defined. Typical midpoint categories of environmental impacts which are used in most agricultural LCA studies are climate change, acidification, eutrophication, toxicity, land use, and, increasingly, water use (e.g. Dijkman et al., 2018, Nemecek et al., 2011, Dalgaard et al., 2007). These are common 'midpoint' categories, which taken further along the cause–effect chain would group into 'endpoint' categories. Endpoint categories represent damage to human health, ecosystems, and natural resources.

Selecting between midpoint and endpoint LCIA methods requires reflection on the goal and scope of the study, including its intended audience and applications. The endpoint approach enables approximation to actual damages (e.g. human health) and aggregation across impact categories, which may render LCA results more accessible and relevant to lay audiences. However, the calculation process is more complicated and the results are of considerably less precision compared to the midpoint approach (De Jong, 2013).

One of the main problems in LCIA is due to the exclusion of time- and space-specific impacts of environmental interventions. In this regard, selecting the appropriate impact category, LCIA method, spatial variation, and environmental dynamics are discussed. The first challenging issue is impact category selection, and since there is not a standard practice here, different practitioners reflect their own, often unjustified, preferences for some impact categories over others (Finnveden, 2000). In the case of damage indicators, there is disagreement between researchers.

4.1 Climate change

Agriculture both contributes to and is affected by climate change. Depending on the region of the world, agriculture may be impacted by climate change as a result of changes in water availability or in crop productivity. The increased frequency and severity of extreme weather events will likely also have pronounced negative impacts, although in some regions agricultural outcomes may actually improve as a result of climate change.

Agricultural activities directly contribute to nearly 14% of global annual GHG emissions, while the indirect impact of land use change accounts for another 4–8% (Turral et al., 2011). Taken as a whole, agri-food supply chains globally contribute as much as 30% of GHG emissions. Carbon dioxide (CO_2), nitrous oxide (N_2O), and methane (CH_4) are the most important agricultural GHGs. CO_2 is generated by activities such as production and burning of fossil fuels for cultivation, pumping irrigation water, transport, application and production of chemical fertilizers and production of other agricultural inputs, as well as from land use change. N_2O from agricultural sources is largely associated with the use of nitrogen fertilizer and with animal manure. Most CH_4 is generated during animal production (in particular, via enteric fermentation in ruminants and decomposition of manure under anaerobic conditions) and during flooded rice cultivation (about 35%) (Turral et al., 2011). Despite the large contribution of the global agricultural sector to climate change, agricultural land use also exhibits a strong potential for climate change mitigation.

The Intergovernmental Panel on Climate Change (IPCC) provides good information on the climate change impact pathway (IPCC, 2013, 2014). Figure 1 shows the impact pathway of climate change. At the first step, GHG emissions impact on radiative forcing. The radiative forcing causes an increase in the earth temperature and changes in rainfall and sea level. These changes cause perturbation of agricultural yields, ecosystem disturbances, disease spread, flooding and so on. These effects lead to damages on human health and ecosystem quality. The impact pathway of climate change is classified to shorter-term and long-term climate change. Some of the long-term climate change impacts are sea-level rise, polar ice caps melting, coral bleaching, and changing biomes and the shorter-term impacts are movement of species (ecosystem adaptation), heat stress, malnutrition (human adaptation), extreme weather events and so on. (Levasseur et al., 2017).

The European Commission's Joint Research Centre (EC-JRC) (Manfredi et al., 2012) and IPCC (2013) recommend global warming potentials (GWPs) (kg $CO_{2\text{-eq}}$) for the characterization of climate forcers. GWPs have been the

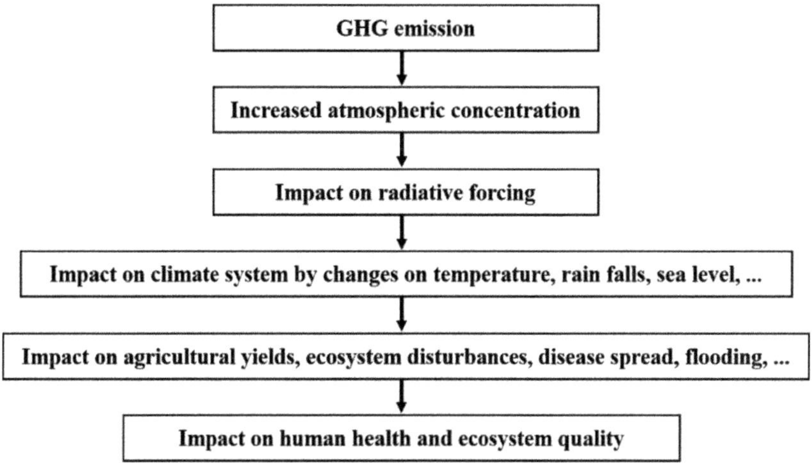

Figure 1 The impact pathway of climate change.

default characterization factors for GHGs and are defined as the integrated relative forcing of a specific gas over a specific time horizon. However, since there are many non-linear interactions between GWP and global temperature or any other climate variable, the United Nations Environment Programme (UNEP) suggested global temperature potential (GTP) as a more general characterization factor for environmental impacts of GHGs (Levasseur et al., 2017). GTP is an instantaneous normalized metric and is defined as an indicator to show the global average increase in atmospheric temperature due to GHG emissions (Levasseur et al., 2017). Both GWP and GTP represent the environmental impacts in terms of kg $CO_{2\text{-eq}}$; however, GTP is closer to actual impacts than radiative forcing (Levasseur et al., 2017).

4.2 Acidification

Acidification happens when acidifying substances are converted in a chemical reaction into acid substances. The burning of fossil fuels, transport, power generation, and heating are the most important sources of acidification. There are a number of buffer reactions in an ecosystem which can neutralize acidifying substances (Blaser et al., 1999). Acidifying substances deposition leads to changes in the chemical composition of the soil and water. These compounds release hydrogen ions. Ecosystems are affected when the hydrogen ion concentration increases (pH decreases). The metabolism of plants is dependent on optimal pH levels. In addition to ecosystem impacts, acidification has different impacts on human health, such as diseases. Different kinds of cancers can be transmitted to humans when fish intoxicated with

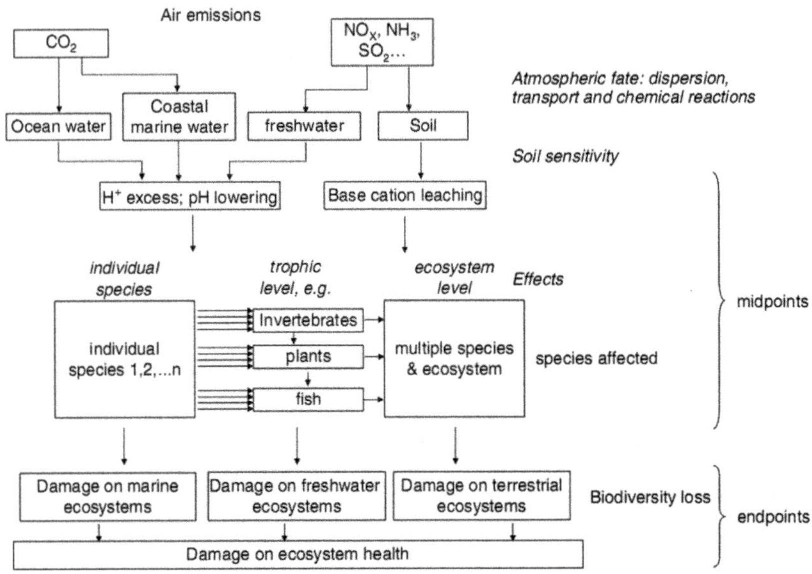

Figure 2 The impact pathway for acidification. Source: adapted from EC-JRC (2011).

higher sulphur concentration are consumed. Fig. 2 shows the impact pathway for acidification.

Proton release in terrestrial or aquatic ecosystems may lead to acidification in low buffering capacity environments. According to Stranddorf et al. (2005), important acidifying compounds include SO_x, NO_x, and ammonia (NH_3). SO_x emissions in agriculture primarily originate from combustion of sulphur-containing fuel (e.g. coal and oil), NO_x from combustion processes in agricultural machineries, and NH_3 from animal manure (EEA, 1998). Volatilization of ammonia from organic and mineral N fertilizers, elemental S fertilizer, and cultivation of legumes (i.e. exudation of acids during plant growth and nutrient uptake in the rhizosphere) may also be important sources of acidifying emissions (Goulding, 2016; Brentrup et al., 2000).

Acidification potential is determined by deposition patterns and the sensitivity of the receiving environment (e.g. buffer capacity of soils and surface waters). Different methods and models have addressed acidification, namely Eco-indicator 99 (Goedkoop and Spriensma, 2001), EPS (Steen, 1999), CML 2002 (Guinée et al., 2002), EDIP 2003 (Potting et al., 1998), TRACI (Norris, 2002), LIME (Hayashi et al., 2004), Accumulated Exceedance (Seppälä et al., 2006), Azevedo (Azevedo et al., 2013) and ReCiPe (Huijbregts et al., 2017). Specifications of each method have been discussed in van Zelm et al. (2015).

Acidification potentials are expressed for midpoint as kg $SO_{2\text{-eq}}$ (Norris, 2002; Huijbregts et al., 2017; Guinée et al., 2002; Hayashi et al., 2004), m^2-UES (Potting et al., 1998), Eq per year (Seppälä et al., 2006), mol $SO_{2\text{-eq}}$ (Steen, 1999),

Table 1 Acidification and terrestrial eutrophication impact categories: regionalized factors for Europe

	Acidification potential (in kg $SO_{2\text{-eq}}$ per kg emission)			Terrestrial eutrophication potential (in kg $NO_{x\text{-eq}}$ per kg emission)	
Emission sources region	SO_2	NO_x	NH_3	NO_x	NH_3
Western Europe (average)	0.79	0.41	1.30	0.99	3.70
Eastern Europe (average)	1.60	0.70	1.80	1.70	5.00
Europe (average)	1.20	0.5	1.60	1.20	4.30

Source: adapted from Huijbregts et al. (2001b).

and for endpoint as m^2.year (Azevedo et al., 2013; Goedkoop and Spriensma, 2001; Huijbregts et al., 2017), kg.year (Hayashi et al., 2004). The EC-JRC in the product environmental footprint (PEF) guide recommends the use of the accumulated exceedance model (mol H^+_{eq}) for the characterization of acidifying substance flows (Manfredi et al., 2012) (details can found in Posch et al., 2008; Seppälä et al., 2006).

It is important to apply region- and time-specific acidification and terrestrial eutrophication potentials. Huijbregts et al. (2001b) applied the RAINS (Regional Air Pollution INformation and Simulation)-LCA models to calculate the CFs of acidification potential for NH_3, nitrogen oxide (NO), and sulphur dioxide (SO_2) for different years, countries, and regions in Europe. Table 1 provides the region-specific and average acidification and terrestrial eutrophication potentials for SO_2, NO_x, and NH_3 emissions in Europe. It should be considered that the region-specific CFs are calculated based on the conditions in Switzerland, while the recalculated CFs were estimated by using the total region-specific emissions related to 1995 and 2010. In other words, the emission-equivalents in Table 1 are relative to the impact from the emissions when occurring in Switzerland.

4.3 Eutrophication

Eutrophication results when nutrients released into water stimulate primary production at levels sufficient to create anoxic conditions. Leaching of phosphate and nitrate from fertilizers and animal manures are major drivers of eutrophication of surface and ground waters.

Eutrophication potential (EP) is commonly expressed as kg phosphate equivalents ($PO_{4\text{-eq}}$) (Dahlgren et al., 2015), but kg P_{eq} or kg N_{eq} (Struijs et al., 2009) are also sometimes used. Since terrestrial and aquatic plants often respond differently to additional nutrients, this impact category can be divided into two sub-categories:

1 *Terrestrial eutrophication*: Nitrogen is the major limiting nutrient in terrestrial ecosystems. NO_x and, in particular, NH_3 depositions are

typically the most important contributors to terrestrial eutrophication (Finnveden and Potting, 1999). The decomposition of urea in animal wastes and uric acid in poultry wastes is a major source of NH_3 emissions.

2 *Aquatic eutrophication*: Emissions of N and P to surface waters are classified into: (1) NO_x emissions from combustion processes and NH_3 from N fertilizers and animal manures; (2) leaching of nitrate (from fertilizer application and animal manures) (Beusen et al., 1995); and (3) direct deposition of N and P (from municipality, industry treatment plants), which is unrelated to agriculture.

The main pathway for N emissions from soil to aquatic ecosystems is via nitrate (NO_3) leaching. However, there is no linear relationship between the level of N input to the soil (e.g. as fertilizer) and nitrate loading of ground and surface water. Fertilization rates and strategies, N removal with harvested crops, and site-specific soil and climate conditions will all influence potential rates of nitrate losses to ground and surface water via leaching (Brentrup et al., 2000).

The direct human impacts including algal blooms may be toxic to humans and result in beach closures (Anderson, 1989; Paerl et al., 2001). The excess nutrients and organic matter would affect cell synthesis and energy supply. Eutrophication pathways can be discussed for terrestrial eutrophication and aquatic eutrophication.

Since terrestrial eutrophication is the results of excess nitrogen, the structure and function of terrestrial, N-limited ecosystems by favouring a (typically) limited number of N-adapted species would change. This changes the tolerance rate against disease or other stressors (e.g. drought, frost), resulting in overall biodiversity and productivity changes. A flow diagram of a eutrophication pathway is illustrated in Fig. 3.

When it comes to the aquatic environment, the increase of nitrogen and phosphorus in aquatic ecosystems can make changes to the structure and function of plants and animals by involving the organisms in the sea bottoms and water columns.

Several competing models for quantifying the eutrophication potential are available. The CML method calculates both terrestrial and aquatic eutrophication in a single indicator and ignores that only a fraction of emissions will have an impact on the aquatic environment (Struijs et al., 2009). ReCiPe 2016 calculates the eutrophication impacts in terms of freshwater eutrophication only (Huijbregts et al., 2017), while ReCiPe 2008 includes both marine and freshwater eutrophication potential indicators (Struijs et al., 2009). The EC's PEF guide recommends the use of the accumulated exceedance model (mol H^+_{eq}) for the characterization of terrestrial eutrophication substance flows (Manfredi et al., 2012), and the use of the EUTREND model (kg P_{eq}; kg N_{eq}) for

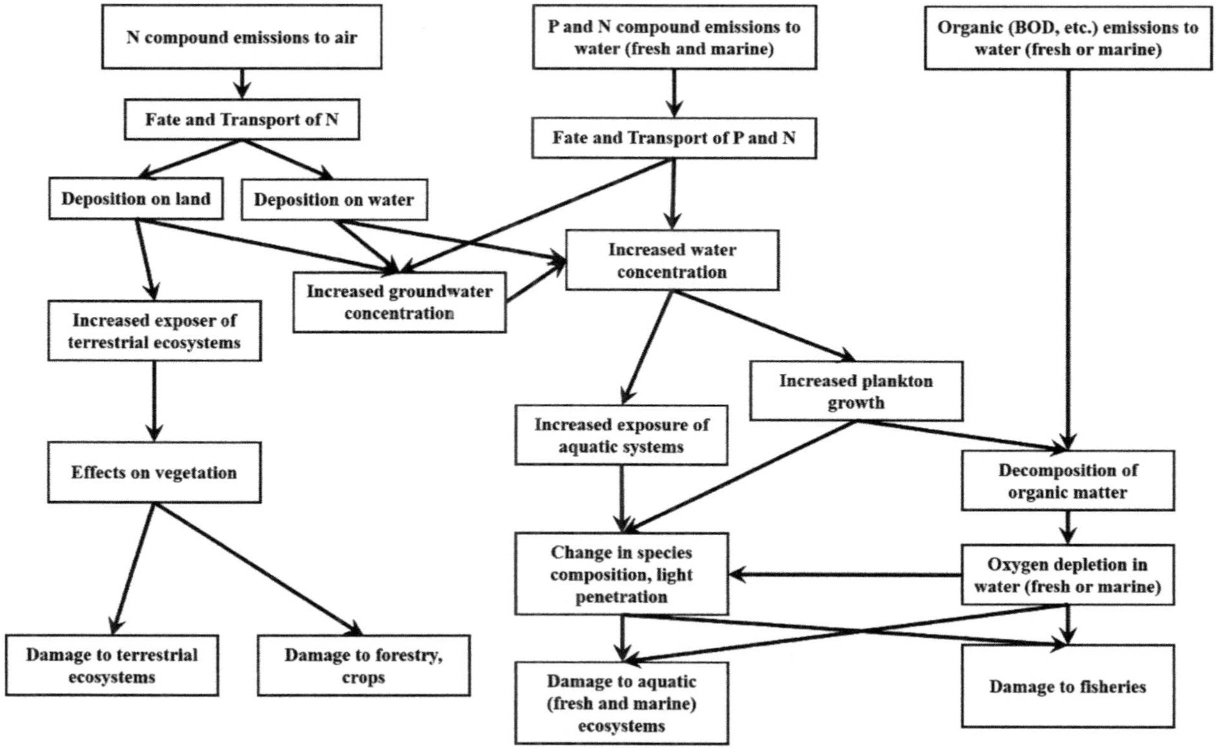

Figure 3 Flow diagram of eutrophication impact pathway (EC-JRC, 2010). The flow diagram also includes the impact pathway for eutrophication-independent oxygen depletion.

the characterization of aquatic eutrophication substance flows (Manfredi et al., 2012), as implemented in ReCiPe (2008). See Struijs et al. (2009).

4.4 Toxicity

The toxicity impact category addresses the effects of pesticides and heavy metals on human health, and both aquatic and soil ecosystems (Wang et al., 2014). Potentially toxic emissions released from agricultural activities include inorganic air pollutants like NH_3, SO_2, and NO_x, plant protection products (chemical pesticides, fungicides, and herbicides), and heavy metals from mineral (in particular, phosphate) fertilizers, slurry, sewage sludge, or compost. A variety of models have been developed to estimate the rate of unintended emissions of toxic substances, their fate in the environment, and their final effects on ecosystems and humans (Goedkoop and Spriensma, 2001; Guinée, 1996; Huijbregts, 2001; Jolliet and Crettaz, 1997). However, there is a good consensus on the USEtox model. The EC-JCR's PEF Guide recommends the use of the USEtox model, expressed in the comparative toxic unit for humans (CTUh) for the characterization of toxic substance flows (Manfredi et al., 2012; Rosenbaum et al., 2008).

Toxicity impacts are different between conventional and organic farming systems related to the use of agrochemicals in conventional farming systems (Moudrý and Moudrý, 2014). Since the product life cycle generates chemicals into the environment and releases chemical mass to air, water, biota, and soil, humans are exposed to these impacts directly by breathing the air and drinking the water or indirectly by accumulation in herbal textures of human food (Huijbregts et al., 2005; Rosenbaum et al., 2011). Moreover, agricultural workers may be exposed to direct contact of chemicals with their skin. When chemicals distribute inside the human body, different organs may be damaged and various diseases occur and cause disability or loss of life. Figure 4 illustrates the pathway of human toxicity impacts.

4.5 Land use impacts on climate change, ecosystem services, and biodiversity

Land use refers to how an area is used for human purposes (e.g. urban activities, agriculture, forestry, etc.) (Mattila et al., 2011). As the world population and the amount of food, feed, fuels, and fibre required to support it continues to grow, the limited availability of productive agricultural land is of increasing concern. While agriculture represents the largest user of land, other economic sectors such as forestry are also important users and competitors for available land. Moreover, land provides habitats for other species, regulates the storage and flow of surface and ground water, and provides a variety of other essential ecosystem services.

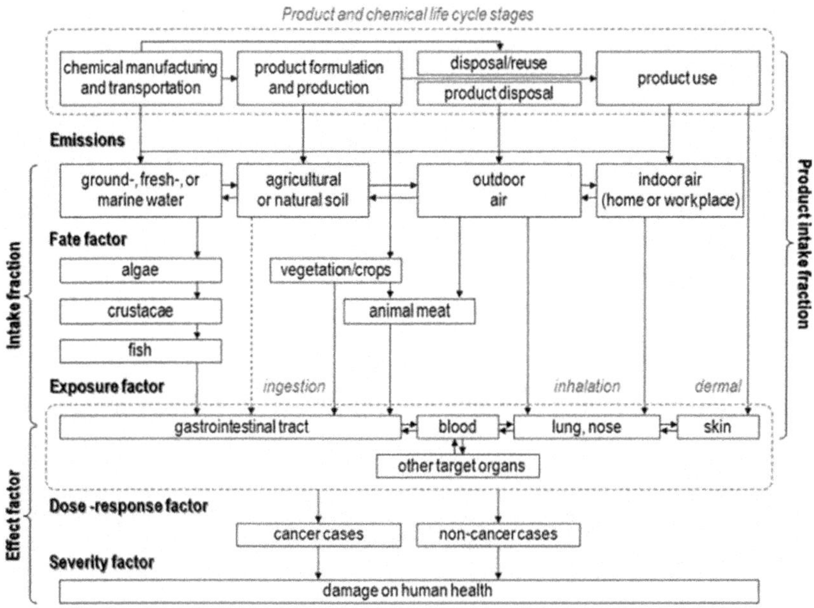

Figure 4 Pathway of human toxicity impacts.

Land use impact is the environmental impact caused by human activities on land, ranging from changes in species composition and abundance to the disruption of ecosystem processes. This impact affects human health, ecosystem quality, and natural resources. Land use impacts take several pathways, including physical (compaction, erosion, etc.), chemical (pH, salts composition, toxicity), or biological (vegetation cover, species composition, etc.). These alterations affect land-based processes (e.g. albedo; water cycle), which are linked to all areas of protection (Koellner et al., 2013). Due to the multi-functionality of land, land use impact assessment requires modelling of several impact pathways. This includes impacts on habitats, resource availability, diversity of wildlife species, and soil quality (EEA, 1998). Figure 5 illustrates the pathways of land use impacts on biodiversity and ecosystem services.

In LCA, land use interventions may be characterized as either land occupation or land transformation (Mattila et al., 2011). Land occupation is defined as the continual use of land for a specific purpose such as forestry, agriculture, or buildings. Conversely, land transformation, which is also known as land use change (LUC), refers to the change from one land use type (e.g. forestry to agriculture) to another. LUC incurs environmental impacts. For example, forest clearing leads to the release of carbon from biomass and soils to the atmosphere, which contributes to climate change. In addition, changes in ecosystem functioning due to LUC lead to biodiversity decline, directly and indirectly, via for example changes in the hydrological cycle by river

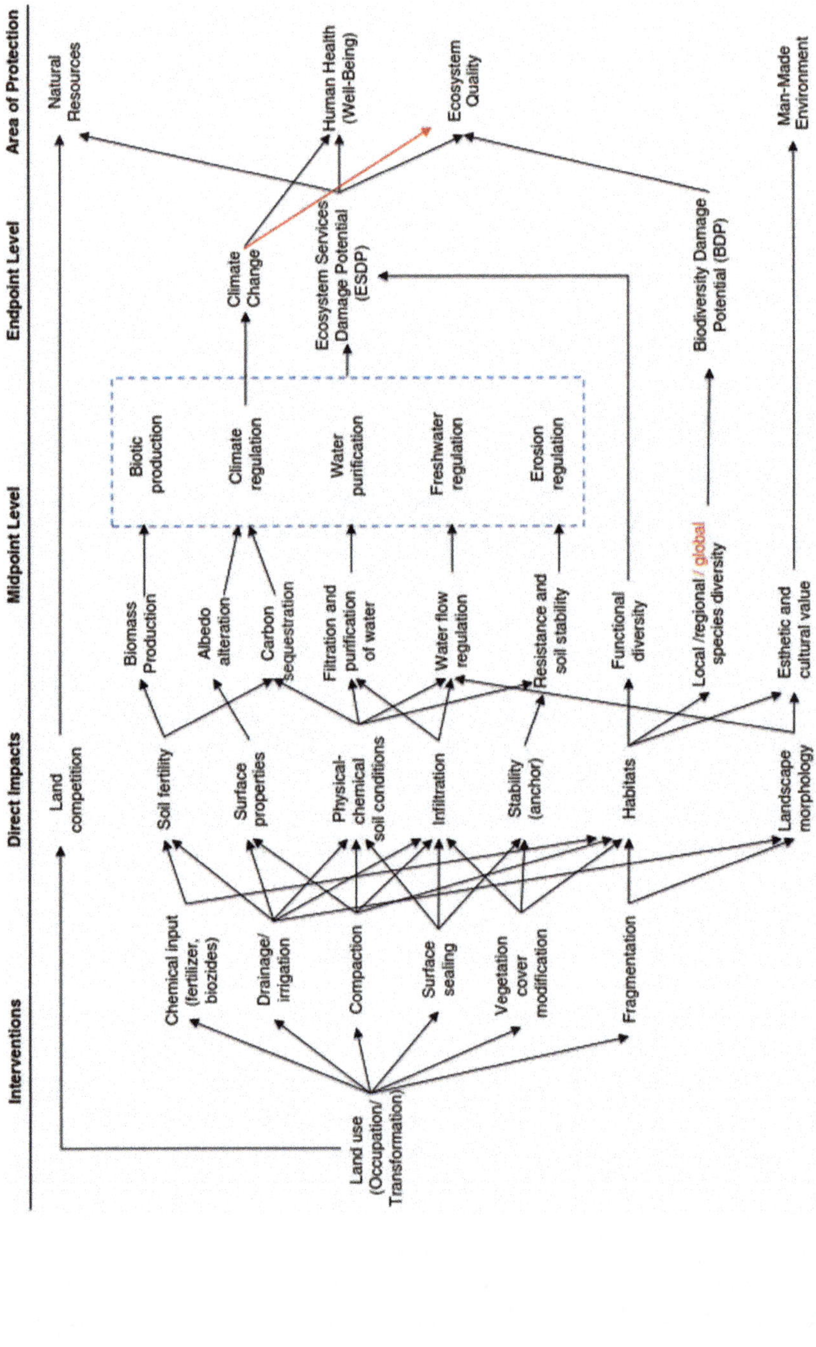

Figure 5 Impact pathways from land use. Source: based on Koellner et al. (2013).

diversion, change of evapotranspiration flows, and by modifying the portion of precipitation into run-off (Rosenbaum et al., 2018). Some of the most important agricultural activities have impacts on soil quality, e.g. compaction and erosion, which affects aquifer and natural soil capacity to remove pollutants (Rosenbaum et al., 2018). The Millennium Ecosystem Assessment (Finlayson et al., 2005) identifies some ecological impacts of land use as: biodiversity decline, regional climate regulation (e.g. changes in land cover changing the precipitation rate), food production decline due to soil erosion, drought and flood risks due to loss of forests and wetlands, and changes in the hydrological cycle.

LUC can be further sub-divided into direct land use change (dLUC) and indirect land use change (iLUC) (Vellinga et al., 2013). The dLUC is defined as the change in land use from a previous land use activity to an alternative use (e.g. forest to agriculture). The iLUC refers to the land transformation that occurs indirectly, but as a result of the studied product system due to its land competition effects (Schmidt et al., 2015). For example, a land use change from food crop to biofuel crop production may cause indirect land use effects since a constant food supply would require that the foregone food be produced elsewhere, and this may involve an indirect land use change from, for example, forest to agricultural cropland elsewhere (Wicke et al., 2012).

Methods for quantifying emissions related to land use have received increased attention due to the pressure that the demand for biofuels places on land (Edenhofer et al., 2011; Finkbeiner, 2014; Warner et al., 2014). The majority of LCA studies have focussed on dLUC effects related to the production of agricultural products and, in particular, bioenergy and biofuel products (Warner et al., 2014). For example, Mattila et al. (2011) present theoretical framework for dLUC and report a comparative land use impact assessment for beer and wine products. More recently, iLUC has also been taken into account by several authors (Verstegen et al., 2016; Plevin et al., 2015; Searchinger et al., 2008).

Some international and national guidelines used to estimate GHG emissions from land use are the EU Renewable Energy Directive (RED), PAS 2050 (carbon footprint specification by BSI), California Low Carbon Fuel Standard (LCFS), and IPCC guidelines on land use in agriculture, forestry, and land use (AFOLU) sector. Land use is also addressed in the context of LCA guidance documents including those of the UNEP/SETAC Life Cycle Initiative, the ILCD Handbook 2010, and several leading LCIA methodologies, such as ReCiPe, CML, and LIME (Mattila et al., 2011). Three types of models – biophysical, economic, and rule-based models – have been used to estimate emissions from iLUC (Schmidt et al., 2015). Searchinger et al. (2008) used the FAPRI[1] partial equilibrium model to estimate GHG emissions from iLUC due to increased biofuel demand.

1 http://www.fapri.iastate.edu/

Kløverpris et al. (2010) modified the GTAP[2] Model to estimate global land use changes due to increased wheat demand. Hertel et al. (2010) used the GTAP-BIO[3] model to assess impacts of maize ethanol on global land use and GHG emissions. To characterize the land use flows for iLUC, we recommend the use of the characterization model by Schmidt et al. (2015). One of the newest guidelines for assessing global land use impacts on biodiversity and ecosystem services in LCA was developed by UNEP-SETAC (Koellner et al., 2013). These guidelines give recommendations on the creation of a spatial model, inventory data collection, and land use impact assessment calculations.

One of the main impacts of livestock production systems is climate change via LUC and associated losses of biomass carbon and soil organic carbon (SOC) (Steinfeld et al., 2006), largely driven by land clearing for producing high-protein feeds, such as soybean and maize (Wassenaar et al., 2007; Garnett, 2009). However, improved diets for cattle can reduce enteric methane emissions, which may offset emissions from LUC (Schils et al., 2007; Beauchemin et al., 2008). Similar trade-offs may be observed with respect to LUC for the production of biofuel crops, such as soybean and sugarcane in Brazil, oil palm in South East Asia, and corn in the USA (Vellinga and Hoving, 2011). While iLUC from biofuels production may result in large losses of SOC in a short period, mitigating environmental effects may occur continuously over a longer period of time (Gibbs et al., 2008).

The EC-JRC's PEF guide recommends the use of the soil organic matter (SOM) model, expressed in kg deficit for the characterization of land use flows (Manfredi et al., 2012; Milà i Canals et al., 2007). We recommend an improvement to that LCIA model for impacts on ecosystem services (Brandão and Milà i Canals, 2013). For biodiversity impacts, we recommend the use of Koellner et al. (2013).

Additional aspects affect most of the above impacts of agricultural systems. Different regions may react differently to the same environmental interventions on the basis of some local characteristics. Modelling local-specific impacts requires spatial data that can represent geographical specificities and their sensitiveness to global impact categories such as global warming. One of the most important limitations of modelling the LCA of agricultural products is that agricultural LCAs do not always address the spatial variability based on regional differences (Potting and Hauschild, 2006; Sedlbauer et al., 2007). The most important LCIA approaches, such as IMPACT 2002+, ReCiPe, TRACI, LUCAS, and LIME are based on the assumption that the emissions are released where the methodology has been developed, i.e. in Europe, the US, Canada, and Japan. This is an obvious and important limitation of the LCIA

2 https://www.gtap.agecon.purdue.edu/
3 https://www.gtap.agecon.purdue.edu/resources/download/3939.pdf

methodology. Therefore, there is still a need for developing some important location-related impact categories, such as water and land use, to have regional impact categories (Margni and Curran, 2012).

5 Conclusion and future trends

The application of LCA to the agricultural sector in support of the sustainable development goals has grown quickly over the last decades. Over the same time, data and methods for agricultural LCAs have been developed, supporting producers to improve their environmental performance and consumers to choose consciously. In addition to supporting the shift to more resource-efficient on-farm activities with fewer emissions, a life cycle perspective identifies ways of abating environmental impacts throughout the value chains involved in the provision of inputs, transportation of inputs to the farm, processing, distribution, consumption, and waste management.

Furthermore, identification of trade-offs among environmental, economic, and social concerns can be included in LCA, which provides a quantitative framework to understand the potential sustainability impacts incurred along product supply chains. Assessing such trade-offs has been reinforced in recent years in agricultural LCAs.

This chapter described the main methodological issues encountered in agricultural LCA studies. Identifying the appropriate methodological parameters (such as FU, allocation method, selection of relevant impact categories) in modelling of agricultural LCAs not only affects the obtained results, but also the interpretation of results. Identifying the appropriate FU, which is one of the key parameters that reflects the relative purpose and the characteristics of an LCA study, is crucial. In other words, incorrect definition of the FU would lead to misleading results. For example, in this chapter, we showed how comparing organic and conventional farming systems are affected by the various FUs. By defining the FU based on 1 ha, organic farming represents a superior alternative to conventional farming, while a mass-based FU prioritizes the conventional farming. Thus, it is recommended to ensure that the FU is identified in accordance with the research question asked.

Data acquisition on inputs and outputs is another fundamental issue for any LCA study, in particular for agricultural systems. IO-based and process-based approaches have different pros and cons based on their completeness, precision, and data requirement levels. Hybrid LCA bears the completeness and precision of process-based LCA and the less data requirement of IO-based LCA; however, it is still under examination for its higher accuracy level.

Another important and controversial issue about agricultural LCAs is partitioning of environmental burdens to the co-products. Different methods are recommended by ISO standards and researchers including system expansion and allocation. As it was demonstrated by Weidema and Schmidt (2010), system expansion preserves all mass and energy balances, while any other form of allocation does not; however, application of system expansion necessarily requires more detailed information from other alternative systems. The approach recently taken for modelling livestock production systems is biophysical allocation, which is not based on the causal relationships between inputs and outputs. This method has been claimed (not proven) to be superior for accuracy despite its obvious limitations. It was shown why some biophysical allocation models are sensitive to the economic value of co-products and how it influences the ultimate share of total environmental burdens. We also mentioned about the dependency of this method on various parameters such as characteristics of the production system (breed of animals, variety of crops, etc.) and time and place of assessment which influence the accuracy of the results. As we discussed, this type of allocation does not explain the causal physical mechanisms of the physical parameters properly. Allocation methods based on mass, energy content, economic value are the most common allocation methods with less accuracy and in some cases with defects in proper allocation of environmental burdens. For instance, in economic allocation, the economic value of animal manure (in livestock production systems) and straw (in crop production system) is overlooked where it does not have any economic value; however, it is necessary to be considered as a by-product. Moreover, economic value of co-products is highly variable over time and space.

CLCA solves the multi-functionality problems by substitution, while ALCA usually deals with multi-functionality via allocation. CLCA identifies the environmental consequences of a change in demand for the FU, which is arguably more relevant for decision makers, whereas ALCA delimits the product system with reference to a physical or economic property. On this basis, ALCA results give an insight into the hotspots that can be used to improve the production system. Moreover, ALCA results can be used for comparing the environmental impacts of different production systems.

In agricultural LCAs, identifying the major impact categories and implementation of their corresponding results to the relevant production system are challenging issues. The outcomes of LCIA consist of a wide range of environmental burdens that need to be categorized into smaller groups with a limited number of environmental impact indicators. The most popular midpoint impact categories associated with agricultural production systems are climate change, acidification, eutrophication, toxicity, and land use. These are common midpoint categories, which further are classified into endpoint categories.

Ultimately, LCA results are influenced by the uncertainty inherently associated with the methodological choices and assumptions adopted, which is important to note when using LCA as a tool for decision making. However, LCA can illuminate the pros and cons of each alternative and thereby elucidate the trade-offs of choosing one alternative over another, by clarifying the understanding of the environmental impacts associated with the alternatives under assessment. We can recommend that the common practice in agricultural LCA be extended to include economic and social impacts as well as more indirect mechanisms (such as rebound, behaviour, price effects, and other dynamics) to overcome the shortcomings of existing LCA studies in the fields of agriculture. If so, the prospects of LCA in supporting sustainable agriculture are promising.

6 Abbreviations

AFOLU	Agriculture, Forestry and Land Use
NH_3	Ammonia
ALCA	Attributional Life Cycle Assessment
CO_2	Carbon Dioxide
CLCA	Consequential Life Cycle Assessment
dLUC	Direct Land Use Change
ECM	Energy Corrected Milk
EC-JRC	European Commission's Joint Research Centre
EP	Eutrophication Potential
FPCM	Fat and Protein Corrected Milk
FU	Functional Unit
GTP	Global Temperature Potential
GWP	Global Warming Potential
GHG	Greenhouse Gas
GDP	Gross Domestic Product
iLUC	Indirect Land Use Change
IO	Input-Output
IPCC	Intergovernmental Panel On Climate Change
IDF	International Dairy Federation
ISO	International Organisation For Standardisation
LUC	Land Use Change
LCA	Life Cycle Assessment
LCC	Life Cycle Costing
LCIA	Life Cycle Impact Assessment
LCI	Life Cycle Inventory
LCSA	Life Cycle Social Assessment
LCFS	Low Carbon Fuel Standard
CH_4	Methane
NO_3	Nitrate

NO	Nitrogen Oxide
N_2O	Nitrous Oxide
PEF	Product Environmental Footprint
RED	Renewable Energy Directive
SOC	Soil Organic Carbon
SOM	Soil Organic Matter
SO_2	Sulphur Dioxide
UNEP	United Nations Environment Programme

7 Where to look for further information

The chapter has been compiled in accordance with standards, databases, and guidelines. Major inputs for further information are:

- Existing and upcoming international standards on life cycle assessment, environmental labels and declarations, and eco-design (e.g. ISO 14040/14044, ISO/TS 14067, ISO1402X, ISO/TR 14062).
- IPCC Guidelines for national greenhouse gas inventories.
- International Reference Life Cycle Data System (ILCD) Handbook.
- Emerging national standards and technical agreements (e.g. PAS 2050, AFNOR BP X 30-323, GHG Protocol).
- Environmental Assessment of Food and Drink Protocol (ENVIFOOD).
- European Product Environmental Footprint (PEF).
- Livestock Environmental Assessment and Performance Partnership (LEAP), developed by FAO.
- Global Livestock Environmental Assessment Model (GLEAM) developed by FAO.
- Agri-footprint, developed by Blonk Consultants.

There are many organizations involved in the field of LCA in agriculture and food production. Some of them are listed as follows:

- ISO (https://www.iso.org/standards.html)
- FAO livestock and environment (http://www.fao.org/sustainability/en/)
- IPCC (https://www.ipcc.ch)
- International Dairy Federation (IDF) (https://www.fil-idf.org)
- The United Nations Environment Programme (UNEP) (https://www.unenvironment.org)
- Society of Environmental Toxicology and Chemistry (SETAC) (https://www.setac.org)
- European Commission Joint Research Center (https://ec.europa.eu/info/departments/joint-research-centre_en)

8 References

Acquaye, A. A., Wiedmann, T., Feng, K., Crawford, R. H., Barrett, J., Kuylenstierna, J., Duffy, A. P., Koh, S. C. L. and Mcqueen-Mason, S. 2011. Identification of 'carbon hot-spots' and quantification of GHG intensities in the biodiesel supply chain using hybrid LCA and structural path analysis. *Environmental Science and Technology* 45(6), 2471–8. doi:10.1021/es103410q.

Anderson, D. M. 1989. Toxic algal blooms and red tides: a global perspective. In: Okaichi, T., Anderson, D. M. and Nemoto, T. (Eds), *Red Tides: Biology, Environmental Science and Toxicology*. Elsevier, New York, pp. 11–16.

Angarita, E. E. Y., Lora, E. E. S., Da Costa, R. E. and Torres, E. A. 2009. The energy balance in the palm oil-derived methyl ester (PME) life cycle for the cases in Brazil and Colombia. *Renewable Energy* 34(12), 2905–13. doi:10.1016/j.renene.2009.05.007.

Anton, A., Montero, J. I., Munoz, P. and Castells, F. 2005. LCA and tomato production in Mediterranean greenhouses. *International Journal of Agricultural Resources, Governance and Ecology* 4(2), 102–12. doi:10.1504/IJARGE.2005.007192.

Ardente, F. and Cellura, M. 2012. Economic allocation in life cycle assessment: the state of the art and discussion of examples. *Journal of Industrial Ecology* 16(3), 387–98. doi:10.1111/j.1530-9290.2011.00434.x.

Ardente, F., Beccali, G., Cellura, M. and Marvuglia, A. 2006. POEMS: a case study of an Italian wine-producing firm. *Environmental Management* 38(3), 350–64. doi:10.1007/s00267-005-0103-8.

Audsley, E., Alber, S., Clift, R., Cowell, S., Crettaz, P., Gaillard, G., Hausheer, J., Jolliet, O., Kleijn, R. and Mortensen, B. 1997. Harmonisation of environmental life cycle assessment for agriculture. Final Report, Concerted Action AIR3-CT94-2028. European Commission, DG VI Agriculture, p. 139.

Ayer, N. W., Tyedmers, P. H., Pelletier, N. L., Sonesson, U. and Scholz, A. 2007. Co-product allocation in life cycle assessments of seafood production systems: review of problems and strategies. *The International Journal of Life Cycle Assessment* 12(7), 480–7. doi:10.1065/lca2006.11.284.

Azevedo, L. B., Van Zelm, R., Hendriks, A. J., Bobbink, R. and Huijbregts, M. A. J. 2013. Quantitative effects of soil pH on plant species richness: a global assessment. *Environmental Pollution* 174, 10–5. doi:10.1016/j.envpol.2012.11.001.

Basset-Mens, C. and Van Der Werf, H. M. G. 2005. Scenario-based environmental assessment of farming systems: the case of pig production in France. *Agriculture, Ecosystems and Environment* 105(1–2), 127–44. doi:10.1016/j.agee.2004.05.007.

Basset-Mens, C., Ledgard, S. and Boyes, M. 2009. Eco-efficiency of intensification scenarios for milk production in New Zealand. *Ecological Economics* 68(6), 1615–25. doi:10.1016/j.ecolecon.2007.11.017.

Beauchemin, K. A., Kreuzer, M., O'mara, F. and Mcallister, T. A. 2008. Nutritional management for enteric methane abatement: a review. *Australian Journal of Experimental Agriculture* 48(2), 21–7. doi:10.1071/EA07199.

Beusen, A. H. W., Klepper, O. and Meinardi, C. R. 1995. Modelling the flow of nitrogen and phosphorus in Europe: from loads to coastal seas. *Water Science and Technology* 31(8), 141–5. doi:10.2166/wst.1995.0281.

Blaser, P., Zysset, M., Zimmermann, S. and Luster, J. 1999. Soil acidification in southern Switzerland between 1987 and 1997: a case study based on the critical load concept. *Environmental Science and Technology* 33(14), 2383–9. doi:10.1021/es9808144.

Botha, T. and Von Blottnitz, H. 2006. A comparison of the environmental benefits of bagasse-derived electricity and fuel ethanol on a life-cycle basis. *Energy Policy* 34(17), 2654–61. doi:10.1016/j.enpol.2004.12.017.

Brandão, M. 2012. Food, feed, fuel, timber or carbon sink? Towards sustainable land use: a consequential life cycle approach. PhD thesis. University of Surrey, UK, 246p. Appendices, 541p.

Brandão, M. and Milà I Canals, L. 2013. Global characterisation factors to assess land use impacts on biotic production. *The International Journal of Life Cycle Assessment* 18(6), 1243–52. doi:10.1007/s11367-012-0381-3.

Brentrup, F., Küsters, J., Lammel, J. and Kuhlmann, H. 2000. Methods to estimate on-field nitrogen emissions from crop production as an input to LCA studies in the agricultural sector. *The International Journal of Life Cycle Assessment* 5(6), 349–57. doi:10.1007/BF02978670.

Bright, R. M., Strømman, A. H. and Hawkins, T. R. 2010. Environmental assessment of wood-based biofuel production and consumption scenarios in Norway. *Journal of Industrial Ecology* 14(3), 422–39. doi:10.1111/j.1530-9290.2010.00249.x.

Caffrey, K. R. and Veal, M. W. 2013a. Conducting an agricultural life cycle assessment: challenges and perspectives. *The Scientific World Journal* 2013, 472431. doi:10.1155/2013/472431.

Caffrey, K. R. and Veal, M. W. 2013b. Conducting an agricultural life cycle assessment: challenges and perspectives. *The Scientific World Journal* 2013, Article ID 472431, 1–13.

Casey, J. W. and Holden, N. M. 2005. Analysis of greenhouse gas emissions from the average Irish milk production system. *Agricultural Systems* 86(1), 97–114. doi:10.1016/j.agsy.2004.09.006.

Casey, J. W. and Holden, N. M. 2006. Greenhouse gas emissions from conventional, agri-environmental scheme, and organic Irish suckler-beef units. *Journal of Environmental Quality* 35(1), 231-9. doi:10.2134/jeq2005.0121.

Cavalett, O. and Ortega, E. 2010. Integrated environmental assessment of biodiesel production from soybean in Brazil. *Journal of Cleaner Production* 18(1), 55–70. doi:10.1016/j.jclepro.2009.09.008.

Cederberg, C. and Darelius, K. 2002. Using LCA methodology to assess the potential environmental impact of intensive beef and pork production. PhD thesis. Department of Applied Environmental Science, Göteborg University, Sweden.

Cederberg, C. and Flysjö, A. 2004. *Life Cycle Inventory of 23 Dairy Farms in South-Western Sweden*. SIK Institutet för livsmedel och bioteknik - The Swedish Institute for Food and Biotechnology, Göteborg, Sweden.

Cederberg, C. and Mattsson, B. 2000. Life cycle assessment of milk production–a comparison of conventional and organic farming. *Journal of Cleaner Production* 8(1), 49–60. doi:10.1016/S0959-6526(99)00311-X.

Cederberg, C. and Stadig, M. 2003. System expansion and allocation in life cycle assessment of milk and beef production. *The International Journal of Life Cycle Assessment* 8(6), 350–6. doi:10.1007/BF02978508.

Cellura, M., Ardente, F. and Longo, S. 2012. From the LCA of food products to the environmental assessment of protected crops districts: a case-study in the south of Italy. *Journal of Environmental Management* 93(1), 194–208. doi:10.1016/j.jenvman.2011.08.019.

Cerutti, A. K., Bruun, S., Donno, D., Beccaro, G. L. and Bounous, G. 2013. Environmental sustainability of traditional foods: the case of ancient apple cultivars in Northern Italy assessed by multifunctional LCA. *Journal of Cleaner Production* 52, 245-52. doi:10.1016/j.jclepro.2013.03.029.

Cherubini, E., Zanghelini, G. M., Alvarenga, R. A. F., Franco, D. and Soares, S. R. 2015. Life cycle assessment of swine production in Brazil: a comparison of four manure management systems. *Journal of Cleaner Production* 87, 68-77. doi:10.1016/j.jclepro.2014.10.035.

Chiaramonti, D. and Recchia, L. 2010. Is life cycle assessment (LCA) a suitable method for quantitative CO_2 saving estimations? The impact of field input on the LCA results for a pure vegetable oil chain. *Biomass and Bioenergy* 34(5), 787-97. doi:10.1016/j.biombioe.2010.01.022.

Chobtang, J., McLaren, S. J., Ledgard, S. F. and Donaghy, D. J. 2017. Consequential life cycle assessment of pasture-based milk production: a case study in the Waikato Region, New Zealand. *Journal of Industrial Ecology* 21(5), 1139-52. doi:10.1111/jiec.12484.

Crawford, R. H., Bontinck, P. A., Stephan, A., Wiedmann, T. and Yu, M. 2018. Hybrid life cycle inventory methods - a review. *Journal of Cleaner Production* 172, 1273-88. doi:10.1016/j.jclepro.2017.10.176.

Curran, M. A. 2015. *Life Cycle Assessment Student Handbook*. John Wiley & Sons, UK.

Dahlgren, L., Stripple, H. and Oliveira, F. 2015. Life cycle assessment, Comparative study of virgin fibre based packaging products with competing plastic materials. Report number: U 5052. IVL Swedish Environmental Research Institute Ltd.

Dalgaard, R., Halberg, N. and Hermansen, J. E. 2007. Danish pork production: an environmental assessment. DJF Animal Science No. 82. University of Aarhus, Tjele.

Dalgaard, R., Schmidt, J., Halberg, N., Christensen, P., Thrane, M. and Pengue, W. A. 2008. LCA of soybean meal. *The International Journal of Life Cycle Assessment* 13(3), 240-54. doi:10.1065/lca2007.06.342.

De Jong, P. 2013. *Sustainable Dairy Production* (1st edn.). John Wiley & Sons, Ltd, UK. doi:10.1002/9781118489451.

De Luca, A. I., Iofrida, N., Strano, A., Falcone, G. and Gulisano, G. 2015. Social life cycle assessment and participatory approaches: a methodological proposal applied to citrus farming in Southern Italy. *Integrated Environmental Assessment and Management* 11(3), 383-96. doi:10.1002/ieam.1611.

De Vries, M. and De Boer, I. J. M. 2010. Comparing environmental impacts for livestock products: a review of life cycle assessments. *Livestock Science* 128(1-3), 1-11. doi:10.1016/j.livsci.2009.11.007.

Dijkman, T. J., Basset-Mens, C., Antón, A. and Núñez, M. 2018. LCA of food and agriculture. In: Hauschild, M., Rosenbaum, R. and Olsen, S. (Eds), *Life Cycle Assessment*. Springer, Cham.

Dollé, J. B. and Gac, A. 2012. Milk and meat biophysical allocation in dairy farms. In: *8th International Conference on LCA in the Agri-Food Sector*, 2012, Rennes, France. INRA, pp. 665-7.

Dreyer, L., Hauschild, M. and Schierbeck, J. 2006. A framework for social life cycle impact assessment (10 pp). *The International Journal of Life Cycle Assessment* 11(2), 88-97. doi:10.1065/lca2005.08.223.

EC-JRC. 2010. Analysis of existing environmental impact assessment methodologies for use in life cycle assessment. In: *ILCD Handbook – International Reference Life Cycle Data System*. European Commission Joint Research Center, Italy. Available at: https://eplca.jrc.ec.europa.eu/uploads/ILCD-Handbook-LCIA-Background-analysis-online-12March2010.pdf (accessed on 01 February 2019).

EC-JRC. 2011. Recommendations for life cycle impact assessment in the European context. In: *ILCD Handbook - International Reference Life Cycle Data System*. European Commission Joint Research Center, Italy. Available at: https://eplca.jrc.ec.europa.eu/uploads/ILCD-Recommendation-of-methods-for-LCIA-def.pdf (accessed on 01 February 2019).

Edenhofer, O., Pichs-Madruga, R., Sokona, Y., Seyboth, K., Matschoss, P., Kadner, S., Zwickel, T., Eickemeier, P., Hansen, G. and Schlömer, S. 2011. IPCC special report on renewable energy sources and climate change mitigation. Prepared By *Working Group III of the Intergovernmental Panel on Climate Change*. Cambridge University Press, Cambridge, UK.

EEA. 1998. *Europe's Environment: the Second Assessment: An Overview*. Office for Official Publications of the European Communities, European Environment Agency (EEA).

Eide, M. H. 2002. Life cycle assessment (LCA) of industrial milk production. *The International Journal of Life Cycle Assessment* 7(2), 115–26. doi:10.1007/BF02978855.

Ekvall, T. and Finnveden, G. 2001. Allocation in ISO 14041—a critical review. *Journal of Cleaner Production* 9(3), 197–208. doi:10.1016/S0959-6526(00)00052-4.

Ekvall, T. and Weidema, B. P. 2004. System boundaries and input data in consequential life cycle inventory analysis. *The International Journal of Life Cycle Assessment* 9(3), 161–71. doi:10.1007/BF02994190.

Eriksson, I. S., Elmquist, H., Stern, S. and Nybrant, T. 2005. Environmental systems analysis of pig production-the impact of feed choice (12 pp). *The International Journal of Life Cycle Assessment* 10(2), 143–54. doi:10.1065/lca2004.06.160.

Eriksson, O., Finnveden, G., Ekvall, T. and Björklund, A. 2007. Life cycle assessment of fuels for district heating: a comparison of waste incineration, biomass-and natural gas combustion. *Energy Policy* 35(2), 1346–62. doi:10.1016/j.enpol.2006.04.005.

Fantin, V., Righi, S., Rondini, I. and Masoni, P. 2017. Environmental assessment of wheat and maize production in an Italian farmers' cooperative. *Journal of Cleaner Production* 140, 631–43. doi:10.1016/j.jclepro.2016.06.136.

FAO. 2006. *Livestock's Long Shadow*. Food and Agriculture Organization of the United Nations. Available at: http://www.fao.org/docrep/010/a0701e/a0701e.pdf (accessed on 01 February 2019).

FAO. 2014. Greenhouse gas emissions and fossil energy demand from poultry supply chains. Online. Available at: http://www.fao.org/3/a-mj752e.pdf (accessed on 01 February 2019).

Fazio, S. and Monti, A. 2011. Life cycle assessment of different bioenergy production systems including perennial and annual crops. *Biomass and Bioenergy* 35(12), 4868–78. doi:10.1016/j.biombioe.2011.10.014.

Ferng, J. J. 2003. Allocating the responsibility of CO_2 over-emissions from the perspectives of benefit principle and ecological deficit. *Ecological Economics* 46(1), 121–41. doi:10.1016/S0921-8009(03)00104-6.

Feschet, P., Macombe, C., Garrabé, M., Loeillet, D., Saez, A. R. and Benhmad, F. 2013. Social impact assessment in LCA using the Preston pathway. *The International Journal of Life Cycle Assessment* 18(2), 490–503. doi:10.1007/s11367-012-0490-z.

Finkbeiner, M. 2014. Indirect land use change - Help beyond the hype? *Biomass and Bioenergy* 62, 218-21. doi:10.1016/j.biombioe.2014.01.024.

Finkbeiner, M. 2016. *Special Types of Life Cycle Assessment*. Springer, Dordrecht. doi:10.1007/978-94-017-7610-3.

Finkbeiner, M., Schau, E. M., Lehmann, A. and Traverso, M. 2010. Towards life cycle sustainability assessment. *Sustainability* 2(10), 3309-22. doi:10.3390/su2103309.

Finlayson, M., Cruz, R. D., Davidson, N., Alder, J., Cork, S., De Groot, R. S., Lévêque, C., Milton, G. R., Peterson, G. and Pritchard, D. 2005. *Millennium Ecosystem Assessment: Ecosystems and Human Well-Being: Wetlands and Water Synthesis*. Island Press, Washington DC.

Finnveden, G. 2000. On the limitations of life cycle assessment and environmental systems analysis tools in general. *The International Journal of Life Cycle Assessment* 5(4), 229-38. doi:10.1007/BF02979365.

Finnveden, G. and Potting, J. 1999. Eutrophication as an impact category. State of the art and research needs. *The International Journal of Life Cycle Assessment* 4(6), 311-4. doi:10.1007/BF02978518.

Finnveden, G., Hauschild, M. Z., Ekvall, T., Guinée, J., Heijungs, R., Hellweg, S., Koehler, A., Pennington, D. and Suh, S. 2009. Recent developments in life cycle assessment. *Journal of Environmental Management* 91(1), 1-21. doi:10.1016/j.jenvman.2009.06.018.

Franze, J. and Ciroth, A. 2011. A comparison of cut roses from Ecuador and the Netherlands. *The International Journal of Life Cycle Assessment* 16(4), 366-79. doi:10.1007/s11367-011-0266-x.

Frischknecht, R. 1998. Life cycle inventory analysis for decision-making. *The International Journal of Life Cycle Assessment* 3(2), 67-. doi:10.1007/BF02978487.

Frischknecht, R. 2000. Allocation in life cycle inventory analysis for joint production. *The International Journal of Life Cycle Assessment* 5(2), 85-95. doi:10.1007/BF02979729.

Garnett, T. 2009. Livestock-related greenhouse gas emissions: impacts and options for policy makers. *Environmental Science and Policy* 12(4), 491-503. doi:10.1016/j.envsci.2009.01.006.

Gerber, P., Vellinga, T., Opio, C. and Steinfeld, H. 2011. Productivity gains and greenhouse gas emissions intensity in dairy systems. *Livestock Science* 139(1-2), 100-8. doi:10.1016/j.livsci.2011.03.012.

Gibbs, H. K., Johnston, M., Foley, J. A., Holloway, T., Monfreda, C., Ramankutty, N. and Zaks, D. 2008. Carbon payback times for crop-based biofuel expansion in the tropics: the effects of changing yield and technology. *Environmental Research Letters* 3(3). doi:10.1088/1748-9326/3/3/034001.

Gibon, T. and Schaubroeck, T. 2017. Lifting the fog on characteristics and limitations of hybrid LCA – a reply to "Does hybrid LCA with a complete system boundary yield adequate results for product promotion?" by Yi Yang (Int J Life Cycle Assess 22 (3): 456-406). *The International Journal of Life Cycle Assessment* 22, 1005-8.

Goedkoop, M. and Spriensma, R. 2001. The Eco-indicator 99: a damage oriented method for life cycle impact assessment. PRé Consultants. Available at: https://www.pre-sustainability.com/download/EI99_annexe_v3.pdf (accessed on 01 February 2019).

Goulding, K. W. T. 2016. Soil acidification and the importance of liming agricultural soils with particular reference to the United Kingdom. *Soil Use and Management* 32(3), 390-9. doi:10.1111/sum.12270.

Guinée, J. 1996. Data for the normalization step within life cycle assessment of products (revised version). CML paper, 14, CML (Centre of Environmental Science), Leiden.

Guinée, J. B., Gorrée, M., Heijungs, R., Huppes, G., Kleijn, R., De Koning, A., Van Oers, L., Wegener Sleeswijk, A., Suh, S., Udo De Haes, H. A., J.A., D. B., Van Duin, R. and Huijbregts, M. A. J. 2002. *Handbook on Life Cycle Assessment: Operational Guide to the ISO Standards*. Springer, Dordrecht, pp. 1-692. doi:10.1007/0-306-48055-7.

Haas, G., Wetterich, F. and Geier, U. 2000. Life cycle assessment framework in agriculture on the farm level. *The International Journal of Life Cycle Assessment* 5(6), 345–8. doi:10.1007/BF02978669.

Haas, G., Wetterich, F. and Köpke, U. 2001. Comparing intensive, extensified and organic grassland farming in southern Germany by process life cycle assessment. *Agriculture, Ecosystems and Environment* 83(1–2), 43–53. doi:10.1016/S0167-8809(00)00160-2.

Hanegraaf, M. C., Biewinga, E. E. and van derBijl, G. 1998. Assessing the ecological and economic sustainability of energy crops. *Biomass and Bioenergy* 15(4–5), 345–55. doi:10.1016/S0961-9534(98)00042-7.

Harding, K. G., Dennis, J. S., Von Blottnitz, H. and Harrison, S. T. L. 2008. A life-cycle comparison between inorganic and biological catalysis for the production of biodiesel. *Journal of Cleaner Production* 16(13), 1368–78. doi:10.1016/j.jclepro.2007.07.003.

Harris, S. and Narayanaswamy, V. 2009. *A Literature Review of Life Cycle Assessment in Agriculture*. Publication No. 09/029. RIRDC, pp. 1–45.

Hauschild, M. Z. and Huijbregts, M. A. J. 2015. Introducing life cycle impact assessment. In: Hauschild, M. Z. and Huijbregts, M. A. J. (Eds), *Life Cycle Impact Assessment*. Springer, Dordrecht, pp. 1–16. doi:10.1007/978-94-017-9744-3.

Hayashi, K. 2005. Environmental indicators for agricultural management: the problem of integration. In: Kolowrocki, K. (Ed.), *Advances in Safety and Reliability: ESREL 2005*. Taylor & Francis Group, London, pp. 821–30.

Hayashi, K., Okazaki, M., Itsubo, N. and Inaba, A. 2004. Development of damage function of acidification for terrestrial ecosystems based on the effect of aluminum toxicity on net primary production. *The International Journal of Life Cycle Assessment* 9(1), 13–22. doi:10.1007/BF02978532.

Heijungs, R. 1997. *Economic Drama and the Environmental Stage: Formal Derivation of Algorithmic Tools for Environmental Analysis and Decision-Support from a Unified Epistemological Principle*. Centre of Environmental Science, Leiden University, Leiden.

Heijungs, R. and Suh, S. 2002. *The Computational Structure of Life Cycle Assessment*. Springer Science & Business Media, Kluwer Academic Publishers, Dordrecht.

Heller, M. C., Keoleian, G. A. and Volk, T. A. 2003. Life cycle assessment of a willow bioenergy cropping system. *Biomass and Bioenergy* 25(2), 147–65. doi:10.1016/S0961-9534(02)00190-3.

Hertel, T. W., Tyner, W. E. and Birur, D. K. 2010. The global impacts of biofuel mandates. *The Energy Journal* 31(1), 75–100. doi:10.5547/ISSN0195-6574-EJ-Vol31-No1-4.

Holme, R. and Watts, P. 2000. *Corporate Social Responsibility: Making Good Business Sense*. The World Business Council for Sustainable Development, Geneva.

Hospido, A. and Sonesson, U. 2005. The environmental impact of mastitis: a case study of dairy herds. *The Science of the Total Environment* 343(1–3), 71–82. doi:10.1016/j.scitotenv.2004.10.006.

Hospido, A., Moreira, M. T. and Feijoo, G. 2003. Simplified life cycle assessment of Galician milk production. *International Dairy Journal* 13(10), 783–96. doi:10.1016/S0958-6946(03)00100-6.

Hu, Z., Tan, P., Yan, X. and Lou, D. 2008. Life cycle energy, environment and economic assessment of soybean-based biodiesel as an alternative automotive fuel in China. *Energy* 33(11), 1654–8. doi:10.1016/j.energy.2008.06.004.

Huguenin-Elie, O. and Nemecek, T. 2014. Life cycle assessment of roughage production. Land use systems in grassland dominated regions. In: *Proceedings of the 20th General Meeting of the European Grassland Federation*, Luzern, Switzerland, 21–24 June 2004. vdf Hochschulverlag AG an der ETH Zurich, pp. 772–4.

Huijbregts, M. 2001. Uncertainty and variability in environmental life-cycle assessment. PhD thesis. University of Amsterdam, Amsterdam.

Huijbregts, M. A. J., Norris, G., Bretz, R., Ciroth, A., Maurice, B., Von Bahr, B., Weidema, B. and De Beaufort, A. S. H. 2001a. Framework for modelling data uncertainty in life cycle inventories. *The International Journal of Life Cycle Assessment* 6(3), 127–32. doi:10.1007/BF02978728.

Huijbregts, M. A. J., Schöpp, W., Verkuijlen, E., Heijungs, R. and Reijnders, L. 2001b. Spatially explicit characterization of acidifying and eutrophying air pollution in life-cycle assessment. *Journal of Industrial Ecology* 4(3), 75–92. doi:10.1162/108819800300106393.

Huijbregts, M. A. J., Struijs, J., Goedkoop, M., Heijungs, R., Hendriks, A. J. and Van De Meent, D. 2005. Human population intake fractions and environmental fate factors of toxic pollutants in life cycle impact assessment. *Chemosphere* 61(10), 1495–504. doi:10.1016/j.chemosphere.2005.04.046.

Huijbregts, M. A. J., Steinmann, Z. J. N., Elshout, P. M. F., Stam, G., Verones, F., Vieira, M., Zijp, M., Hollander, A. and Van Zelm, R. 2017. ReCiPe2016: a harmonised life cycle impact assessment method at midpoint and endpoint level. *The International Journal of Life Cycle Assessment* 22(2), 138–47. doi:10.1007/s11367-016-1246-y.

IDF. 2010. A common carbon footprint approach for dairy - the IDF guide to standard lifecycle assessment methodology. Bulletin of the International Dairy Federation 445/2010. International Dairy Federation (IDF), p. 445. Available at: http://www.idf-lca-guide.org/Files/media/Documents/445-2010-A-common-carbon-footprint-approach-for-dairy.pdf (accessed on 08 June 2018).

IPCC. 2013. *Climate Change 2013: The Physical Science Basis. Contribution of Working Group I to the fifth Assessment Report of the Intergovernmental Panel on Climate Change*. Stocker, T. F., Qin, D., Plattner, G. K., Tignor, M., Allen, S. K., Boschung, J., Nauels, A., Xia, Y., Bex, V. and Midgley, P. M. (Eds). Cambridge University Press, Cambridge, UK and New York, NY.

IPCC. 2014. *Climate Change 2014: Impacts, Adaptation, and Vulnerability. Part A: Global and Sectoral Aspects. Contribution of Working Group II to the Fifth Assessment Report of the Intergovernmental Panel on Climate Change*. Field, C. B., Barros, V. R., Dokken, D. J., Mach, K. J., Mastrandrea, M. D., Bilir, T. E., Chatterjee, M., Ebi, K. L., Estrada, Y. O., Genova, R. C., Girma, B., Kissel, E. S., Levy, A. N., MacCracken, S., Mastrandrea, P. R. and White, L. L. (Eds). Cambridge University Press, Cambridge, UK and New York, NY.

ISO. 2006a. *ISO 14040:2006 Environmental Management - Life Cycle Assessment - Principles and Framework* (2nd edn). Technical Committee of International Organization for Standardization.

ISO. 2006b. *ISO 14044:2006 Environmental management - Life Cycle Assessment - Requirements and Guidelines*. Technical Committee of International Organization for Standardization.

ISO. 2010. *ISO 26000:2010 Guidance on Social Responsibility*. Technical Committee of International Organization for Standardization.

Jansen, K. and Vellema, S. 2004. *Agribusiness and Society: Corporate Responses to Environmentalism, Market Opportunities and Public Regulation*. Zed Books, London, UK and New York, NY.

Jolliet, O. and Crettaz, P. 1997. Fate coefficients for the toxicity assessment of air pollutants. *The International Journal of Life Cycle Assessment* 2(2), 104–10. doi:10.1007/BF02978768.

Joshi, S. 1999. Product environmental life-cycle assessment using input-output techniques. *Journal of Industrial Ecology* 3(2–3), 95–120. doi:10.1162/108819899569449.

Karlsson, M. and Sund, L. 2016. Life cycle assessments of arable land use options and protein feeds: a comparative study investigating the climate impact from different scenarios in the agricultural sector. Master thesis at Department of Management and Engineering, Linköping University, Sweden.

Katajajuuri, J. M. 2007. Experiences and improvement possibilities–LCA case study of broiler chicken production. In: *3rd International Conference on Life Cycle Management*, Zurich, Switzerland, 2007, pp. 27–9.

Kim, S. and Dale, B. E. 2008. Life cycle assessment of fuel ethanol derived from corn grain via dry milling. *Bioresource Technology* 99(12), 5250–60. doi:10.1016/j.biortech.2007.09.034.

Klöpffer, W. and Grahl, B. 2014. *Life Cycle Assessment (LCA): a Guide to Best Practice*. Wiley-VCH, Weinheim,Germany.

Klöpffer, W. and Renner, I. 2008. Life-cycle based sustainability assessment of products. In: Schaltegger, S., Bennett, M., Burritt, R. L. and Jasch, C. (Eds), *Environmental Management Accounting for Cleaner Production*. Springer, Dordrecht, pp. 91–102. doi:10.1007/978-1-4020-8913-8.

Kløverpris, J. H., Baltzer, K. and Nielsen, P. H. 2010. Life cycle inventory modelling of land use induced by crop consumption. Part 2: example of wheat consumption in Brazil, China, Denmark and the USA. *International Journal of Life Cycle Assessment* 15, 90–103.

Koellner, T., De Baan, L., Beck, T., Brandão, M., Civit, B., Margni, M., Canals, L. M. I., Saad, R., De Souza, D. M. and Müller-Wenk, R. 2013. UNEP-SETAC guideline on global land use impact assessment on biodiversity and ecosystem services in LCA. *The International Journal of Life Cycle Assessment* 18(6), 1188–202. doi:10.1007/s11367-013-0579-z.

Lam, M. K., Lee, K. T. and Mohamed, A. R. 2009. Life cycle assessment for the production of biodiesel: a case study in Malaysia for palm oil versus jatropha oil. *Biofuels, Bioproducts and Biorefining: Innovation for a Sustainable Economy* 3(6), 601–12. doi:10.1002/bbb.182.

Lardon, L., Helias, A., Sialve, B., Steyer, J. P. and Bernard, O. 2009. Life-Cycle Assessment of biodiesel production from microalgae. *Environmental Science and Technology* 43(17), 6475–81.

Lee, K.-M. and Inaba, A. 2004. *Life Cycle Assessment: Best Practices of ISO 14040 Series*. Center for Ecodesign and LCA (CEL), Ajou University, Korea.

Lenzen, M. and Crawford, R. 2009. The path exchange method for hybrid LCA. *Environmental Science and Technology* 43(21), 8251–6. doi:10.1021/es902090z.

Lenzen, M., Murray, J., Sack, F. and Wiedmann, T. 2007. Shared producer and consumer responsibility – theory and practice. *Ecological Economics* 61(1), 27–42. doi:10.1016/j.ecolecon.2006.05.018.

Levasseur, A., De Schryver, A., Hauschild, M., Kabe, Y., Sahnoune, A., Tanaka, K. and Cherubini, F. 2017. Greenhouse gas emissions and climate change impacts. In: Frischknecht, R. and Jolliet, O. (Eds), *Global Guidance for Life Cycle Assessment Indicators*. United Nations Environment Programme, pp. 58–75.

Mackenzie, S. G., Leinonen, I. and Kyriazakis, I. 2017. The need for co-product allocation in the life cycle assessment of agricultural systems–is "biophysical" allocation progress? *The International Journal of Life Cycle Assessment* 22(2), 128–37. doi:10.1007/s11367-016-1161-2.

Manfredi, S., Allacker, K., Pelletier, N., Chomkhamsri, K. and De Souza, D. M. 2012. Product environmental footprint (PEF) guide. Ref. Ares(2012)873782. European Commission Joint Research Center (EC-JRC), Institute for Environment and Sustainability (IES).

Mankiw, G. and Taylor, M. P. 2006. *Economics*. Thomson Learning, London.

Margni, M. and Curran, M. A. 2012. Life cycle impact assessment. In: Curran, M. A. (Ed.), *Life Cycle Assessment Handbook: a Guide for Environmentally Sustainable Products*. Scrivener Publishing, Wiley, pp. 67–104.

Mattila, T. J. 2018. Use of input-output analysis in LCA. In: Hauschild, M. Z., Rosenbaum, R. K. and Olsen, S. I. (Eds), *Life Cycle Assessment*. Springer, Cham, pp. 349–72. doi:10.1007/978-3-319-56475-3.

Mattila, T. J., Helin, T., Antikainen, R., Soimakallio, S., Pingoud, K. and Wessman, H. 2011. Landuse in life cycle assessment. The Finnish Environment Institute 24. Available at: https://helda.helsinki.fi/bitstream/handle/10138/37049/FE_24_2011.pdf?sequence=3&isAllowed=y.

Mazur-Wierzbicka, E. 2015. The application of corporate social responsibility in European agriculture. *Miscellanea Geographica* 19(1), 19–23. doi:10.1515/mgrsd-2015-0001.

Milà i Canals, L., Romanyà, J. and Cowell, S. J. 2007. Method for assessing impacts on life support functions (LSF) related to the use of 'fertile land' in Life Cycle Assessment (LCA). *Journal of Cleaner Production* 15(15), 1426–40. doi:10.1016/j.jclepro.2006.05.005.

Mollenhorst, H., Berentsen, P. B. and De Boer, I. J. 2006. On-farm quantification of sustainability indicators: an application to egg production systems. *British Poultry Science* 47(4), 405–17. doi:10.1080/00071660600829282.

Monti, A., Fazio, S. and Venturi, G. 2009. Cradle-to-farm gate life cycle assessment in perennial energy crops. *European Journal of Agronomy* 31(2), 77–84. doi:10.1016/j.eja.2009.04.001.

Moudrý, J. J. and Moudrý, J. 2014. Environmental aspects of organic farming. In: Pilipavicius, V. (Ed.), *Organic Agriculture Towards Sustainability*. IntechOpen.

Mouron, P., Nemecek, T., Scholz, R. W. and Weber, O. 2006a. Management influence on environmental impacts in an apple production system on Swiss fruit farms: combining life cycle assessment with statistical risk assessment. *Agriculture, Ecosystems and Environment* 114(2–4), 311–22. doi:10.1016/j.agee.2005.11.020.

Mouron, P., Scholz, R. W., Nemecek, T. and Weber, O. 2006b. Life cycle management on Swiss fruit farms: relating environmental and income indicators for apple-growing. *Ecological Economics* 58(3), 561–78. doi:10.1016/j.ecolecon.2005.08.007.

Ndong, R., Montrejaud-Vignoles, M., Saint Girons, O., Gabrielle, B., Pirot, R., Domergue, M. and Sablayrolles, C. 2009. Life cycle assessment of biofuels from *Jatropha curcas* in West Africa: a field study. *GCB Bioenergy* 1(3), 197–210. doi:10.1111/j.1757-1707.2009.01014.x.

Nemecek, T., Frick, C., Dubois, D. and Gaillard, G. 2001. Comparing farming systems at crop rotation level by LCA. In: *Proceedings of the International Conference on LCA in Foods*, Gothenburg. SIK, VITO, Gothenburg, pp. 65–9.

Nemecek, T., Dubois, D., Huguenin-Elie, O. and Gaillard, G. 2011. Life cycle assessment of Swiss farming systems: I. Integrated and organic farming. *Agricultural Systems* 104(3), 217–32. doi:10.1016/j.agsy.2010.10.002.

Nienhuis, I. J. K. and De Vreede, I. P. J. A. 1996. Utility of the environmental life cycle assessment method in horticulture. In: *XIII International Symposium on Horticultural Economics 429*, 1996, pp. 531–8. doi:10.17660/ActaHortic.1996.429.69.

Norris, G. A. 2002. Impact characterization in the tool for the reduction and assessment of chemical and other environmental impacts: methods for acidification, eutrophication, and ozone formation. *Journal of Industrial Ecology* 6(3–4), 79–101. doi:10.1162/108819802766269548.

Paerl, H. W., Fulton, R. S., Moisander, P. H. and Dyble, J. 2001. Harmful freshwater algal blooms, with an emphasis on cyanobacteria. *The Scientific World Journal* 1, 76–113. doi:10.1100/tsw.2001.16.

Palma-Rojas, S., Caldeira-Pires, A. and Nogueira, J. M. 2017. Environmental and economic hybrid life cycle assessment of bagasse-derived ethanol produced in Brazil. *The International Journal of Life Cycle Assessment* 22(3), 317–27. doi:10.1007/s11367-015-0892-9.

Pedersen, O. G. and De Haan, M. 2006. The system of environmental and economic accounts–2003 and the economic relevance of physical flow accounting. *Journal of Industrial Ecology* 10(1–2), 19–42. doi:10.1162/108819806775545466.

Pelletier, N. 2018. Social sustainability assessment of Canadian egg production facilities: methods, analysis, and recommendations. *Sustainability* 10(5), 1601. doi:10.3390/su10051601.

Pelletier, N. and Tyedmers, P. 2011. An ecological economic critique of the use of market information in life cycle assessment research. *Journal of Industrial Ecology* 15(3), 342–54. doi:10.1111/j.1530-9290.2011.00337.x.

Pelletier, N., Pirog, R. and Rasmussen, R. 2010. Comparative life cycle environmental impacts of three beef production strategies in the Upper Midwestern United States. *Agricultural Systems* 103(6), 380–9. doi:10.1016/j.agsy.2010.03.009.

Pelletier, N., Ardente, F., Brandão, M., De Camillis, C. and Pennington, D. 2015. Rationales for and limitations of preferred solutions for multi-functionality problems in LCA: is increased consistency possible? *The International Journal of Life Cycle Assessment* 20(1), 74–86. doi:10.1007/s11367-014-0812-4.

Peters, G. M., Rowley, H. V., Wiedemann, S. G., Tucker, R. W., Short, M. D. and Schulz, M. 2010a. Red meat production in Australia: life cycle assessment and comparison with overseas studies. *Environmental Science and Technology* 44(4), 1327–32. doi:10.1021/es901131e.

Peters, G. M., Wiedemann, S. G., Rowley, H. V. and Tucker, R. W. 2010b. Accounting for water use in Australian red meat production. *The International Journal of Life Cycle Assessment* 15(3), 311–20. doi:10.1007/s11367-010-0161-x.

Petti, L., Ramirez, P. K. S., Traverso, M. and Ugaya, C. M. L. 2018. An Italian tomato "Cuore di Bue" case study: challenges and benefits using subcategory assessment method for social life cycle assessment. *The International Journal of Life Cycle Assessment* 23(3), 569–80. doi:10.1007/s11367-016-1175-9.

Pirlo, G. and Lolli, S. 2019. Environmental impact of milk production from samples of organic and conventional farms in Lombardy (Italy). *Journal of Cleaner Production* 211, 962–71. doi:10.1016/j.jclepro.2018.11.070.

Pishgar-Komleh, S. H., Omid, M. and Heidari, M. D. 2013. On the study of energy use and GHG (greenhouse gas) emissions in greenhouse cucumber production in Yazd Province. *Energy* 59, 63–71. doi:10.1016/j.energy.2013.07.037.

Pishgar-Komleh, S. H., Akram, A., Keyhani, A., Raei, M., Elshout, P. M. F., Huijbregts, M. A. J. and Van Zelm, R. 2017a. Variability in the carbon footprint of open-field tomato production in Iran-A case study of Alborz and East-Azerbaijan provinces. *Journal of Cleaner Production* 142, 1510–7. doi:10.1016/j.jclepro.2016.11.154.

Pishgar-Komleh, S. H., Akram, A., Keyhani, A. and Van Zelm, R. 2017b. Life cycle energy use, costs, and greenhouse gas emission of broiler farms in different production systems in Iran–a case study of Alborz Province. *Environmental Science and Pollution Research* 24(19), 16041–9. doi:10.1007/s11356-017-9255-3.

Pizzol, M., Laurent, A., Sala, S., Weidema, B., Verones, F. and Koffler, C. 2017. Normalisation and weighting in life cycle assessment: quo vadis? *The International Journal of Life Cycle Assessment* 22(6), 853–66. doi:10.1007/s11367-016-1199-1.

Plevin, R. J., Beckman, J., Golub, A. A., Witcover, J. and O'hare, M. 2015. Carbon accounting and economic model uncertainty of emissions from biofuels-induced land use change. *Environmental Science and Technology* 49(5), 2656–64. doi:10.1021/es505481d.

Pomponi, F. and Lenzen, M. 2018. Hybrid life cycle assessment (LCA) will likely yield more accurate results than process-based LCA. *Journal of Cleaner Production* 176, 210–5. doi:10.1016/j.jclepro.2017.12.119.

Posch, M., Seppälä, J., Hettelingh, J.-P., Johansson, M., Margni, M. and Jolliet, O. 2008. The role of atmospheric dispersion models and ecosystem sensitivity in the determination of characterisation factors for acidifying and eutrophying emissions in LCIA. *The International Journal of Life Cycle Assessment* 13(6), 477–86. doi:10.1007/s11367-008-0025-9.

Potting, J. and Hauschild, M. Z. 2006. Spatial differentiation in life cycle impact assessment: a decade of method development to increase the environmental realism of LCIA. *The International Journal of Life Cycle Assessment* 11(S1), 11–3. doi:10.1065/lca2006.04.005.

Potting, J., Schöpp, W., Blok, K. and Hauschild, M. 1998. Site-dependent life-cycle impact assessment of acidification. *Journal of Industrial Ecology* 2(2), 63–87. doi:10.1162/jiec.1998.2.2.63.

Rebitzer, G., Ekvall, T., Frischknecht, R., Hunkeler, D., Norris, G., Rydberg, T., Schmidt, W. P., Suh, S., Weidema, B. P. and Pennington, D. W. 2004. Life cycle assessment: part 1: framework, goal and scope definition, inventory analysis, and applications. *Environment International* 30(5), 701–20. doi:10.1016/j.envint.2003.11.005.

Reckmann, K. 2013. Life Cycle Assessment of pork especially emphasising feed and pig production. PhD thesis. Selbstverl. des Inst. für Tierzucht und Tierhaltung der Christian-Albrechts-Univ. zu Kiel.

Revéret, J. P., Couture, J. M. and Parent, J. 2015. Socioeconomic LCA of milk production in Canada. In: Muthu, S. S. (Ed.), *Social Life Cycle Assessment*. Springer, Singapore, pp. 25–69. doi:10.1007/978-981-287-296-8.

Rodrigues, J., Domingos, T., Giljum, S. and Schneider, F. 2006. Designing an indicator of environmental responsibility. *Ecological Economics* 59(3), 256–66. doi:10.1016/j.ecolecon.2005.10.002.

Rosenbaum, R. K., Bachmann, T. M., Gold, L. S., Huijbregts, M. A. J., Jolliet, O., Juraske, R., Koehler, A., Larsen, H. F., Macleod, M., Margni, M., McKone, T. E., Payet, J., Schuhmacher, M., van de Meent, D. and Hauschild, M. Z. 2008. USEtox–the UNEP-SETAC toxicity model: recommended characterisation factors for human toxicity and freshwater ecotoxicity in life cycle impact assessment. *The International Journal of Life Cycle Assessment* 13(7), 532–46. doi:10.1007/s11367-008-0038-4.

Rosenbaum, R. K., Huijbregts, M. A. J., Henderson, A. D., Margni, M., Mckone, T. E., Van De Meent, D., Hauschild, M. Z., Shaked, S., Li, D. S., Gold, L. S. and Jolliet, O. 2011. USEtox human exposure and toxicity factors for comparative assessment of toxic emissions in life cycle analysis: sensitivity to key chemical properties. *The International Journal of Life Cycle Assessment* 16(8), 710–27. doi:10.1007/s11367-011-0316-4.

Rosenbaum, R. K., Hauschild, M. Z., Boulay, A.-M., Fantke, P., Laurent, A., Núñez, M. and Vieira, M. 2018. Life cycle impact assessment. In: Hauschild, M. Z., Rosenbaum, R. K. and Olsen, S. I. (Eds), *Life Cycle Assessment*. Springer, Cham, pp. 167–270. doi:10.1007/978-3-319-56475-3.

Sánchez, J. A. G., Martínez, J. M. L., Martín, J. L. and Holgado, M. N. F. 2012. Comparison of life cycle energy consumption and GHG emissions of natural gas, biodiesel and diesel buses of the Madrid transportation system. *Energy* 47(1), 174–98. doi:10.1016/j.energy.2012.09.052.

Schaubroeck, T. and Gibon, T. 2017. Outlining reasons to apply hybrid LCA--a reply to "rethinking system boundary in LCA" by Yi Yang (2017). *The International Journal of Life Cycle Assessment* 22(6), 1012–3. doi:10.1007/s11367-017-1311-1.

Schils, R. L. M., Olesen, J. E., Del Prado, A. and Soussana, J. F. 2007. A review of farm level modelling approaches for mitigating greenhouse gas emissions from ruminant livestock systems. *Livestock Science* 112(3), 240–51. doi:10.1016/j.livsci.2007.09.005.

Schmidt, J. H. and Weidema, B. P. 2008. Shift in the marginal supply of vegetable oil. *The International Journal of Life Cycle Assessment* 13(3), 235–9. doi:10.1065/lca2007.07.351.

Schmidt, J. H., Weidema, B. P. and Brandão, M. 2015. A framework for modelling indirect land use changes in life cycle assessment. *Journal of Cleaner Production* 99, 230–8. doi:10.1016/j.jclepro.2015.03.013.

Searchinger, T., Heimlich, R., Houghton, R. A., Dong, F., Elobeid, A., Fabiosa, J., Tokgoz, S., Hayes, D. and Yu, T. H. 2008. Use of US croplands for biofuels increases greenhouse gases through emissions from land-use change. *Science* 319(5867), 1238–40. doi:10.1126/science.1151861.

Searcy, E. and Flynn, P. C. 2008. Processing of straw/corn stover: comparison of life cycle emissions. *International Journal of Green Energy* 5(6), 423–37. doi:10.1080/15435070802498010.

Sedlbauer, V. K., Braune, A., Humbert, S., Margni, M., Schuller, O. and Fischer, M. 2007. Spatial differentiation in LCA moving forward to more operational sustainability. *Technikfolgenabschätzung. Theorie und Praxis* 3, 24–31.

Seppälä, J., Posch, M., Johansson, M. and Hettelingh, J. P. 2006. Country-dependent characterisation factors for acidification and terrestrial eutrophication based on accumulated exceedance as an impact category indicator (14 pp). *The International Journal of Life Cycle Assessment* 11(6), 403–16. doi:10.1065/lca2005.06.215.

Spatari, S., Bagley, D. M. and Maclean, H. L. 2010. Life cycle evaluation of emerging lignocellulosic ethanol conversion technologies. *Bioresource Technology* 101(2), 654–67. doi:10.1016/j.biortech.2009.08.067.

Steen, B. 1999. A systematic approach to environmental priority strategies in product development (EPS): version 2000 - models and data of the default method. CPM report. Chalmers University of Technology, Göteborg, Sweden.

Steinfeld, H., Gerber, P., Wassenaar, T., Castel, V., Rosales, M. and De Haan, C. 2006. *Livestock's Long Shadow*. FAO, Rome.

Stoessel, F., Juraske, R., Pfister, S. and Hellweg, S. 2012. Life cycle inventory and carbon and water foodprint of fruits and vegetables: application to a Swiss retailer. *Environmental Science and Technology* 46(6), 3253–62. doi:10.1021/es2030577.

Stranddorf, H., Hoffmann, L. and Schmidt, A. 2005. Update on impact categories, normalisation and weighting in LCA-selected EDIP97 data. Environmental Project Nr. 995. Danish Environmental Protection Agency, Copenhagen.

Struijs, J., Beusen, A., Van Jaarsveld, H. and Huijbregts, M. A. J. 2009. Aquatic eutrophication. In: Goedkoop, M., Heijungs, R., Huijbregts, M. A. J., De Schryver, A., Struijs, J. and Van Zelm, R. (Eds), *ReCiPe 2008, a Life Cycle Impact Assessment Method Which Comprises Harmonised Category Indicators at the Midpoint and the Endpoint Level*. The Ministry of Housing, Spatial Planning and the Environment, the Netherlands, pp. 59–67. Chapter 6.

Suh, S. and Huppes, G. 2005. Methods for life cycle inventory of a product. *Journal of Cleaner Production* 13(7), 687–97. doi:10.1016/j.jclepro.2003.04.001.

Suh, S., Weidema, B., Schmidt, J. H. and Heijungs, R. 2010. Generalized make and use framework for allocation in life cycle assessment. *Journal of Industrial Ecology* 14(2), 335–53. doi:10.1111/j.1530-9290.2010.00235.x.

Teh, S. H., Wiedmann, T. and Moore, S. 2018. Mixed-unit hybrid life cycle assessment applied to the recycling of construction materials. *Journal of Economic Structures* 7(1), 13. doi:10.1186/s40008-018-0112-4.

Thoma, G., Jolliet, O. and Wang, Y. 2013a. A biophysical approach to allocation of life cycle environmental burdens for fluid milk supply chain analysis. *International Dairy Journal* 31, S41–9. doi:10.1016/j.idairyj.2012.08.012.

Thoma, G., Popp, J., Nutter, D., Shonnard, D., Ulrich, R., Matlock, M., Kim, D. S., Neiderman, Z., Kemper, N., East, C. and Adom, F. 2013b. Greenhouse gas emissions from milk production and consumption in the United States: A cradle-to-grave life cycle assessment circa 2008. *International Dairy Journal* 31, S3–S14. doi:10.1016/j.idairyj.2012.08.013.

Thomassen, M. A., Dalgaard, R., Heijungs, R. and De Boer, I. J. M. 2008a. Attributional and consequential LCA of milk production. *The International Journal of Life Cycle Assessment* 13(4), 339–49. doi:10.1007/s11367-008-0007-y.

Thomassen, M. A., Van Calker, K. J., Smits, M. C. J., Iepema, G. L. and De Boer, I. J. M. 2008b. Life cycle assessment of conventional and organic milk production in the Netherlands. *Agricultural Systems* 96(1–3), 95–107. doi:10.1016/j.agsy.2007.06.001.

Thomassen, M. A., Dolman, M. A., Van Calker, K. J. and De Boer, I. J. M. 2009. Relating life cycle assessment indicators to gross value added for Dutch dairy farms. *Ecological Economics* 68(8–9), 2278–84. doi:10.1016/j.ecolecon.2009.02.011.

Tillman, A. M. 2000. Significance of decision-making for LCA methodology. *Environmental Impact Assessment Review* 20(1), 113–23. doi:10.1016/S0195-9255(99)00035-9.

Tillman, A. M., Baumann, H., Eriksson, E. and Rydberg, T. 1991. *Life-Cycle Analysis of Packaging Materials: Calculation of Environmental Load*. Chalmers Industri Teknik, Gothemburg.

Tuomisto, H. L., Hodge, I. D., Riordan, P. and Macdonald, D. W. 2012. Does organic farming reduce environmental impacts?-A meta-analysis of European research. *Journal of Environmental Management* 112, 309–20. doi:10.1016/j.jenvman.2012.08.018.

Turral, H., Burke, J. J. and Faurès, J. M. 2011. *Climate Change, Water and Food Security*. Food and Agriculture Organization of the United Nations, Rome.

Uihlein, A., Ehrenberger, S. and Schebek, L. 2008. Utilisation options of renewable resources: a life cycle assessment of selected products. *Journal of Cleaner Production* 16(12), 1306–20. doi:10.1016/j.jclepro.2007.06.009.

Van Zanten, H. H. E., Mollenhorst, H., De Vries, J. W., Van Middelaar, C. E., Van Kernebeek, H. R. J. and De Boer, I. J. M. 2014. Assessing environmental consequences of using co-products in animal feed. *The International Journal of Life Cycle Assessment* 19(1), 79–88. doi:10.1007/s11367-013-0633-x.

Van Zanten, H. H. E., Bikker, P., Meerburg, B. G. and De Boer, I. J. M. 2018. Attributional versus consequential life cycle assessment and feed optimization: alternative protein sources in pig diets. *The International Journal of Life Cycle Assessment* 23(1), 1–11. doi:10.1007/s11367-017-1299-6.

Van Zelm, R., Roy, P. O., Hauschild, M. Z. and Huijbregts, M. A. J. 2015. Acidification. In: Hauschild, M. Z. and Huijbregts, M. A. J. (Eds), *Life Cycle Impact Assessment*. Springer, the Netherlands, pp. 163–76. doi:10.1007/978-94-017-9744-3.

Vellinga, T. V. and Hoving, I. E. 2011. Maize silage for dairy cows: mitigation of methane emissions can be offset by land use change. *Nutrient Cycling in Agroecosystems* 89(3), 413–26. doi:10.1007/s10705-010-9405-1.

Vellinga, T. V., Blonk, H., Marinussen, M., Van Zeist, W. J. and Starmans, D. A. J. 2013. Methodology used in feedprint: a tool quantifying greenhouse gas emissions of feed production and utilization. Wageningen UR Livestock Research.

Verstegen, J. A., Van Der Hilst, F., Woltjer, G., Karssenberg, D., De Jong, S. M. and Faaij, A. P. C. 2016. What can and can't we say about indirect land-use change in Brazil using an integrated economic-land-use change model? *GCB Bioenergy* 8(3), 561–78. doi:10.1111/gcbb.12270.

Vigon, B. W., Tolle, D. A., Cornaby, B. W., Latham, H. C., Harrison, C. L., Boguski, T. L., Hunt, R. G. and Sellers, J. D. 1993. *Life-Cycle Assessment: Inventory Guidelines and Principles*. United States Environmental Protection Agency, Office of Research and Development (EPA/600/R-92/245), Washington DC.

Virtanen, Y., Kurppa, S., Saarinen, M., Katajajuuri, J. M., Usva, K., Mäenpää, I., Mäkelä, J., Grönroos, J. and Nissinen, A. 2011. Carbon footprint of food-approaches from national input-output statistics and a LCA of a food portion. *Journal of Cleaner Production* 19(16), 1849–56. doi:10.1016/j.jclepro.2011.07.001.

Wang, C., Li, X., Gong, T. and Zhang, H. 2014. Life cycle assessment of wheat-maize rotation system emphasizing high crop yield and high resource use efficiency in Quzhou County. *Journal of Cleaner Production* 68, 56–63. doi:10.1016/j.jclepro.2014.01.018.

Wangel, A. 2014. Social LCA scenarios: engaging producers and consumers in new domestic oyster value chains in Denmark. Social LCA in progress–4th SocSem, Fruitrop Thema. Presented at the *4th International Seminar on Social LCA*. CIRAD, Montpellier.

Warner, E., Zhang, Y., Inman, D. and Heath, G. 2014. Challenges in the estimation of greenhouse gas emissions from biofuel-induced global land-use change. *Biofuels, Bioproducts and Biorefining* 8(1), 114–25. doi:10.1002/bbb.1434.

Wassenaar, T., Gerber, P., Verburg, P. H., Rosales, M., Ibrahim, M. and Steinfeld, H. 2007. Projecting land use changes in the Neotropics: the geography of pasture expansion into forest. *Global Environmental Change* 17(1), 86–104. doi:10.1016/j.gloenvcha.2006.03.007.

Watanabe, M. D. B., Chagas, M. F., Cavalett, O., Guilhoto, J. J. M., Griffin, W. M., Cunha, M. P. and Bonomi, A. 2016. Hybrid input-output life cycle assessment of first-and second-generation ethanol production technologies in Brazil. *Journal of Industrial Ecology* 20(4), 764–74. doi:10.1111/jiec.12325.

Weidema, B. 2000. Avoiding co-product allocation in life-cycle assessment. *Journal of Industrial Ecology* 4(3), 11–33. doi:10.1162/108819800300106366.

Weidema, B. P. 2003. Market information in life cycle assessment. Environmental Project No. 863, Miljøprojekt. The Danish Environmental Protection Agency.

Weidema, B. P. 2018. In search of a consistent solution to allocation of joint production. *Journal of Industrial Ecology* 22(2), 252–62. doi:10.1111/jiec.12571.

Weidema, B. P. and Schmidt, J. H. 2010. Avoiding allocation in life cycle assessment revisited. *Journal of Industrial Ecology* 14(2), 192–5. doi:10.1111/j.1530-9290.2010.00236.x.

Weidema, B., Wenzel, H., Petersen, C. and Hansen, K. 2004. The product, functional unit and reference flows in LCA. *Environmental News* 70, 1–46.

Weidema, B. P., Pizzol, M., Schmidt, J. and Thoma, G. 2018. Attributional or consequential Life Cycle Assessment: a matter of social responsibility. *Journal of Cleaner Production* 174, 305–14. doi:10.1016/j.jclepro.2017.10.340.

Werner, F. and Richter, K. 2000. Economic allocation in LCA: a case study about aluminium window frames. *The International Journal of Life Cycle Assessment* 5(2), 79–83. doi:10.1007/BF02979727.

Wicke, B., Verweij, P., Van Meijl, H., Van Vuuren, D. P. and Faaij, A. P. C. 2012. Indirect land use change: review of existing models and strategies for mitigation. *Biofuels* 3(1), 87–100. doi:10.4155/bfs.11.154.

Wiedemann, S. G., Simmons, A., Watson, K. J. L. and Biggs, L. 2019. Effect of methodological choice on the estimated impacts of wool production and the significance for LCA-based rating systems. *The International Journal of Life Cycle Assessment* 24(5), 848–55. doi:10.1007/s11367-018-1538-5.

Williams, A., Audsley, E. and Sandars, D. 2006. Determining the environmental burdens and resource use in the production of agricultural and horticultural commodities: Defra project report IS0205. Available at: http://randd.defra.gov.uk/Default.aspx.

Xiao, G., Zhao, Z., Liang, L., Meng, F., Wu, W. and Guo, Y. 2019. Improving nitrogen and water use efficiency in a wheat-maize rotation system in the North China Plain using optimized farming practices. *Agricultural Water Management* 212, 172–80. doi:10.1016/j.agwat.2018.09.011.

Yan, M.-J., Humphreys, J. and Holden, N. M. 2013. Evaluation of process and input-output-based life-cycle assessment of Irish milk production. *The Journal of Agricultural Science* 151(5), 701–13. doi:10.1017/S0021859613000257.

Yang, Y. 2017a. Does hybrid LCA with a complete system boundary yield adequate results for product promotion? *The International Journal of Life Cycle Assessment* 22(3), 456–60. doi:10.1007/s11367-016-1256-9.

Yang, Y. 2017b. Rethinking system boundary in LCA–reply to "Lifting the fog on the characteristics and limitations of hybrid LCA" by Thomas Gibon and Thomas Schaubroeck (2017). *The International Journal of Life Cycle Assessment* 22(6), 1009–11. doi:10.1007/s11367-017-1295-x.

Yang, Y., Heijungs, R. and Brandão, M. 2017. Hybrid life cycle assessment (LCA) does not necessarily yield more accurate results than process-based LCA. *Journal of Cleaner Production* 150, 237–42. doi:10.1016/j.jclepro.2017.03.006.

Zhu, X. and Van Ierland, E. C. 2004. Protein chains and environmental pressures: a comparison of pork and novel protein foods. *Environmental Sciences* 1(3), 254–76. doi:10.1080/15693430412331291652.

Chapter 2

Key challenges in modelling of agricultural activities and their environmental impact

Evelyne A. Groen, Wageningen University, The Netherlands

1 Introduction

1.1 Modelling environmental impact of agricultural activities

Agriculture is one of the main drivers of environmental pressure (Steinfeld et al., 2006; Bajželj et al., 2014; Tilman and Clark, 2014). Mitigation strategies, both to reduce environmental impact and moderate depletion of natural resources, are needed to avert irreversible ecosystem damage (Tilman and Clark, 2014; Asseng et al., 2015) and the development of such strategies requires a thorough understanding of the processes at play. Modelling the environmental impacts of livestock production, such as climate change, is of interest because of the relatively large contribution of livestock to the global anthropogenic greenhouse gas emissions (Gerber et al., 2013; Groen, 2016). Regarding crop production, modelling of nutrient losses are of interest because they can contribute to eutrophication, acidification, emission of greenhouse gas emissions (Sutton et al., 2013) and depletion of fossil phosphorus (Suh and Yee, 2011; Groen, 2016).

Modelling, specifically environmental impact of agricultural activities, supports three needs: quantification of current practices, testing the effect of changes and support of decision-making (adapted from: Marion, 2008). From the perspective of the environmental impact modeller (and for the purposes

http://dx.doi.org/10.19103/AS.2018.0044.02

of this chapter), three types of models are identified (adapted from: Wrisberg et al., 2002; Marion, 2008; Moore et al., 2014; Jones et al., 2017a):

- *Chain-analysis models* are models that quantify product or substance flows of a production chain (Wrisberg et al., 2002). For example, quantification of greenhouse gas emissions of milk consumption over the entire production chain or nitrogen losses at dairy farms. Examples of tools or methods used in chain-analysis models are life cycle assessment (LCA) and nutrient balance approaches.
- *Optimization models* are models that optimize a cost function under exterior constraints (Heckelei, 2003). For example, the minimization of greenhouse gas emissions while maintaining a healthy diet, or maximization of farm profit, given constraints on on-site nitrogen production. A typical example of an optimization method is linear programming.
- *Simulation models* are models that focus on simulating an environment at a specific scale under certain conditions (Moore et al., 2014; Jones et al., 2017a). For example, crop models that simulate crop yield over time, taking into account the nitrogen dynamics in the soil. Examples are mechanistic models or biophysical models (e.g. based on chemical reactions or differential equations), integrated assessment models or agent-based modelling. Integrated assessment models describe interactions between human development and the natural environment (Stehfest et al., 2014), such as (long-term) environmental policies and their effects on the climate, for example IPCC assessment reports (IPCC, 2007). Agent-based modelling is a type of modelling that can be used to simulate (animal) interactions.

Some researchers combined different types of models to improve the quality of the calculated emissions. For example, Helin et al. (2013) provided an overview of studies of dynamic forest (simulation) models that can be used to include carbon flows in LCA. van Middelaar et al. (2013) included a mechanistic model to calculate emissions due to enteric fermentation to calculate the emissions of milk production over the entire production chain. Gabrielle and Gagnaire (2008) used a biophysical model to calculate, for example, nitrate leaching and ammonia emissions that were subsequently used in an LCA that quantified emissions from bioethanol. In another LCA study, van Zanten et al. (2018) calculated the optimal diet for pigs using linear programming. Schmidt et al. (2015) used a biophysical model that includes indirect land use change in LCA.

Decisions to incorporate more fine-grained calculations for specific parts of a model to improve model quality is only one of the challenges a researcher faces. Other aspects, such as setting the model boundaries (e.g. excluding parts of the production chain), obtaining high-quality data or dealing with uncertainty, all have implications for model quality.

1.2 Key challenges: modelling assumptions, obtaining data and dealing with uncertainty

How do model choices influence results? Do finer-grained production processes improve the accuracy of the results? What are the implications of low-quality data? How can you determine if uncertainties around input variables contribute to model output variance?

Model quality of agricultural activities and their environmental impact depends on assessing the following three key challenges: (1) addressing modelling assumptions, such as choosing the appropriate aggregation level or system boundary; (2) obtaining (high-quality) data from, for example, databases, IPCC tier 1 emission factors, empirical data collection or estimations and (3) dealing with uncertainty of input data. Uncertainty in environmental modelling is inevitable, since model input data varies due to seasonal changes, geographical conditions and socio-economic factors. Additionally, epistemic uncertainties, such as approximations and measurement errors, exist around input variables like field emissions and technical data such as farm inputs (Walker et al., 2003; Groen, 2016).

The next section starts with an overview of model characteristics, followed by a section on obtaining data. Both sections combined will lay the foundation for methods that can help to gain more insight into model quality.

2 Addressing key challenges

2.1 Model characteristics

Model aspects of the chain-analysis class, to which LCAs belong, are described below in more detail. Chain-analysis models can be divided into two categories (Wrisberg et al., 2002; Groen, 2016):

- *Function-oriented:* the functional unit of the product (a measure of the function of the product under study, to which inputs and outputs can be related) defines the system boundary, ideally from cradle-to-grave. Function-oriented approaches can be time dependent, but often consider the production of the final product for a specific year, and are therefore considered as static. Examples of function-oriented approaches are (1) LCA and (2) cumulative energy requirement analysis. LCA, which quantifies relevant resource use and emissions of a product over the entire production chain, is a function-oriented approach that is widely used in agricultural studies (Groen, 2016). Usually, an LCA is modelled as a set of linear equations. In matrix-based LCA, all technical variables are represented by the **A**-matrix, and the emissions in the **B**-matrix. To produce the functional unit **f**, the scaling vector of the **A**-matrix is calculated and multiplied with the emissions in the **B**-matrix and the characterization matrix **Q**, which contains the characterization factors (e.g. CO_2 equivalents for greenhouse gases such as N_2O and CH_4) (Heijungs and Suh, 2002). In most LCA models, input variables do not contain uncertainties (deterministic), in contrast to stochastic models, where input variables are represented by distribution functions. LCA can be used to quantify different types of environmental impacts, for example eutrophication, climate change or eco-toxicity (Finnveden et al., 2009; Hellweg and Milà i Canals, 2014).

- *Region-oriented:* a specific area and time interval define the system boundary. Usually, a region-oriented approach is based on national statistics and considers a specific country, or on a much smaller scale all processes controlled by a single company, such as farms (Groen, 2016). Even though they concern a specific year, they are time-independent (static). Examples of region-oriented approaches are (Wrisberg et al., 2002) (1) material flow analysis (e.g. plastic flows); (2) substance flow analysis (e.g. nitrogen flows); (3) nutrient balance analysis, concerning input and output of nutrients or substances through a region or a farm; and (4) environmental input–output analysis, focusing on pollution of industrial trade of a country (Groen, 2016). Nutrient balance analysis, computed as the difference between nutrients entering and leaving the system, is a region-oriented approach that is widely used in agriculture (Aarts et al., 1992; Schröder et al., 2003; Lesschen et al., 2011).

To sum up, agricultural models can have several characteristics, which will be highlighted here (Fig. 1). First, models can be deterministic (each calculation results in the same outcome, typically used in mass balance equations) or stochastic (e.g. input data are described by probability distributions, each model run will lead to a different outcome) (property A; Fig. 1). Second, models can be linear, such as models based on linear equations or mass balances (e.g. models based on LCA and nutrient balances), or non-linear, such as simulation models (property B; Fig. 1). An example of non-linearity in simulation models is the *LPJmL* model, which simulates non-linear changes of replacing natural vegetation by agriculture (Bondeau et al., 2007). A third property (property C; Fig. 1) is time-dependence. Although chain perspective models usually consider a specific time frame, the calculations are time-independent (or static). Time-dependent (or dynamic) models are, for example, models that simulate crop growth over time (including time steps for each calculation), such as the *MONICA* model, which simulates nitrogen and carbon dynamics of crops (Nendel, 2014).

Furthermore, configuration settings such as spatial scale, initial conditions and constraints influence model outcome (setting I–III; Fig. 1). Scale (or aggregation level) refers to the granularity of individual production processes.

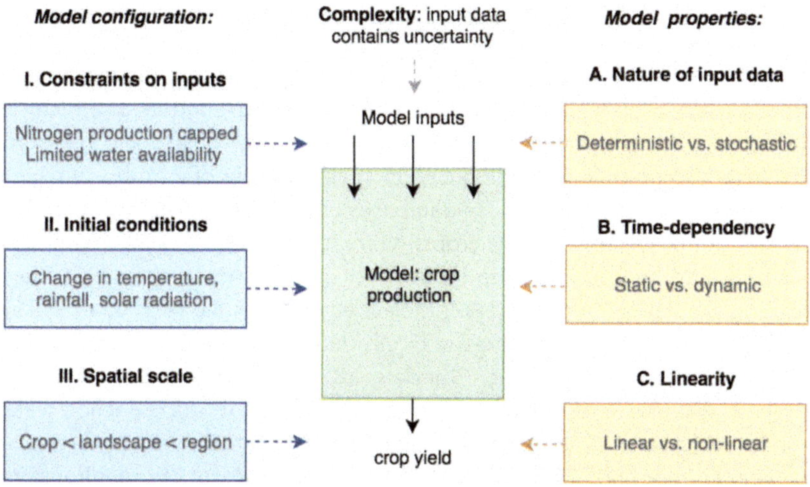

Figure 1 Aspects of modelling agricultural activities and their environmental impacts: examples of configuration settings and model properties; crop production is taken as an example.

For example, crops can be modelled on a regional scale as a single production process, where, for example, fertilizers and fuel for harvesting are averaged over all farms within that region. Alternatively, crops can be modelled on a farm level, expanding the model to the number of farms present in that region, each farm containing specific inputs for fertilizer and fuel use. The chain perspective is essentially a steady-state: given a set of model parameters, determined by a choice for a certain crop or region, the model provides an unambiguous and time-independent result. For dynamic models this is no longer true: at each moment in time the results of the model feed back into the model's conditions. Constraints, such as maximizing crop yields given constraints on water and nitrogen availability, have a strong presence in linear programming models. However, they also appear in mass balances, for example when nitrogen losses are the closing item of the balance, they can be a constraint on negative values.

An additional complexity, apart from setting model configurations and properties, concerns the nature of the input data. Input data of environmental models inherently contain uncertainty (Fig. 1). When this uncertainty is included in the model, the input data is commonly described by distribution functions, which makes the model stochastic. However in most LCA studies, uncertainty is ignored (Finnveden et al., 2009).

There are methods available, which will be discussed in Section 2.3, that, depending on model properties like linearity and the incorporation of uncertainty, can assess model quality. In some cases, it is also possible to assess the effect of constraints, initial conditions and scale, on the model output.

2.2 Data collection and uncertainty

When modelling agriculture and its environmental impact, several sources of data are available for data collection (Fig. 2). In LCA, inputs of agricultural production systems such as crop production are based on measured, empirical data, public or commercial databases or rule-of-thumb estimations. In addition to these inputs, these databases commonly also contain emissions of agricultural systems such as crops. Alternatively, values for these emissions can be found by applying crop input-specific equations (such as IPCC tier 1) or using simulation models (van Middelaar et al., 2012; Richards et al., 2016).

Combining the data sources of Fig. 2, data are coming from (1) empirical data collection; (2) simulation models; (3) empirically based equations; (4) databases and (5) estimations. Examples of empirical data collection of farm data can be found in, for example, Lee et al. (2017) (feed for dairy farms over a 6-year time period), Richards et al. (2016) (measurement of greenhouse gas emissions in eight countries) and Henriksson et al. (2015a) (200 aquaculture farms in four Asian countries). Examples of simulation models can be found in Nendel (2014) and Holzworth et al. (2014) (both models simulate crops) and Moore et al. (2014) (simulate farm management). Empirically based equations, for example for emission factors of agricultural emissions, including uncertainty ranges, can be found in IPCC (2006). Another example of empirically based equations are the indicators to measure biodiversity impact, as mentioned in de Baan et al. (2013). Examples of public national and international databases (e.g.

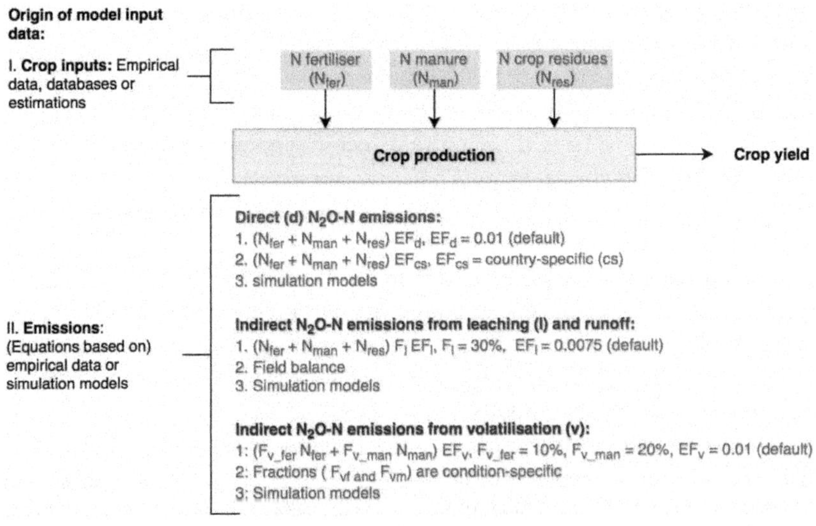

Figure 2 Three ways of calculating N_2O emissions of crop production, based on fertilizer, manure and crop residues (adapted from: van Middelaar et al., 2012).

Sweden, Japan and Australia), and internal business (steel, copper) databases for LCA can be found in Finnveden et al. (2009). Other examples of (industry) databases can be found in Rebitzer et al. (2004).

Although it is not possible to make a general statement regarding the quality of each of those sources, empirical data collection probably leads to a lower uncertainty range than estimating input variables. Also, empirical data might be prone to measurement errors, while empirically based equations, such as the IPCC tier 1 emission factors, are based on averages and therefore inherently contain natural variability:

- *Natural variability* relates to observable variation, which means it is inherent to the system and therefore cannot be reduced (Walker et al., 2003). An example of natural variability in agriculture is variation between annual crop yields of wheat across years and across countries or soil types (Groen, 2016). Examples of studies that include natural variability of dairy farms are Lee et al. (2017) (Korea), Henriksson et al. (2011) (Sweden) and Chen and Corson (2014) (France).
- *Epistemic uncertainty* originates from a lack of knowledge and can be reduced by gaining more or better data (Walker et al., 2003). Epistemic uncertainty includes errors during measurements or estimations (Groen, 2016). Examples of studies that quantified epistemic uncertainty are Zehetmeier et al. (2014), who based epistemic uncertainty ranges on IPCC (2006), Clavreul et al. (2013), who based the epistemic uncertainty ranges on a previous study of bioenergy from energy crops and Oenema et al. (2015). Oenema et al. (2015) performed an uncertainty analysis on nitrogen flows on a dairy farm and based the ranges of epistemic uncertainty on expert panel judgement.

However, in many studies, these uncertainties are ignored or overlooked. A recent study showed that out of all publications that made use of modelling (over all disciplines, physics, environmental science, economics etc.) less than 1% of those studies included the effect of input uncertainties on the model outcome (Ferretti et al., 2016). Also LCA studies often ignore uncertainties (Heijungs and Kleijn, 2001; Finnveden et al., 2009). While the result in the form of a point value[1] may be easy to interpret and used in policy-making, point values are sometimes insufficient to accurately describe model inputs and may even lead to a misleading output. For example, when comparing two product alternatives a deterministic result will favour one of the two alternatives that might not have been found if the actual uncertainty ranges were included in the comparison (Heijungs, 1996; Heijungs and Lenzen, 2014).

1 Each input variable is represented by a single scalar value as opposed to a distribution function.

Including epistemic uncertainty and natural variability in modelling reflects the conditions under which input data was collected, and often results in more reliable model outcomes. Furthermore, approaches like uncertainty analysis and sensitivity analysis can improve model quality by quantifying confidence intervals or finding those variables that largely determine output variance. In addition, uncertainty and sensitivity analyses can help to reduce the efforts for data collection, support the development of mitigation strategies and improve overall reliability, leading to more informed decision-making.

2.3 Uncertainty and sensitivity analysis

In Sections 2.1 and 2.2, the key challenges of modelling environmental impacts were illustrated: the model configuration (such as the scale and constraints on the inputs), the model properties (such as time-dependence and linearity), the sourcing of data and the incorporation of uncertainty (due to variability or measurement errors). Keeping those challenges in mind, one is now in a position to, or at least to try to, 'thoroughly corroborate' the model that is used (Saltelli et al., 2008).

Uncertainty analysis and sensitivity analysis can be applied regardless of which model is used (Helton et al., 2005; Saltelli et al., 2008). Uncertainty analyses are used to determine the model output uncertainty (Saltelli et al., 2008). Sensitivity analyses, on the other hand, can be used to find those variables that dominantly influence model output, or indicate parts of the model that are over-parameterized and could be simplified. In relation to these issues, three types of analysis exist:

- *Local sensitivity analysis* addresses what happens to the model output when input variables are changed around their central values, that is the intrinsic model behaviour of a variable. The variables that have the largest effect on the model output are referred to as the most locally sensitive or influential variables (Heijungs, 2010; Groen, 2016). A common approach (in LCA) is to change each input variable with the same amount, for example 1% or 5%. For an analytical model, this is equal to calculating the partial derivatives of the model output with respect to model inputs.
- *Uncertainty propagation* refers to the estimation of the uncertainty of model output using uncertainty attributes of the model inputs (Groen, 2016). A common approach is a Monte Carlo simulation, in which values are drawn from the joint distribution functions of all variables.
- *Global sensitivity analysis* is the study of how the uncertainty in the output of a model can be apportioned to the different sources of uncertainty in the model input (Saltelli et al., 2008). For global sensitivity analysis many methods are available, and the suitability of the methods depends on the linearity of the model.

2.3.1 Local sensitivity analysis

Local sensitivity analysis in environmental impact assessment models, such as LCA and nutrient balance analysis, are generally performed using a one-at-a-time approach (Huijbregts et al., 2001; Suh and Yee, 2011; Yang et al., 2011; Clavreul et al., 2012; van Middelaar et al., 2012; van Zanten et al., 2015). A one-at-a-time approach takes a variable and makes a small change, for example 5%, and quantifies the effect on the model output. A typical question that can be addressed during a local sensitivity analysis is: does a small change around an input variable (e.g. the emission factor for enteric fermentation) influence the model outcome?

A scenario analysis can be seen as an extension of a local sensitivity analysis, where several parameters are changed to represent, for example, different management practices or soil conditions. For simulation of farming models, a typical approach to dealing with uncertainty of input parameters is to calculate ensemble of models simulations with slightly perturbed model inputs (Moore et al., 2014). Holzworth et al. (2014) simulated the crop yield, animal husbandry and cash balance of farms in Vietnam under different livestock management conditions.

One-at-a-time approaches in modelling usually consider a subset of all available input variables, based on expert judgement or perhaps a quantitative criterion, such as the contribution of the individual production processes to the total environmental impact (sometimes referred to as a hotspot analysis, Thomassen et al., 2008; Hellweg and Milà i Canals, 2014). Since a 'small change' in the analysis can be arbitrary, calculating the partial derivatives of the model output with respect to each input variable is a more systematic approach. When an LCA model is parameterized in the matrix-based equation for LCA (Heijungs and Suh, 2002) the multiplier method can be applied (Heijungs, 2002, 2010; Sakai and Yokoyama, 2002; Groen et al., 2016). The multipliers are equal to the normalized partial derivatives for each variable in the LCA model. For non-LCA models, the feasibility of calculating the partial derivatives depends on the linear behaviour of the model. In case of models that behave non-linearly, the first order partial derivatives give very little insight into model results when actual uncertainties are taken into account (Saltelli et al., 2008). Table 1 gives an overview of the required data, examples of typically used methods and the goal of applying the sensitivity or uncertainty analysis.

A local sensitivity analysis reflects the scale of the model (Fig. 1): production processes that are aggregated (i.e. based on a regional average) are more influential than processes modelled at a more fine-grained manner or level of detail (i.e. each farm in the region is modelled as an individual process). The results of a local sensitivity analysis can therefore help with determining the appropriate aggregation level: if a production process is not influential (i.e. increasing crop yield with 5% did not significantly change emissions), creating

Table 1 Data requirements of local sensitivity analysis, uncertainty propagation and global sensitivity analysis, including examples of methods and goals

	Local sensitivity analysis	Uncertainty propagation	Global sensitivity analysis
Data	Point values	Distribution functions	Distribution functions
Methods	(Normalized) partial derivatives	Monte Carlo simulation	Squared standardized regression coefficients
Goal	Assess the effect of small changes around input variables	Uncertainty analysis, statistical inference	Explain model output variance

a more fine-grained production process will not give more insight into (the sensitivity of) this particular production process. In addition, a local sensitivity analysis can help to set the appropriate system boundaries: if a production process is influential (i.e. increasing crop yield with 5% reduced the emissions by 5%), broadening the system boundary of that process (e.g. including fuel for transportation) will probably affect the result. These conclusions only hold for models behaving (approximately) linearly, such as LCA and nutrient balances.

More formally, a local sensitivity analysis quantifies the change of a variable x with some value (Δx) and calculating the relative effect on the model output y equals:

$$\frac{x}{y}\frac{\left(y\left(x+\Delta x\right)-y\left(x\right)\right)}{\left(\Delta x\right)}$$

and is therefore an approximation of the normalized partial derivative:

$$S_x^{local} = \frac{x}{y}\frac{\partial y}{\partial x} \tag{1}$$

Equation (1) is also referred to as a local sensitivity index. These formal expressions will be used again when discussing global sensitivity analysis.

When applying a local sensitivity analysis, under most circumstances, the change in output will be at most the change in input of the most influential variable. For example, if the most influential variable is changed with 5%, the output is expected to change at most with 5%. Typically, only a few variables with such a high sensitivity will be found. This behaviour is a result from the fact that (most) LCA and nutrient balance models behave linearly. Examples can be found in Groen et al. (2016) and Wolf et al. (2017). The (local) sensitivity indices, as defined by eq. (1), can both be positive or negative (Heijungs, 2010); this can be understood by recalling that the one-at-a-time approach changes one input variable while all others remain fixed. For example, a slight increase in the use of fuel will scale the production process of fuel accordingly and therefore increase emissions, and therefore has a positive local sensitivity index. A slight increase in crop yield will reduce emissions per kg product and has a negative local sensitivity index. The indices can be interpreted as a kind of efficiency measure.

Some LCA and nutrient balance studies (Kim and Dale, 2002; Sayagh et al., 2010; Huang et al., 2013; Mendoza Beltran et al., 2016) use a one-at-a-time approach to refer to the effect of modelling decisions on the model output, such as the effect of changing allocation techniques or characterization factors. A local sensitivity analysis can also be used to indicate parameters that can be used in a scenario analysis later on.

2.3.2 Uncertainty analysis

Uncertainty analysis consists of two phases. The first is uncertainty propagation, which quantifies how variabilities around input variables propagate through the model, and determine the output uncertainty (Saltelli et al., 2008; Heijungs and Lenzen, 2014) (Fig. 1). After that, the output uncertainties have to be interpreted (Helton et al., 2005; Mendoza Beltran et al., 2018). Possible approaches are for example, determining the output variance, determining a confidence interval or significant differences between two alternatives (Table 1; Fig. 3). The output statistics, such as the output variance or the confidence interval, can then be used to, for example, verify if improvement options or trends are robust under input uncertainties or if alternatives still differ significantly from each other if uncertainties are included.

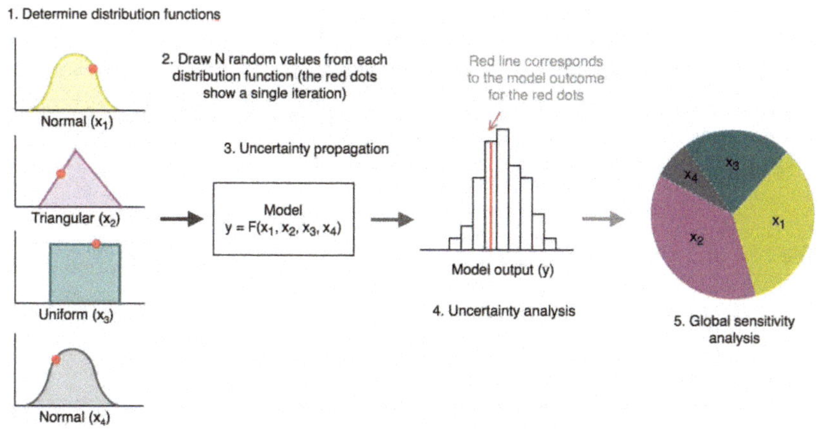

Figure 3 Four steps of a Monte Carlo simulation: (1) defining input uncertainty, (2) drawing random values from each input, (3) using those values to calculate the output uncertainty and (4) performing uncertainty analysis on the generated model output. The global sensitivity analysis is illustrated as well, but is not part of the Monte Carlo simulation.

Because Monte Carlo simulation is such an important tool for both uncertainty analysis and global sensitivity analysis and is the most common method applied for uncertainty propagation in LCA (Lloyd and Ries, 2007; Igos et al., 2018), the method will be explained here in brief. To apply Monte Carlo simulation, the first of each (uncertain) variable in the model a distribution function should be determined (step 1). Our assumption, which is not a prerequisite for Monte Carlo simulation in general, but which is valid within this domain of application, is that these distribution functions are statistically independent. This means that random samples can be drawn from each distribution function independently (step 2). The results of run 1–3 and run N, for an artificial example, are given in Table 2. Advanced sampling methods can be applied, such as Latin hypercube sampling or quasi-Monte Carlo sampling (Basson and Petrie, 2007; Lloyd and Ries, 2007; Subramanyan et al., 2008; Groen et al., 2014) or bootstrap methods (Lee et al., 2017), but normal random sampling is used most often (Igos et al., 2018). Examples of studies that use Monte Carlo simulation in LCA are Zehetmeier et al. (2014) (milk production), Bisinella et al. (2016) and Sonnemann et al. (2003) (waste treatment), Wei et al. (2014) (liquid ammonia production), de Koning et al. (2010) (detergents) and Geisler et al. (2005) (plant-protection products).

After each run, the model output is calculated based on the values drawn for each input variable (step 3). In Table 2, the model outputs based on the values drawn for each run are given in the bottom row. For example, during run two in Table 2, the model output results in 20.5 kg CO_2/kg product. After N runs, or when the output sample converges, for example the mean and standard deviation are not affected when the sample size is increased, the simulation is finished and further statistical analysis can be applied on the sampled model output (step 4). Examples of statistical analysis are calculating the mean value and standard deviation, but also applying statistical test when comparing the distribution functions of two products. An overview and detailed examples of methods can be found in Heijungs et al. (2005) and Mendoza Beltran et al. (2018). Steps 1 to 4 are visualized in Fig. 3.

Table 2 Random sampling of four distribution functions (fuel, electricity, CO_2 and CH_4)

	Run 1	Run 2	Run 3	. . .	Run N
Fuel (L)	12.1	10.8	11.3		11.9
Electricity (kWh)	4.56	3.55	4.02		3.49
CO_2 (kg)	1.23	1.22	1.39		1.23
CH_4 (kg)	3.45	3.47	3.21		3.43
Model output (kg CO_2-eq)	23.1	20.5	22.4	. . .	23.9

Results are shown for the first three runs and the last run (N). Based on the values drawn in each run, the model output is calculated (bottom row).

Another method for uncertainty propagation in LCA is analytical uncertainty propagation, which uses a Taylor series to approximate the output variance. The advantage of this method is that no sampling is required (and is therefore a lot faster than Monte Carlo simulation). A disadvantage is that only the output variance is calculated; uncertainty analysis methods that require distribution functions cannot be applied. Examples can be found in, for example, Heijungs and Lenzen (2014) and von Pfingsten et al. (2017).

It should be noted that increasing the number of uncertain variables does not necessarily increase the output uncertainty. In addition, increasing the variance of the input variables does not necessarily increase the output uncertainty. The reason for this behaviour will be explained in Section 2.3.3.

While most studies that perform Monte Carlo simulation in LCA assume independence between inputs (Lloyd and Ries, 2007; Igos et al., 2018), as was done earlier in this section, there are several things to consider when generating samples: (1) correlation between input variables, (2) maintaining a mass or energy balance and (3) dependent sampling when comparing alternatives.

Correlation between input variables (1): including a correlation coefficient can improve the amount of realistic model representations during a sampling. Correlations between input variables occur, for example, when the distribution functions used during uncertainty propagation are based on empirical data over time or (geographical) locations. For example, Lee et al. (2017) performed an uncertainty analysis of greenhouse gas emission of feedstuffs and energy consumption of dairy cows over a 6-year time span. Correlation was included between several feedstuffs and energy use on a dairy farm, based on the Spearman rank correlation coefficient found between the input variables within a year. The results showed that the output variance increased when correlations were included. Lesschen et al. (2007) implemented cross-correlation coefficients between the inputs and outputs of a soil (mass) balance of three types of nutrients (N, P and K) based on spatially explicit modelling of nutrient flows within Burkina Faso. Results showed extremely strong correlation between nitrogen flows of gaseous losses and mineral fertilizer input, harvested products and crop residue removal, and symbiotic N-fixation and crop production. Kelliher et al. (2017) performed an uncertainty analysis of nitrous oxide emissions of grasslands in New Zealand. For six different soil types the correlation between N input and the emission factor of N_2O was determined. Results showed that including correlation increased the output variance. Other examples of including correlation in agricultural LCA studies are, for example, Bojacá and Schrevens (2010) who performed an uncertainty analysis of potato production. Correlation coefficients between inputs such as urea and potassium chloride (KCL) and urea and potato yield based on empirical data that was collected every 2 weeks from 33 production fields between 2006 and 2008 in Peru. Results showed that including correlation reduced the output variance.

In an uncertainty analysis of carbon footprints of milk at Swedish dairy farms, Henriksson et al. (2011) found strong correlation between energy corrected milk and feed dry matter intake across 1051 farms in 2005.

The effect of including correlation during uncertainty propagation is slightly more complex. Including correlations between input variables will only change the output variance when both have a moderate contribution to the output variance compared to the other variables (Smith et al., 1992; Groen and Heijungs, 2017). Counter-intuitively, including correlation can both increase and decrease the output variance, depending on the model details. More details can be found in Groen and Heijungs (2017). The approach of including correlations during uncertainty propagation will be illustrated with a case study in Section 3.1.

Maintaining a mass or energy balance (2): in models that are based on mass balances, such as nutrient use efficiency calculations, maintaining a mass balance during uncertainty propagation is a must. For example, consider a very simple mass balance with one input and two outputs. If all three variables are described by a distribution function and could vary independently from each other during uncertainty propagation, that could result in either the input exceeding the output, or the outputs exceeding the input. During random sampling of all three variables, maintaining a mass balance actually only occurs by chance. To maintain the mass balance when performing sampling, the closing item of the balance should depend on the random values drawn for other items in the mass balance (Mu et al., 2017).

Dependent sampling when comparing alternatives (3): consider for example the case study of Henriksson et al. (2015b) who compared the carbon footprint of catfish (*Pangasius*) from two different aquaculture farms in Vietnam. It is important to note that both farms, although they differ in size, rely on the same production processes. For example, both alternatives include electricity production and transportation. To arrive at a fair comparison, the two farms and their shared processes are modelled as a single extended system: when sampling values for these shared processes, one should take the same random values when comparing (also referred to as dependent sampling). When dependent sampling was included, a significant difference was found between the farms, while no significant difference was found when normal sampling was applied. Details of this approach are found in Henriksson et al. (2015b) and Mendoza Beltran et al. (2018).

2.3.3 Global sensitivity analysis

A global sensitivity analysis is applied to find out which input variables explain the output variance. A global sensitivity analysis uses the model output variance and is therefore performed after propagation of input uncertainties

(Table 1; Fig. 3). For example, in Chen and Corson (2014), (squared) correlation coefficients were used to find out which emission factor contributed most to the variance around the mean greenhouse gas emissions of dairy farms in France. Results showed that epistemic uncertainty around the emission factor of manure deposition on pasture explained most of the output variance (67% in conventional farming systems), while all other emission factors contributed less than 10%. In a study on the same topic, Basset-Mens et al. (2009) calculated the effect of uncertainty around emission factors on the variance of greenhouse gas emissions of (average) dairy farms in New Zealand. Using regression coefficients, the uncertainty around the emission factor of manure deposition was the second most important variable (it accounted for 30% of the output variance), and on-farm dry matter intake explained most of the output variance (49%).

Specka et al. (2015) applied a global sensitivity analysis to the crop model MONICA, and determined which model parameter contributed most in the output variance for six crops (winter wheat, spring barley, silage maize, sugar beet, clover grass ley and winter rape) at three different sites in Germany. Results showed that out of the 117 model parameters analysed, only 28 contributed to the output variance for each of the six crops. Moreover, it was shown that the maximum assimilation rate (related to the plants' photosynthesis efficiency) showed up as the most important parameter in each crop. The importance of the other model parameters, like nitrogen concentration (mainly only important for sugar beet) and day length (important for e.g. maize) differed over different crops, but were similar for each crop over the three different sites.

More formally, the global sensitivity index (for linear models) for each model parameter x can be written as follows:

$$S_x^{global} = \frac{\mathrm{var}(\hat{x})}{\mathrm{var}(\hat{y})}\hat{a}^2 \tag{2}$$

where $\mathrm{var}(\hat{x})$ equals the variance of the model input parameter, $\mathrm{var}(\hat{y})$ equals the model output variance and $\hat{a}$ equals, for example, the regression coefficient as applied in Basset-Mens et al. (2009) or the correlation coefficient[2] as applied in Chen and Corson (2014)[3,4] in the examples introduced previously above. The global sensitivity index in eq. (2) combines the effect of the variance of each input variable ($\mathrm{var}(\hat{x})$) with the local sensitivity of each model variable ($\hat{a}$). An overview of methods for global sensitivity in LCA can be found in Groen et al. (2017) and Igos et al. (2018). Most commonly applied are correlation-based

2 Not to be confused with correlation between input variables: correlation-based methods used for global sensitivity analysis refers to correlation between the input and the output variables.
3 When applying key Issue analysis (Heijungs, 1996; Heijungs and Kleijn, 2001), $\hat{a}$ is the partial derivative, equal to $\partial y/\partial x$ of the local sensitivity index in eq. (1).
4 When applying regression-based sensitivity analysis, $\hat{a}$ is a regression coefficient and the global sensitivity index in eq. (2) is equal to the standardized squared regression coefficient. When applying a correlation-based method, eq. (2) is equal to the squared correlation coefficient.

methods (e.g. Geisler et al., 2005; Heijungs and Lenzen, 2014; Onat et al., 2014) or regression-based methods (e.g. Sugiyama et al., 2005; Basset-Mens et al., 2009; Aktas and Bilec, 2012; Vigne et al., 2012).

It should be noted that usually only a few variables dominate the explained output variance (Saltelli et al., 2008; Specka et al., 2015). In addition, variables that were found to be locally sensitive do not necessarily show up as globally sensitive. This is understood from looking at eq. (2): both the local sensitivity of the model input parameter captured by $\hat{a}$ and the variance of the model input parameter should be large enough to show up in the global sensitivity analysis.

The result of a combined local and global sensitivity analysis is illustrated in Fig. 4.

The horizontal axis shows the (relative) uncertainty of the model inputs, the vertical axis their (absolute) local sensitivity index within the model. Each dot represents a model input parameter. Four types of parameters are shown.

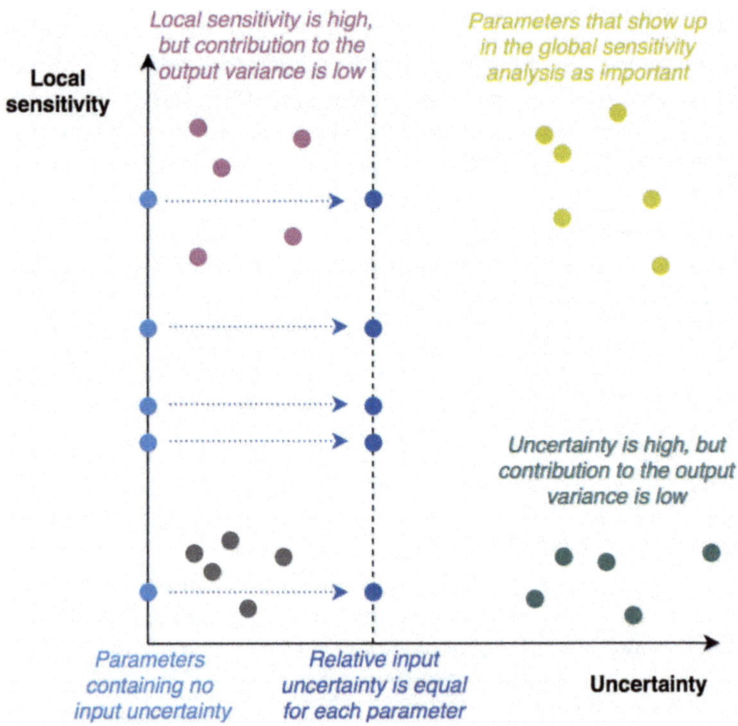

Figure 4 Combined results of a local and global sensitivity analysis, each dot represents a model input parameter. The parameters that show up as most important are represented by the yellow dots in the upper right corner. This figure is based on the PhD thesis of Valentina Bisinella (Bisinella, 2017), especially the idea of showing the results of varying each variable with the same (relative) uncertainty, and on the work of Reinout Heijungs (e.g. Heijungs, 1996).

The grey dots in the lower left corner are parameters that do not show up in the global sensitivity analysis: changing their mean value or uncertainty will not change the result of the analysis. The pink dots in the upper left corner are parameters that are locally sensitive. Increasing the uncertainty of those parameters can make them more important in the result of the global sensitivity analysis (moving them towards the upper right corner). The yellow dots in the upper right corner are the parameters that explain most of the output variance (hence are the most important). Efforts to reduce the output variance should focus on those parameters, as reducing the input uncertainty of those parameters will have the most effect. The green dots in the lower right corner are parameters of which the input uncertainty is high, but which nevertheless do not contribute significantly to the output variance.

Furthermore, two extreme cases are shown. First, the blue dots on the vertical axis represent model inputs that do not contain uncertainty, only their local sensitivity index can be determined according to eq. (1). Second, if each parameter is varied with the same relative amount (e.g. all model inputs are normally distributed where the relative standard deviation $\sigma/|\mu|$ equals 10%) the importance of the parameter in the global sensitivity analysis is the same as the outcome of the local sensitivity analysis (this should be kept in mind when assuming distribution functions when input data is only limitedly available).

Because only a handful of variables typically explain the output variance, one now also understands why increasing the variance of some variables (green dots in Fig. 4) does not necessarily increase their contribution to the output variance: those variables are not locally sensitive (Fig. 4) and changing their input uncertainty cannot significantly affect the output variance. Additionally, the pink and yellow dots represent parameters for which changing the mean values will result in the highest increase in environmental performance. The approach of combining a local and global sensitivity analysis will be illustrated with a case study in Section 3.2.

The large uncertainty ranges around the IPCC emission factors for agricultural activities illustrate the use of applying a global sensitivity analysis. Although it is recommended to take country-specific or modelled data, many research papers rely on the default equations and their ranges (e.g. for volatilization: 0.01 (0.002–0.05) kg $N\text{-}N_2O$/kg NH_3-N + NO_x-N (IPCC, 2006)), even though the emission factors are based on empirical data from northwest Europe and are less reliable for countries outside Europe or the specific crops under study. The relevance of the uncertainty ranges is highlighted by, for example, Richards et al. (2016), who showed that for individual crops modelled throughout Africa, the default factors from IPCC almost overestimated the measured outcome by a factor of more than two. In another study, Ruser et al. (2017) recommended a new rapeseed-specific N_2O emission factor of 0.6%

instead of the IPCC tier 1 factor of 1.0%, based on measurements on five sites in Germany.

What do these results mean for global sensitivity analysis of agricultural models using those emissions factors? In general, assuming that crop production is only a part of the production chain or simulation model, so, for example, the study concerns the environmental impact of a diet (of which only a few components are modelled via the IPCC tier 1 emissions) only a few specific feed ingredients are considered while simulating farm management strategies. If uncertainty is included in the IPCC emission factors, the global sensitivity analysis will show whether the uncertainty ranges of the emission factors contribute to the output variance or not. A conclusion of a global sensitivity analysis can be that the production processes modelled with the IPCC tier 1 emissions, taking into account their uncertainty, do not contribute to the output variance, and can therefore remain fixed. Alternatively, if those processes do show up as important, it can be decided to collect better data, or use more elaborate simulation models to improve the accuracy of the emissions and therefore model quality.

Applying a global sensitivity analysis can lead to the following iterative actions:

- *Factor fixing*: a variable whose variance does not (or very little) contribute to the output variance (low S_x^{global})[5] can be held fixed. This means that in future model evaluations of the distribution function of that variable can be fixed to a point value, without changing the model output variance by much. This saves time on data collection, and in the case of more complex models, can reduce calculation time.
- *Factor prioritizing*: a variable whose variance contributes considerably to the output variance (high S_x^{global}). This means that in future model evaluations, the distribution function of that variable should be defined accurately, to give the best estimate for the model output variance.

When simulation models are included within LCA studies to model specific parts, such as soil dynamics, and that model does not behave linearly anymore, more advanced methods should be applied to determine the global sensitivity indexes. At that point, a local sensitivity analysis will also be less insightful. In that case, methods based on the conditional variance, such as Sobol' indices or methods using Fourier transformations are more useful, as was applied in Specka et al. (2015). A general introduction to higher order global sensitivity indices can be found in Saltelli et al. (2008), adapted to LCA in, for example, Groen et al. (2017), Cucurachi et al. (2015) and Mutel et al. (2013).

5 In the case of non-linear models, variance-based global sensitivity analysis methods such as the Sobol' indices should be applied and factor fixing should be based on low total sensitivity indices.

3 Case studies

Two case studies will be discussed: first a case study is shown that demonstrates how to quantify the effect of correlation on the output variance. Second, a case study is shown that combines a local and a global sensitivity analysis, showing the power of combining the two approaches.

3.1 The effect of correlation on output variance

In LCA, when working with average data based on empirical datasets, uncertainty distributions can be defined based on the original data points. For example, for a production process describing crop production, the use of mineral fertilizer is based on 100 farms in one region producing the same crop. The input of mineral fertilizer can be described as a distribution function, which now includes the natural variability that was present between the farms from the empirical dataset. In addition, the yield of that crop is also derived from the same dataset, and depicted as a distribution function as well. However, unless a mass balance or other equation between fertilizer and yield for each farm is explicitly included within the production process in the model, the relation between these quantities is lost. Commonly, such relations are neglected. A way of implementing this (assumingly linear) relation would be via a correlation coefficient, which describes the relation between fertilizer and yield over the different farms. For example, a positive correlation coefficient would describe that farms using more fertilizer have a higher yield.

The effect of including correlations during uncertainty propagation will be illustrated with a case study based on the data available in Lee et al. (2017), using Monte Carlo simulation and analytical uncertainty propagation. Lee et al. (2017) performed an uncertainty analysis on the greenhouse gas emissions of a dairy farm in Korea. Input data on feedstuffs and energy consumption was collected during 6 years on a monthly basis. In addition, correlation coefficients between the input parameters were defined to maintain the original relation between the data points that was found over the years. For illustrative purposes, some assumptions were slightly simplified or adjusted. For example, it was assumed that all distributions were normally distributed (the original data contained logistic and lognormal distributions as well), correlation was only included if the correlation coefficient was bigger than 0.3 (the original dataset contained 15 correlation coefficients) and the standard deviations were rescaled with a factor $1/\sqrt{n}$ (n equals 12, the number of months the standard deviation was based on). An overview of the data, as used in this example, can be found in Table 3.

From the data available in Lee et al. (2017), a matrix-based LCA model was created, as described in Section 2.1. The **A**-matrix is based on the mean value in Table 3, the **B**-matrix is given in Table 3 in the column emissions, based on

Table 3 Input data based on Lee et al. (2017), only correlations >0.3 were considered here

Parameter	Mean (kg/ cow-month)	Standard deviation (kg/cow-month)	Emissions (kg CO_2-eq)	Correlated with
Feed for dry cows	173	30.7	65.6	-
Feed for lactating cows	421	58.3	270	Soybean meal (0.374), electricity (0.547), diesel (0.418)
Alfalfa	27.7	2.46	9.14	-
Straw	27.4	3.84	26.0	Electricity (0.423)
Oats	25.2	5.14	14.6	-
Corn silage	141	21.2	10.8	-
Rape seed	22.9	1.99	9.85	-
Bagasse	22.2	5.11	0.466	-
Soybean meal	17.4	2.85	12.4	Diesel (0.389), feed for lactating cows (0.374)
Electricity	130	17.1	65.2	Feed for lactating cows (0.547), straw (0.423)
Diesel	7.30	1.55	24.1	Feed for lactating cows (0.418), soybean meal (0.389)

In this case study, all input parameters are normally distributed. The units of the feedstuffs and diesel are in kg/cow-month except for electricity, which is expressed in kWh/cow-month.

the emissions factors as described in Lee et al. (2017). The total emissions equal 508 kg CO_2-eq per cow per month (which can also be found by summing the column under emissions) (Table 4).

To study the uncertainty propagation using Monte Carlo simulation, a sample size of N = 1000 was generated from 11 normal distributions, their mean and standard deviations given in Table 3. When correlation was not included, the output variance equalled $183 \cdot 10^1$ kg CO_2-eq. Next, a multivariate sample of N = 1000 is generated that includes the correlations as described in Table 3.

Table 4 Comparison of the mean, variance and standard deviation using a Monte Carlo simulation (MCS; using normal random sampling without correlation and multivariate sampling that included the correlations) and analytical uncertainty propagation (AUP; excluding the covariance term thus ignoring correlation and including the covariance term)

	Mean or baseline (kg CO_2-eq)	Standard deviation	Output variance
MCS (normal random sampling)	518	42.7	$183 \cdot 10^1$
MCS (multivariate sampling)	520	54.1	$292 \cdot 10^1$ (+60%)
AUP (no covariance term)	508	40.8	$166 \cdot 10^1$
AUP (including covariance term)	508	47.6	$226 \cdot 10^1$ (+60%)

When correlation is included the output variance equals 292•10[1] kg CO_2-eq, showing an increase of 60% (Table 4).

For uncertainty propagation using analytical uncertainty propagation, a Taylor approximation of the matrix-based LCA model can be made. The results are shown in Table 4. In a later stage, the covariance terms (Smith et al., 1992; Heijungs, 2010) are included. In this example, five covariance terms determine the change in output variance, because five correlation coefficients are included. The output variance $\text{var}(y)$ including the covariance term is equal to:

$$\text{var}(y) = \text{var}(y)^* + \sum_{i,j} 2\frac{\partial y}{\partial x_i}\frac{\partial y}{\partial x_j}\text{cov}(x_i, x_j) \tag{3}$$

where $\text{var}(y)^*$ is equal to the variance absent correlations. Notice that the change in variance (the second term in eq. (3)) depends on the combined effect of the local sensitivities of the model variable at that point $\partial y/\partial x_i \partial y/\partial x_j$ and the covariance term $\text{cov}(x_i, x_j)$. In this case study, the predicted change was positive and predicted a similar increase (of 60%) in output variance as the multivariate sampling (Table 4).

The observed mean and standard deviation are slightly lower for analytical uncertainty propagation than for the stochastic uncertainty propagation. These kind of deviations have been observed before in for example Lee et al. (2017), Groen et al. (2014) and Heijungs and Lenzen (2014).

Including correlations is necessary to accurately determine the standard deviation and confidence interval of the output variance. Ignoring correlations could underestimate (or overestimate) the output variance (Groen and Heijungs, 2017), as was shown in Table 4. Including correlations is especially important when making comparisons. Ignoring correlation can lead to false positives between two alternatives, for example a significant difference is found, which actually does not exist. Also, ignoring correlation can lead to false negatives comparing two alternatives. Equation (3) can be used in advance, to predict what the effect will be of including a correlation coefficient on the output variance, based on the partial derivatives and the covariance of the input variables, to avoid such false positives and false negatives.

3.2 Combined uncertainty and sensitivity plot

Nitrogen (N) losses at dairy farms contribute to leaching of NO_3^-, emissions of N_2O and NH_3 and lead to eutrophication, acidification and global warming (Whitehead, 1995; Smith et al., 1999). To evaluate or compare N-efficiency of farms, or to identify where losses occur, nutrient use efficiencies are calculated. Nutrient use efficiency of farms is based on a mass balance of all nutrients going in and all nutrients going out (Schröder et al., 2003; Spears et al., 2003; Gourley et al., 2012).

In this case study four European dairy farms are included. Some farms are more intensive than others because of limited land availability and import of nitrogen through purchased feed. Farms 1 and 3 are located on sandy soil, and are intensive in terms of milk production per cow per ha and amount of purchased concentrates. Farms 2 and 4 are located on loamy soil, with relatively many grazing hours compared to farms 1 and 3. Data collection is partly based on farm data, but also includes estimations and is therefore prone to measurement errors and other forms of epistemic uncertainty. Further details and data, including uncertainty ranges, are described in Mu et al. (2017) and Oenema et al. (2015).

Each farm is modelled using a maximum of 16 variables, such as inputs of concentrates, roughage, fertilizer and manure, but also deposition and fixation on on-site crop production was taken into account. Stock changes for imported feed and fertilizer and the replacement of animals was considered as well. All farms produced milk, but some also exported roughage or animals (Mu et al., 2017).

For the local sensitivity analysis, each variable was increased by 1%, and the relative change on the output was calculated. The maximum change is expected to be 1% as well. For the global sensitivity analysis, each variable was distributed according to the data collection described in Mu et al. (2017) and Oenema et al. (2015). Figure 5 shows the result of the combined local and global sensitivity analysis, only values explaining more than 10% of the output variance are labelled. The more intensive farms are represented by the orange (farm 1) and purple (farm 3) labels, the less intensive farms are indicated by blue (farm 2) and grey (farm 4) labels (Fig. 5).

For none of the farm inputs (such as fixation, deposition or manure), stock changes (e.g. concentrates of fertilizers) and outputs (e.g. roughage) play a role in either the local or the global sensitivity analysis, shown as the unlabelled variables in the bottom left corner of Fig. 5.

Figure 5 shows that for all farms, changing the milk output will directly affect the nutrient use efficiency, also since the production of milk is the main goal of a dairy farm those values are usually known with high certainty and contribute less to the output variance. Still, since the values explain about 10–30% of the output variance, it is worth gaining more accurate data. Only for the intensive farms (farms 1 and 3) input of concentrates and roughage show up as important. For farms 2 and 4 getting high-quality data regarding mineral fertilizer improves quality most, since those farms rely on on-site feed-crop production. For farms 2 and 4, the same variables are highlighted: output of milk and animals, and the input of mineral fertilizer.

The results of Fig. 5 are, for example, relevant when benchmarking nutrient use efficiency of farms: high-quality data with low epistemic uncertainty should be collected for the variables depicted in Fig. 5, which can differ from farm

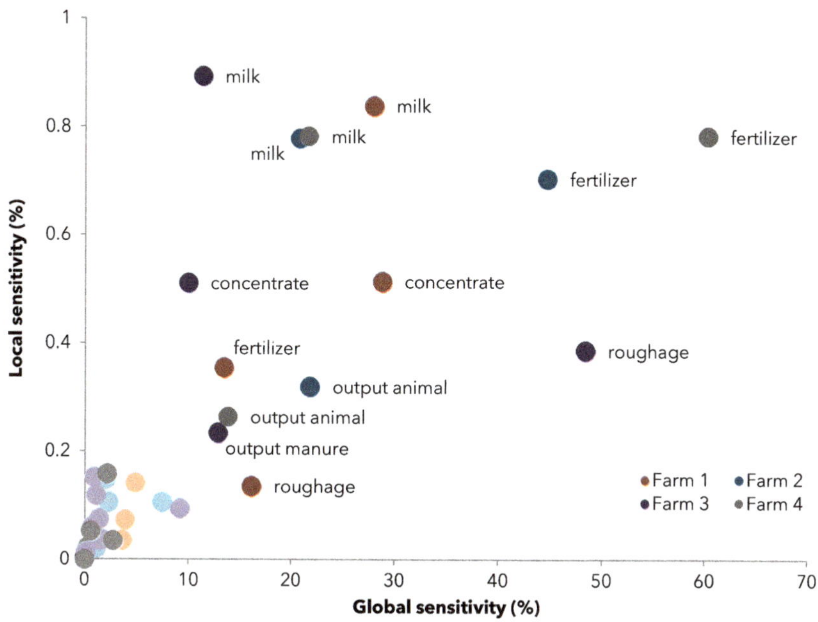

Figure 5 Results of combined local and global sensitivity analysis. In the upper right corner the most import variables are found.

to farm. When improving efficiency on farm 3, decreasing roughage has more effect than fertilizer, which has more effect on farm 1. When comparing N-efficiency, reducing the uncertainty around mineral fertilizer on farms 2 and 4 can reduce the output variance maximum between 45% and 60% respectively, making for much more reliable results.

4 Conclusion

Model quality of agricultural activities and their environmental impact depends on assessing the following three key challenges: (1) addressing modelling assumptions, such as choosing the appropriate aggregation level or system boundary; (2) obtaining (high-quality) data from, for example, databases, IPCC tier 1 emission factors, empirical data collection or estimations and (3) dealing with uncertainty of input data. Uncertainty in environmental modelling is inevitable, since model input data varies due to seasonal changes, geographical conditions and socio-economic factors and epistemic uncertainties, such as measurement errors and estimations of emission factors and technical data.

Uncertainty and sensitivity analysis were introduced as tools that can help address those challenges. Uncertainty analysis and sensitivity analysis can be applied independently of the model that is used, but their usefulness depends on model properties such as linearity. An uncertainty analysis can be used to determine the model output uncertainty. A sensitivity analysis can be used to find important model variables.

Local sensitivity analysis addresses what happens to the model output when input variables are changed around their central values, that is the intrinsic model behaviour of a variable. A common approach is to change each input variable with the same amount, for example 1% or 5%, or calculating the partial derivatives. When applying a local sensitivity analysis to LCA and nutrient balance models, usually the most locally sensitive variable is as sensitive as the relative change of the input variable.

Uncertainty analysis quantifies the uncertainty of model output using uncertainty attributes of the model inputs. A common approach is a Monte Carlo simulation, which was illustrated in Section 2.3.2. An uncertainty analysis can be used to determine output statistics, such as the output variance and the confidence interval, which in turn can be used to, for example, verify if improvement options or trends are robust under input uncertainties or if alternatives still differ significantly from each other if uncertainties are included.

There are several things to consider when generating samples in Monte Carlo simulations: (1) the presence of correlation between input variables, which was illustrated for a case study in Section 3.1; (2) the addition of constraints, such as maintaining a mass or energy balance; and (3) dependent sampling methods, which are applicable when comparing alternatives that share certain production processes.

Global sensitivity analysis calculates the contribution to the output variance of each input variable. For global sensitivity analysis many methods are available: a commonly applied method uses the squared standardized regression coefficients. When applying a global sensitivity analysis, only a few variables explain the output variance. These are usually ranked in a different order from the variables found in the local sensitivity analysis as illustrated in Section 3.2.

Applying sensitivity analysis allows one to test modelling assumptions such as inspecting the aggregation level and the choice of system boundaries. For example, uncertainty propagation allows a systematic analysis of the effect of input data uncertainties on the model output. Further analysis, such as global sensitivity analysis, can determine the importance of finding high-quality data. Its result can help to reduce uncertainty of the model output by directing data collection efforts towards the most important variables, which contributes to improved model quality.

5 Future trends

As the scope of LCAs deepens by including, for example, crop- and nutrient-dedicated simulation models, LCA is increasingly used as a modelling approach. This shift was already noted by Guinée et al. (2010) who mention that the inventory analysis (data collection) and impact assessment will be merged into one modelling phase. For example, in the data collection phase for the emission factors for direct N_2O emissions (Fig. 2), one can simply resort to establishing a default country value, or go as far as to implement a complete simulation model which includes site- and soil-specific emission factors.

Although LCA studies aim to improve results by including simulation models for some studies, a gap between models dedicated to crop production and LCA calculations still exists. Peter et al. (2017) reported that none of the reviewed tools for LCA calculations could accurately describe the (dynamic) effects of including crop rotation. Within agricultural LCAs, topics for expanding the LCA models are, for example, nutrient cycles (Gabrielle and Gagnaire, 2008) and carbon cycles (Helin et al., 2013), modelling indirect land use (Finnveden et al., 2009; Cherubini and Stromman, 2011; McManus and Taylor, 2015; Schmidt et al., 2015) including impacts of pesticides (Fantke et al., 2011), and reducing large uncertainty around the emission factor of indirect land use change (Cherubini and Stromman, 2011; Mullins et al., 2011). Other areas of interest are impacts of water use and (human) toxicity (Finnveden et al., 2009).

On the other side of the agricultural modelling spectrum, models once dedicated solely to simulating crop yields, developed in a more holistic direction (Payraudeau and Werf, 2005), and include climate change, food security or function as decision-making tools, such as on resource efficiency (Holzworth et al., 2015; Jones et al., 2017a). For example, the APSIM model has evolved from a model describing crop yields to a fully-fledged agro-ecosystem model describing aspects that range from farm management to climate change (Holzworth et al., 2014).

Within the deepening scope of LCA models and the more holistic analysis of crop simulation models, several trends and challenges can be identified:

- A growing need for making models *site-specific*. In a study of Reap et al. (2008), which describes unresolved issues in LCA, a strong argument is made for including local environmental uniqueness (such as site-specific input constraints and boundary conditions, Fig. 1). The need for site-specificity is illustrated by several case studies, for example by a study of Potting et al. (1998) who showed that the impact of acidification differs several orders of magnitude between countries (Potting et al., 1998; Finnveden et al., 2009). Furthermore, researchers of the APSIM model emphasize the need for spatial heterogeneity of root and soil systems (Holzworth et al., 2014) to improve the model implementation of competing resources, such as nutrients and light. In a study of Richards et al. (2016) it was shown that default values for nitrous oxide and methane emissions overestimated country-specific measured greenhouse gas emissions by a factor of two or more.

- The ability to *scale models* (Holzworth et al., 2014). A case study of Yang, Tao and Suh (2018), who compared environmental impacts of several crops across several states in the United States, showed that for the production of cotton, the average GHG emissions are 2.8 ton CO_2-eq, but in the state of South Carolina, the GHG equals 5.4 ton CO_2-eq, almost twice of the average (Yang et al., 2018). Spatial differentiation in LCA can be implemented in matrix-based LCA and improves the accuracy of its results (Yang and Heijungs, 2017). Moreover, these authors emphasize the need for a regionalized quantification of the uncertainty ranges of emissions (Yang et al., 2011). Additionally, it is important to account for interaction effects between farms when going from a farm level to a regional level and reveal underlying variability between farms when scaling down (Payraudeau and Werf, 2005).
- Dealing with missing data and *uncertainty analysis*. Several studies, such as from Finnveden et al. (2009), point out that uncertainty analysis is a crucial part of LCA, since uncertainty is inherently present in LCAs due to, for example, missing data, errors of incompleteness and variability within the data. Evaluation of input uncertainty should be an integral part of the assessment (Payraudeau and Werf, 2005; Moore et al., 2014; Holzworth et al., 2015). Richards et al. (2016) emphasize the importance of including uncertainty estimates in the calculation of, for example, N_2O fluxes of tropical agriculture, while the default values consistently overestimates empirical measurements (Richards et al., 2016). One of the first publicly available (and free) LCA software packages which includes many tools for uncertainty and sensitivity analysis is CMLCA (Heijungs and Suh, 2002; Heijungs, 2004). The CMLCA contains a graphical user interface, but the code is not open-source, making it cumbersome for the user to adapt or include case study-specific scripts or models.
- The availability of *robust data sets for benchmarking*. Jones et al. (2017b) mention a lack of data to evaluate the results of agricultural system models. Data scarcity not only leads to an inability to test or fine-tune models for a range of possible inputs (Jones et al., 2017a). Agricultural models should be benchmarked by reference data, so results arising from new applications can be mutually compared (Janssen et al., 2017). Initiatives such as *openLCA Nexus* (GreenDelta, 2018) provide an overview of public databases, such as *AGRIBALYSE* (Colomb et al., 2015), which provides life cycle inventory data of agricultural products, and *EXIOBASE* (EXIOBASE, 2015), which contains environmentally extended input/output tables of 43 countries by industry. Additionally, standardized databases should include

uncertainty distributions to perform uncertainty analysis later on (Reap et al., 2008).

- *Reusability of code and open-source projects*: Jones et al. (2017b) pointed out that a possible way to deal with data scarcity is to reuse code from other models that can simulate the missing data. Moreover, good coding practices can stimulate interoperability and integration of data and models (Janssen et al., 2017), and between researchers when code is developed under an open-source framework (Holzworth et al., 2014). Open-source projects allow users to adjust and expand the source code, making it easier for users to collaborate and share code. For example, once the developers of the APSIM model made their model open source, the number of users as well as the number of global contributors to the models increased (Holzworth et al., 2014). Examples of free open-source LCA software projects are *Brightway* (Mutel, 2017) and *openLCA* (Ciroth, 2007). *Brightway* is written in the Python programming language and is designed for analysis and data inspection using Jupyter notebooks (Kluyver et al., 2016). *openLCA* is an open-source LCA alternative to commercial LCA software, and comes with a user interface. The source code of both projects is available on *Github* (an online platform to share and reuse code).

Model quality depends on accuracy of modelling decisions and quality of the obtained data. Agricultural system models have grown to be more complex, due to expanding the scope beyond crop yields or greenhouse gas emissions, and are integrated or reused by other models to improve model accuracy. Scalability and site-dependency will improve usability and accuracy of model predictions. Reusable (open-source) code, availability of datasets to fine-tune models and benchmark model outcomes, and uncertainty analysis will improve reliability of future studies in agricultural modelling.

6 Where to look for further information

The application of tools such as uncertainty and sensitivity analysis requires some kind of formal way of describing a model captured by mathematical equations. Most software tools have those equations running in the background when sensitivity analysis is applied. In the case of LCA, a matrix-based formulation is available, which is described in much detail with many examples by Heijungs and Suh (2002). This book also contains examples of methods for uncertainty and sensitivity analysis, and is recommended reading if one wants to apply sensitivity and uncertainty analysis in LCA. Most available software tools only apply basic or limited functionality for uncertainty and sensitivity analysis (Heijungs and Kleijn, 2001; Igos et al., 2018).

A general introduction to sensitivity analysis *Sensitivity Analysis: the Primer* (Saltelli et al., 2008), which is highlighted here due to its broad coverage and well-explained uncertainty and sensitivity methods. Every two years the authors of this book provide a summer school (free of charge for European students) that is held in Italy: *SAMO (Summer School on Sensitivity Analysis of Model Output).* Other recommended papers on sensitivity analysis are Hamby (1994) and Iooss and Lemâitre (2014) (comparison of global sensitivity methods); Campolongo et al. (2007) (screening methods); Xu and Gertner (2008) (including of correlation during uncertainty propagation); and Ferretti et al. (2016) (overview of the use of sensitivity analysis in modelling).

The following books, theses and papers that give an overview of application of uncertainty and sensitivity analysis in LCA are recommended to read: Rosenbaum et al. (2018) (general practices for LCA and introduction to uncertainty analysis); Bisinella (2017) and Clavreul et al. (2013) (application of sensitivity analysis in waste treatment); Mutel et al. (2013) (application of sensitivity analysis to biodiversity impacts from land use of cocoa production); Cucurachi (2014) (impact assessment model of noise, including a protocol for global sensitivity analysis); Mendoza Beltran et al. (2018) and Henriksson et al. (2015b) (methods for uncertainty analysis); and Björklund (2002), Groen et al. (2014, 2017), Igos et al. (2018) and Lloyd and Ries (2007) (review of methods for uncertainty and sensitivity analysis in LCA).

7 References

Aarts, H. F. M., Biewinga, E. E. and van Keulen, H. (1992), 'Dairy farming systems based on efficient nutrient management', *Netherlands Journal of Agricultural Science*, 40, pp. 285–99.

Aktas, C. B. and Bilec, M. M. (2012), 'Impact of lifetime on US residential building LCA results', *International Journal of Life Cycle Assessment*. Springer, 17(3), pp. 337–49.

Asseng, S., Ewert, F., Martre, P., Rötter, R. P., Lobell, D. B., Cammarano, D., Kimball, B. A., Ottman, M. J., Wall, G. W., White, J. W., et al. (2015), 'Rising temperatures reduce global wheat production', *Nature Climate Change*. doi:10.1038/nclimate2470.

de Baan, L., Alkemade, R. and Koellner, T. (2013), 'Land use impacts on biodiversity in LCA: A global approach', *International Journal of Life Cycle Assessment*. doi:10.1007/s11367-012-0412-0.

Bajželj, B., Richards, K. S., Allwood, J. M., Smith, P., Dennis, J. S., Curmi, E. and Gilligan, C. A. (2014), 'Importance of food-demand management for climate mitigation', *Nature Climate Change*. Nature Publishing Group, 4(10), pp. 924–9.

Basset-Mens, C., Kelliher, F. M., Ledgard, S. and Cox, N. (2009), 'Uncertainty of global warming potential for milk production on a New Zealand farm and implications for decision making', *International Journal of Life Cycle Assessment*. Springer, 14(7), pp. 630–8.

Basson, L. and Petrie, J. G. (2007), 'An integrated approach for the consideration of uncertainty in decision making supported by Life Cycle Assessment', *Environmental Modelling & Software*. Elsevier, 22(2), pp. 167–76.

Bisinella, V. (2017), Future scenario development within Life Cycle Assessment of waste management systems, Department of Environmental Engineering, Technical University of Denmark (DTU), Kongens Lyngby.

Bisinella, V., Conradsen, K., Christensen, T. H. and Astrup, T. F. (2016), 'A global approach for sparse representation of uncertainty in Life Cycle Assessments of waste management systems', *International Journal of Life Cycle Assessment*. doi:10.1007/s11367-015-1014-4.

Björklund, A. E. (2002), 'Survey of approaches to improve reliability in LCA', *International Journal of Life Cycle Assessment*. Springer, 7(2), pp. 64–72.

Bojacá, C. R. and Schrevens, E. (2010), 'Parameter uncertainty in LCA: Stochastic sampling under correlation', *International Journal of Life Cycle Assessment*. Springer, 15(3), pp. 238–46.

Bondeau, A., Smith, P. C., Zaehle, S., Schaphoff, S., Lucht, W., Cramer, W., Gerten, D., Lotze-Campen, H., Müller, C., et al. (2007), 'Modelling the role of agriculture for the 20th century global terrestrial carbon balance', *Global Change Biology*. doi:10.1111/j.1365-2486.2006.01305.x.

Campolongo, F., Cariboni, J. and Saltelli, A. (2007), 'An effective screening design for sensitivity analysis of large models', *Environmental Modelling & Software*, 22(10), pp. 1509–18.

Chen, X. and Corson, M. S. (2014), 'Influence of emission-factor uncertainty and farm-characteristic variability in LCA estimates of environmental impacts of French dairy farms', *Journal of Cleaner Production*, 81, pp. 150–7.

Cherubini, F. and Stromman, A. H. (2011), 'Life cycle assessment of bioenergy systems: State of the art and future challenges', *Bioresource Technology*. doi:10.1016/j.biortech.2010.08.010.

Ciroth, A. (2007), 'ICT for environment in life cycle applications openLCA - A new open source software for Life Cycle Assessment', *International Journal of Life Cycle Assessment*. doi:10.1065/lca2007.06.337.

Clavreul, J., Guyonnet, D. and Christensen, T. H. (2012), 'Quantifying uncertainty in LCA-modelling of waste management systems', *Waste Management*. doi:10.1016/j.wasman.2012.07.008.

Clavreul, J., Guyonnet, D., Tonini, D. and Christensen, T. H. (2013), 'Stochastic and epistemic uncertainty propagation in LCA', *International Journal of Life Cycle Assessment*. Springer, 18(7), pp. 1393–403.

Colomb, V., Ait Amar, S., Basset-Mens, C., Gac, A., Gaillard, G., Koch, P., Mousset, J., Salou, T., Tailleur, A. and van der Werf, H. M. G. (2015), 'AGRIBALYSE®, the French LCI Database for agricultural products: High quality data for producers and environmental labelling', *OCL*. doi:10.1051/ocl/20140047.

Cucurachi, S. (2014), 'Impact assessment modelling of matter-less stressors in the context of Life Cycle Assessment', Doctoral thesis, Institute of Environmental Sciences (CML), Industrial Ecology, Faculty of Science, Leiden University.

Cucurachi, S., Borgonovo, E. and Heijungs, R. (2015), 'A protocol for the global sensitivity analysis of impact assessment models in Life Cycle Assessment', *Risk Analysis*. Wiley Online Library, 36(2), pp. 357-77.

EXIOBASE (2015), 'About EXIOBASE', *EXIOBASE*, 14 October. https://www.exiobase.eu/index.php/about-exiobase (accessed on 1 June 2018).

Fantke, P., Juraske, R., Antón, A., Friedrich, R. and Jolliet, O. (2011), 'Dynamic multicrop model to characterize impacts of pesticides in food', *Environmental Science and Technology*. doi:10.1021/es201989d.

Ferretti, F., Saltelli, A. and Tarantola, S. (2016), 'Trends in sensitivity analysis practice in the last decade', *Science of the Total Environment*. doi:10.1016/j.scitotenv.2016.02.133.

Finnveden, G., Hauschild, M. Z., Ekvall, T., Guinée, J., Heijungs, R., Hellweg, S., Koehler, A., Pennington, D. and Suh, S. (2009), 'Recent developments in life cycle assessment', *Journal of Environmental Management*. Elsevier, 91(1), pp. 1–21.

Gabrielle, B. and Gagnaire, N. (2008), 'Life-cycle assessment of straw use in bio-ethanol production: A case study based on biophysical modelling', *Biomass and Bioenergy*. doi:10.1016/j.biombioe.2007.10.017.

Geisler, G., Hellweg, S. and Hungerbühler, K. (2005), 'Uncertainty analysis in life cycle assessment: Case study on plant-protection products and implications for decision making', *International Journal of Life Cycle Assessment*. Springer, 10(3), pp. 184–92.

Gerber, P. J., Steinfeld, H., Henderson, B., Mottet, A., Opio, C., Dijkman, J., Falcucci, A. and Tempio, G. (2013), *Tackling Climate Change through Livestock: A Global Assessment of Emissions and Mitigation Opportunities*, Food and Agriculture Organization, 115p.

Gourley, C. J. P., Aarons, S. R. and Powell, J. M. (2012), 'Nitrogen use efficiency and manure management practices in contrasting dairy production systems', *Agriculture, Ecosystems & Environment*. Elsevier, 147, pp. 73–81.

GreenDelta (2018), *openLCA Nexus*, Available at: nexus.openlca.org/databases.

Groen, E. A. (2016), 'An uncertain climate: The value of uncertainty and sensitivity analysis in environmental impact assessment of food', Dissertation, Wageningen University, Wageningen.

Groen, E. A. and Heijungs, R. (2017), 'Ignoring correlation in uncertainty and sensitivity analysis in life cycle assessment: What is the risk?' *Environmental Impact Assessment Review*. doi:10.1016/j.eiar.2016.10.006.

Groen, E. A., Bokkers, E. A. M., Heijungs, R. and de Boer, I. J. M. (2014), 'Methods for uncertainty propagation in life cycle assessment', *Environmental Modelling & Software*. Elsevier, 62, pp. 316–25.

Groen, E. A., van Zanten, H. H. E., Heijungs, R., Bokkers, E. A. M. and de Boer, I. J. M. (2016), 'Sensitivity analysis of greenhouse gas emissions from a pork production chain', *Journal of Cleaner Production*, 129, pp. 202–11. doi:10.1016/j.jclepro.2016.04.081.

Groen, E. A., Heijungs, R., Bokkers, E. A. M. and de Boer, I. J. M. (2017), 'Methods for global sensitivity analysis in life cycle assessment', *International Journal of Life Cycle Assessment*. doi:10.1007/s11367-016-1217-3.

Guinée, J. B., Heijungs, R., Huppes, G., Zamagni, A., Masoni, P., Buonamici, R., Ekvall, T. and Rydberg, T. (2010), 'Life Cycle Assessment: Past, present, and future?', *Environmental Science & Technology*. ACS Publications, 45(1), pp. 90–6

Hamby, D. M. (1994), 'A review of techniques for parameter sensitivity analysis of environmental models', *Environmental Monitoring and Assessment*. Springer, 32(2), pp. 135–54.

Heckelei, T. (2003), 'Estimation of constrained optimisation models for agricultural supply analysis based on generalised maximum entropy', *European Review of Agriculture Economics*. doi:10.1093/erae/30.1.27.

Heijungs, R. (1996), 'Identification of key issues for further investigation in improving the reliability of life cycle assessments', *Journal of Cleaner Production*. Elsevier, 4(3), pp. 159–66.

Heijungs, R. (2002), 'The use of matrix perturbation theory for addressing sensitivity and uncertainty issues in LCA', in *Proceedings of The Fifth International Conference on*

EcoBalance - practical tools and thoughtful principles for sustainability. The Society of Non-traditional Technology, Tokyo, Japan.
Heijungs, R. (2004), 'A theory of the environment and economic systems', *International Journal of Life Cycle Assessment*. doi:10.1007/BF02978541.
Heijungs, R. (2010), 'Sensitivity coefficients for matrix-based LCA', *International Journal of Life Cycle Assessment*. Springer, 15(5), pp. 511-20.
Heijungs, R. and Kleijn, R. (2001), 'Numerical approaches towards life cycle interpretation five examples', *International Journal of Life Cycle Assessment*. Springer, 6(3), pp. 141-8.
Heijungs, R. and Lenzen, M. (2014), 'Error propagation methods for LCA: A comparison', *International Journal of Life Cycle Assessment*. Springer, 19(7), pp. 1445-61.
Heijungs, R. and Suh, S. (2002), *The Computational Structure of Life Cycle Assessment*, Kluwer Academic Publishers, Dordrecht, the Netherlands.
Heijungs, R., Suh, S. and Kleijn, R. (2005), 'Numerical approaches to life cycle interpretation: The case of the ecoinvent'96 database', *International Journal of Life Cycle Assessment*. Springer, 10(2), pp. 103-12.
Helin, T., Sokka, L., Soimakallio, S., Pingoud, K. and Pajula, T. (2013), 'Approaches for inclusion of forest carbon cycle in life cycle assessment - A review', *GCB Bioenergy*. doi:10.1111/gcbb.12016.
Hellweg, S. and Milà i Canals, L. (2014), 'Emerging approaches, challenges and opportunities in life cycle assessment', *Science (New York, N.Y.)*. doi:10.1126/science.1248361.
Helton, J. C., Davis, F. J. and Johnson, J. D. (2005), 'A comparison of uncertainty and sensitivity analysis results obtained with random and Latin hypercube sampling', *Reliability Engineering & System Safety*, 89(3), pp. 305-30.
Henriksson, M., Flysjö, A., Cederberg, C. and Swensson, C. (2011), 'Variation in carbon footprint of milk due to management differences between Swedish dairy farms', *Animal*. Cambridge University Press, 5(9), pp. 1474-84.
Henriksson, P. J. G., Rico, A., Zhang, W., Ahmad-Al-Nahid, S., Newton, R., Phan, L. T., Zhang, Z., Jaithiang, J., Dao, H. M., Phu, T. M., et al. (2015a), 'Comparison of Asian aquaculture products by use of statistically supported life cycle assessment', *Environmental Science and Technology*. doi:10.1021/acs.est.5b04634.
Henriksson, P. J. G., Heijungs, R., Dao, H. M., Phan, L. T., de Snoo, G. R. and Guinée, J. B. (2015b), 'Product carbon footprints and their uncertainties in comparative decision contexts', *PLoS ONE*. Public Library of Science, 10(3).
Holzworth, D. P., Huth, N. I., de Voil, P. G., Zurcher, E. J., Herrmann, N. I., McLean, G., Chenu, K., van Oosterom, E. J., Snow, V., Murphy, C., et al. (2014), 'APSIM - Evolution towards a new generation of agricultural systems simulation', *Environmental Modelling & Software*. doi:10.1016/j.envsoft.2014.07.009.
Holzworth, D. P., Snow, V., Janssen, S., Athanasiadis, I. N., Donatelli, M., Hoogenboom, G., White, J. W. and Thorburn, P. (2015), 'Agricultural production systems modelling and software: Current status and future prospects', *Environmental Modelling & Software*. doi:10.1016/j.envsoft.2014.12.013.
Huang, Y., Spray, A. and Parry, T. (2013), 'Sensitivity analysis of methodological choices in road pavement LCA', *International Journal of Life Cycle Assessment*. Springer, 18(1), pp. 93-101.
Huijbregts, M. A. J., Norris, G., Bretz, R., Ciroth, A., Maurice, B., von Bahr, B., Weidema, B. and de Beaufort, A. S. H. (2001), 'Framework for modelling data uncertainty in life

cycle inventories', *International Journal of Life Cycle Assessment*. Springer, 6(3), pp. 127–32.

Igos, E., Benetto, E., Meyer, R., Baustert, P. and Othoniel, B. (2018), 'How to treat uncertainties in life cycle assessment studies?', *International Journal of Life Cycle Assessment*. doi:10.1007/s11367-018-1477-1.

Iooss, B. and Lemaître, P. (2014), 'A review on global sensitivity analysis methods', *arXiv preprint arXiv:1404.2405*.

IPCC (2006), IPCC Guidelines for National Greenhouse Gas Inventories, Chapter 11: N_2O emissions from managed soils, and CO_2 emissions from lime and urea application. Guidelines for National Greenhouse Gas Inventories.

IPCC (2007), IPCC Guidelines for National Greenhouse Gas Inventories, Climate change 2007: Synthesis report. Contribution of working groups I, II and III to the fourth assessment report of the Intergovernmental Panel on Climate Change.

Janssen, S. J. C., Porter, C. H., Moore, A. D., Athanasiadis, I. N., Foster, I., Jones, J. W. and Antle, J. M. (2017), 'Towards a new generation of agricultural system data, models and knowledge products: Information and communication technology', *Agricultural Systems*. doi:10.1016/j.agsy.2016.09.017.

Jones, J. W., Antle, J. M., Basso, B., Boote, K. J., Conant, R. T., Foster, I., Godfray, H. C. J., Herrero, M., Howitt, R. E., Janssen, S., et al. (2017a), 'Brief history of agricultural systems modeling', *Agricultural Systems*. doi:10.1016/j.agsy.2016.05.014.

Jones, J. W., Antle, J. M., Basso, B., Boote, K. J., Conant, R. T., Foster, I., Godfray, H. C. J., Herrero, M., Howitt, R. E., Janssen, S., et al. (2017b), 'Toward a new generation of agricultural system data, models, and knowledge products: State of agricultural systems science', *Agricultural Systems*. doi:10.1016/j.agsy.2016.09.021.

Kelliher, F. M., Henderson, H. V. and Cox, N. R. (2017), 'The uncertainty of nitrous oxide emissions from grazed grasslands: A New Zealand case study', *Atmospheric Environment*. doi:10.1016/j.atmosenv.2016.11.005.

Kim, S. and Dale, B. E. (2002), 'Allocation procedure in ethanol production system from corn grain I. system expansion', *International Journal of Life Cycle Assessment*. Springer, 7(4), pp. 237–43.

Kluyver, T., Ragan-Kelley, B., Pérez, F., Granger, B., Bussonnier, M., Frederic, J., Kelley, K., Hamrick, J., Grout, J., Corlay, S., et al. (2016), *Jupyter Notebooks – a publishing format for reproducible computational workflows*, Positioning and Power in Academic Publishing: Players, Agents and Agendas. doi:10.3233/978-1-61499-649-1-87.

de Koning, A., Schowanek, D., Dewaele, J., Weisbrod, A. and Guinée, J. (2010), 'Uncertainties in a carbon footprint model for detergents; quantifying the confidence in a comparative result', *International Journal of Life Cycle Assessment*. Springer, 15(1), pp. 79–89.

Lee, M., Lee, J., Lee, J., Kim, Y., Park, Y. and Lee, K. (2017), 'Uncertainty analysis of a GHG emission model output using the block bootstrap and Monte Carlo simulation', *Sustainability*. doi:10.3390/su9091522.

Lesschen, J. P., Stoorvogel, J. J., Smaling, E. M. A., Heuvelink, G. B. M. and Veldkamp, A. (2007), 'A spatially explicit methodology to quantify soil nutrient balances and their uncertainties at the national level', *Nutrient Cycling in Agroecosystems*. doi:10.1007/s10705-006-9078-y.

Lesschen, J. P., van den Berg, M., Westhoek, H. J., Witzke, H. P. and Oenema, O. (2011), 'Greenhouse gas emission profiles of European livestock sectors', *Animal Feed Science and Technology*. Elsevier, 166, pp. 16–28.

Lloyd, S. M. and Ries, R. (2007), 'Characterizing, propagating, and analyzing uncertainty in life cycle assessment: A survey of quantitative approaches', *Journal of Industrial Ecology*. Wiley Online Library, 11(1), pp. 161–79.

Marion, G. (2008), *An Introduction to Mathematical Modelling*, Institution of Bioinformatics and Statistics, Scotland, UK.

McManus, M. C. and Taylor, C. M. (2015), 'The changing nature of life cycle assessment', *Biomass and Bioenergy*. doi:10.1016/j.biombioe.2015.04.024.

Mendoza Beltran, A. M., Heijungs, R., Guinée, J. and Tukker, A. (2016), 'A pseudo-statistical approach to treat choice uncertainty: The example of partitioning allocation methods', *International Journal of Life Cycle Assessment*. Springer, 21(2), pp. 252–64.

Mendoza Beltran, A., Prado, V., Font Vivanco, D., Henriksson, P. J. G., Guinée, J. B. and Heijungs, R. (2018), 'Quantified uncertainties in comparative life cycle assessment: What can be concluded?' *Environmental Science and Technology*. doi:10.1021/acs.est.7b06365.

van Middelaar, C. E., Cederberg, C., Vellinga, T. V., van der Werf, H. M. G. and de Boer, I. J. M. (2012), 'Exploring variability in methods and data sensitivity in carbon footprints of feed ingredients', *International Journal of Life Cycle Assessment*, 18(4), pp. 768–82.

van Middelaar, C. E., Berentsen, P. B. M., Dijkstra, J. and de Boer, I. J. M. (2013), 'Evaluation of a feeding strategy to reduce greenhouse gas emissions from dairy farming: The level of analysis matters', *Agricultural Systems*. doi:10.1016/j.agsy.2013.05.009.

Moore, A. D., Holzworth, D. P., Herrmann, N. I., Brown, H. E., de Voil, P. G., Snow, V. O., Zurcher, E. J. and Huth, N. I. (2014), 'Modelling the manager: Representing rule-based management in farming systems simulation models', *Environmental Modelling and Software*. doi:10.1016/j.envsoft.2014.09.001.

Mu, W., Groen, E. A., van Middelaar, C. E., Bokkers, E. A. M., Hennart, S., Stilmant, D. and de Boer, I. J.M. (2017), 'Benchmarking nutrient use efficiency of dairy farms: The effect of epistemic uncertainty', *Agricultural Systems*. doi:10.1016/j.agsy.2017.04.001.

Mullins, K. A., Griffin, W. M. and Matthews, H. S. (2011), 'Policy implications of uncertainty in modeled life-cycle greenhouse gas emissions of biofuels', *Environmental Science and Technology*. doi:10.1021/es1024993.

Mutel, C. (2017), 'Brightway: An open source framework for Life Cycle Assessment', *The Journal of Open Source Software*. doi:10.21105/joss.00236.

Mutel, C. L., de Baan, L. and Hellweg, S. (2013), 'Two-step sensitivity testing of parametrized and regionalized life cycle assessments: Methodology and case study', *Environmental Science & Technology*, 47(11), pp. 5660–7.

Nendel, C. (2014), 'MONICA: A simulation model for nitrogen and carbon dynamics in agro-ecosystems', in *Environmental Science and Engineering (Subseries: Environmental Science)*. doi:10.1007/978-3-319-01017-5_23.

Oenema, J., Burgers, S., van Keulen, H. and van Ittersum, M. (2015), 'Stochastic uncertainty and sensitivities of nitrogen flows on dairy farms in the Netherlands', *Agricultural Systems*. Elsevier, 137, pp. 126–38.

Onat, N. C., Kucukvar, M. and Tatari, O. (2014), 'Integrating triple bottom line input - output analysis into life cycle sustainability assessment framework: The case for US buildings', *International Journal of Life Cycle Assessment*. Springer, 19(8), pp. 1488–505.

Payraudeau, S. and van Werf, H. M. G. (2005), 'Environmental impact assessment for a farming region: A review of methods', *Agriculture, Ecosystems & Environment*. doi:10.1016/j.agee.2004.12.012.

Peter, C., Helming, K. and Nendel, C. (2017), 'Do greenhouse gas emission calculations from energy crop cultivation reflect actual agricultural management practices? – A review of carbon footprint calculators', *Renewable and Sustainable Energy Reviews*. doi:10.1016/j.rser.2016.09.059.

von Pfingsten, S., Broll, D. O., von der Assen, N. and Bardow, A. (2017), 'Second-order analytical uncertainty analysis in Life Cycle Assessment', *Environmental Science & Technology*. doi:10.1021/acs.est.7b01406.

Potting, J., Schöpp, W., Blok, K. and Hauschild, M. (1998), 'Site-dependent life-cycle impact assessment of acidification', *Journal of Industrial Ecology*. doi:10.1162/jiec.1998.2.2.63.

Reap, J., Roman, F., Duncan, S. and Bras, B. (2008), 'A survey of unresolved problems in life cycle assessment', *International Journal of Life Cycle Assessment*. Springer, 13(5), pp. 374–88.

Rebitzer, G., Ekvall, T., Frischknecht, R., Hunkeler, D., Norris, G., Rydberg, T., Schmidt, W. P., Suh, S., Weidema, B. P. and Pennington, D. W. (2004), 'Life cycle assessment Part 1: Framework, goal and scope definition, inventory analysis, and applications', *Environment International*. doi:10.1016/j.envint.2003.11.005.

Richards, M., Metzel, R., Chirinda, N., Ly, P., Nyamadzawo, G., Duong Vu, Q., de Neergaard, A., Oelofse, M., Wollenberg, E., Keller, E., et al. (2016), 'Limits of agricultural greenhouse gas calculators to predict soil N_2O and CH_4 fluxes in tropical agriculture', *Scientific Reports*. doi:10.1038/srep26279.

Rosenbaum, R. K., Hauschild, M. Z., Boulay, A.-M., Fantke, P., Laurent, A., Núñez, M. and Vieira, M. (2018), 'Life cycle impact assessment', in *Life Cycle Assessment*. doi: 10.1007/978-3-319-56475-3_10.

Ruser, R., Fuß, R., Andres, M., Hegewald, H., Kesenheimer, K., Köbke, S., Räbiger, T., Quinones, T.S., Augustin, J., Christen, O., et al. (2017), 'Nitrous oxide emissions from winter oilseed rape cultivation', *Agriculture, Ecosystems and Environment*. doi:10.1016/j.agee.2017.07.039.

Sakai, S. and Yokoyama, K. (2002), 'Formulation of sensitivity analysis in life cycle assessment using a perturbation method', *Clean Technologies and Environmental Policy*. Springer, 4(2), pp. 72–8.

Saltelli, A., Ratto, M., Andres, T., Campolongo, F., Cariboni, J., Gatelli, D., Saisana, M. and Tarantola, S. (2008), *Global Sensitivity Analysis: The Primer*, John Wiley & Sons.

Sayagh, S., Ventura, A., Hoang, T., François, D. and Jullien, A. (2010), 'Sensitivity of the LCA allocation procedure for BFS recycled into pavement structures', *Resources, Conservation and Recycling*. Elsevier, 54(6), pp. 348–58.

Schmidt, J. H., Weidema, B. P. and Brandão, M. (2015), 'A framework for modelling indirect land use changes in Life Cycle Assessment', *Journal of Cleaner Production*. doi:10.1016/j.jclepro.2015.03.013.

Schröder, J. J., Aarts, H. F. M., ten Berge, H. F. M., van Keulen, H. and Neeteson, J. J. (2003), 'An evaluation of whole-farm nitrogen balances and related indices for efficient nitrogen use', *European Journal of Agronomy*. Elsevier, 20(1), pp. 33–44.

Smith, A. E., Ryan, P. B. and Evans, J. S. (1992), 'The effect of neglecting correlations when propagating uncertainty and estimating the population distribution of risk', *Risk Analysis*. Wiley Online Library, 12(4), pp. 467–74.

Smith, V. H., Tilman, G. D. and Nekola, J. C. (1999), 'Eutrophication: Impacts of excess nutrient inputs on freshwater, marine, and terrestrial ecosystems', *Environmental Pollution*. Elsevier, 100(1), pp. 179–96.

Sonnemann, G. W., Schuhmacher, M. and Castells, F. (2003), 'Uncertainty assessment by a Monte Carlo simulation in a life cycle inventory of electricity produced by a waste incinerator', *Journal of Cleaner Production*. Elsevier, 11(3), pp. 279–92.

Spears, R. A., Kohn, R. A. and Young, A. J. (2003), 'Whole-farm nitrogen balance on western dairy farms', *Journal of Dairy Science*. Elsevier, 86(12), pp. 4178–86.

Specka, X., Nendel, C. and Wieland, R. (2015), 'Analysing the parameter sensitivity of the agro-ecosystem model MONICA for different crops', *European Journal of Agronomy*. doi:10.1016/j.eja.2015.08.004.

Stehfest, E., van Vuuren, D., Bouwman, L. and Kram, T. (2014), *Integrated Assessment of Global Environmental Change with IMAGE 3.0: Model Description and Policy Applications*, Netherlands Environmental Assessment Agency (PBL).

Steinfeld, H., Gerber, P., Wassenaar, T., Castel, V., Rosales, M. and de Haan, C. (2006), *Livestock's Long Shadow: Environmental Issues and Options*, Food and Agriculture Organization of the United Nations (FAO), Rome.

Subramanyan, K., Wu, Y., Diwekar, U. M. and Wang, M. Q. (2008), 'New stochastic simulation capability applied to the GREET model', *International Journal of Life Cycle Assessment*. Springer, 13(3), pp. 278–85.

Sugiyama, H., Fukushima, Y., Hirao, M., Hellweg, S. and Hungerbühler, K. (2005), 'Using standard statistics to consider uncertainty in industry-based life cycle inventory databases', *International Journal of Life Cycle Assessment*. Springer, 10(6), pp. 399–405.

Suh, S. and Yee, S. (2011), 'Phosphorus use-efficiency of agriculture and food system in the US', *Chemosphere*, 84(6), pp. 806–13.

Sutton, M. A., Bleeker, A., Howard, C. M., Bekunda, M., Grizzetti, B., de Vries, W., van Grinsven, H. J. M., Abrol, Y. P., Adhya, T. K., Billen, G., et al. (2013), *Our Nutrient World: The Challenge to Produce More Food and Energy with Less Pollution*, Centre for Ecology and Hydrology (CEH).

Thomassen, M. A., Dalgaard, R., Heijungs, R. and de Boer, I. J. M. (2008), 'Attributional and consequential LCA of milk production', *International Journal of Life Cycle Assessment*. doi:10.1007/s11367-008-0007-y.

Tilman, D. and Clark, M. (2014), 'Global diets link environmental sustainability and human health', *Nature*. Nature Publishing Group, 515(7528), pp. 518–22.

Vigne, M., Martin, O., Faverdin, P. and Peyraud, J.-L. (2012), 'Comparative uncertainty analysis of energy coefficients in energy analysis of dairy farms from two French territories', *Journal of Cleaner Production*. Elsevier, 37, pp. 185–91.

Walker, W. E., Harremoës, P., Rotmans, J., van der Sluijs, J. P., van Asselt, M. B. A., Janssen, P. and von Krauss, M. P. (2003), 'Defining uncertainty: A conceptual basis for uncertainty management in model-based decision support', *Integrated Assessment*. Taylor & Francis, 4(1), pp. 5–17.

Wei, W., Larrey-Lassalle, P., Faure, T., Dumoulin, N., Roux, P. and Mathias, J.-D. (2014), 'How to conduct a proper sensitivity analysis in life cycle assessment: Taking into account correlations within LCI data and interactions within the LCA calculation model', *Environmental Science & Technology*. ACS Publications, 49(1), pp. 377–85.

Whitehead, D. C. (1995), 'Volatilization of ammonia', *in Grassland Nitrogen*, CAB International, Wallingford, UK, pp. 152–79.

Wolf, P., Groen, E. A., Berg, W., Prochnow, A., Bokkers, E. A. M., Heijungs, R. and de Boer, I. J. M. (2017), 'Assessing greenhouse gas emissions of milk production: Which parameters are essential?' *International Journal of Life Cycle Assessment*. doi:10.1007/s11367-016-1165-y.

Wrisberg, N., Udo de Haes, H. A., Triebswetter, U., Eder, P. and Clift, R. (2002), *Analytical Tools for Environmental Design and Management in a Systems Perspective: The Combined Use of Analytical Tools*, Kluwer, Dordrecht, The Netherlands.

Xu, C. and Gertner, G. Z. (2008), 'Uncertainty and sensitivity analysis for models with correlated parameters', *Reliability Engineering & System Safety*. Elsevier, 93(10), pp. 1563–73.

Yang, Y. and Heijungs, R. (2017), 'A generalized computational structure for regional life-cycle assessment', *International Journal of Life Cycle Assessment*. doi:10.1007/s11367-016-1155-0.

Yang, J., Xu, M., Zhang, X., Hu, Q., Sommerfeld, M. and Chen, Y. (2011), 'Life cycle analysis on biodiesel production from micro algae: Water footprint and nutrients balance', *Bioresource Technology*. Elsevier, 102(1), pp. 159–65.

Yang, Y., Tao, M. and Suh, S. (2018), 'Geographic variability of agriculture requires sector-specific uncertainty characterization', *International Journal of Life Cycle Assessment*. doi: 10.1007/s11367-017-1388-6.

van Zanten, H. E. E., Mollenhorst, H., Oonincx, D. G. A. B., Bikker, P., Meerburg, B. G. and de Boer, I. J. M. (2015), 'From environmental nuisance to environmental opportunity: Housefly larvae convert waste to livestock feed', *Journal of Cleaner Production*, 102, pp. 362–9.

van Zanten, H. H. E., Bikker, P., Meerburg, B. G. and de Boer, I. J. M. (2018), 'Attributional versus consequential life cycle assessment and feed optimization: Alternative protein sources in pig diets', *International Journal of Life Cycle Assessment*. doi:10.1007/s11367-017-1299-6.

Zehetmeier, M., Gandorfer, M., Hoffmann, H., Müller, U. K., de Boer, I. J. M. and Heißenhuber, A. (2014), 'The impact of uncertainties on predicted greenhouse gas emissions of dairy cow production systems', *Journal of Cleaner Production*. Elsevier, 73, pp. 116–24.

Chapter 3

The use of farm-level models to assess the environmental impact of livestock production

Alasdair Sykes, Cairistiona Topp and Robert Rees, Scotland's Rural College (SRUC), UK

1 The role of modelling in agricultural greenhouse gas mitigation

1.1 Rationale for environmental modelling in agriculture

We know that our climate is changing, and the international community is committed to taking action to reduce the overall levels of greenhouse gas emissions in order to avoid dangerous climate change. Agriculture will play an important role here given that agriculture and land use are responsible globally for around 24% of greenhouse gas emissions. Climate change has been identified as the most potent threat facing the global economy (World Economic Forum, 2019), and public, scientific and political consensus generally reflects this sentiment (Lorenzoni and Pidgeon, 2006). In line with this, the Paris Agreement formalised a commitment to hold the increase in global average temperature to below 2°C since pre-industrial times (Griscom et al., 2017a). Achieving this commitment is certain to be extremely challenging (Peters et al., 2013) and will require significant reduction in emissions from all sectors. If rising demand for livestock protein is to be met, livestock agriculture therefore faces the dual challenge of reducing emissions while increasing production. This process has already begun; Opio et al. (2011) show the changes which global production has undergone in response to the three decades of population

http://dx.doi.org/10.19103/AS.2018.0044.04

and income growth. However, livestock remains a significant contributor to the global greenhouse gas (GHG) budget; global emissions from livestock production contributed 18% to total annual anthropogenic emissions in the first years of the 21st century (Steinfield et al., 2006). A growing global herd and high per-head emissions (Opio et al., 2013) means that cattle (i.e. beef and dairy production) form a significant proportion of this total, contributing almost three-quarters of total emissions (Caro et al., 2014). Over time, emissions from beef have risen by an estimated 59% over 50 years (Caro et al., 2016).

Production of calories from livestock is inherently a less efficient process than equivalent arable production. This low efficiency is a key factor in rendering livestock production, and its increase a critical environmental concern (Opio et al., 2011). Greenhouse gas emissions per kg of production ('emissions intensity') vary between types of animal-derived protein; ruminant production systems, particularly beef, tend to be the least efficient (Eshel et al., 2014). There is considerable disparity in production efficiency within sectors, which provides opportunity for mitigation of emissions. Qualifying this on a global scale, Smith et al. (2008) identified a number of approaches by which livestock emissions or emission intensities can be reduced. Three types of direct management practices were explored: improved feeding practices, dietary additives and longer-term management and breeding practices to improve the performance potential. In addition, practices relating to management of livestock manures, grazing land and cropland for feed production all had the potential to further reduce the emissions associated with production.

In general terms, livestock agriculture in the developing world is characterised by low-input, low-performance systems (Subak, 1999). Extractive farming practices (e.g. overgrazing) may reduce soil nutrient content and result in low-producing pastures and croplands, as well as direct CO_2 emission from soils (Vågen et al., 2005). Progressive intensification of these low-input systems has occurred over the past century in many western nations and has been shown to improve emission intensity through improvement of animal performance and the offset of enteric methane emissions (e.g. Cardoso et al., 2016). Currently, most beef production is located in developing nations (Kastner et al., 2012), and this is also where the majority of efficiency gains can be made; these factors mean that developed nations, which have undergone historical improvement of their production systems, have a role as practice leaders (e.g. Gerber et al., 2013; Opio et al., 2013).

In addition to potential for efficiency gains, it is worth noting that the global livestock production sector could reduce its contribution to global emissions simply by reducing production; Steinfeld and Gerber (2010) point out that either approach is possible, since the current rapid change in livestock agriculture is driven only in part by growing populations and to a larger extent by a growing middle class. In this sense, demand is manageable. With this in mind, it has been suggested that a move away from Western-level reliance on animal protein may be necessary or inevitable; a number of studies (Cassidy et al., 2013; Davis and D'Odorico, 2015; Davis et al., 2016; Gephart et al., 2016) demonstrate that apportioning human-edible arable crops away from the inherently inefficient livestock sector would be one way of mitigating environmental concerns and providing adequate food for a growing population. Land availability may also impose a limit on global meat consumption; Kastner et al. (2012) show that an area double the size of that currently cultivated would be required to feed a global peak population with an equivalent Western diet. Even if this were physically possible, the environmental damage caused by this shift would likely be severe before this point was reached (e.g. Machovina et al., 2015). Based on this logic, there exists a school of thought which suggests that livestock production practices are inefficient to the extent that they are irreconcilable with sustainable agriculture; more recently, some influential viewpoints in the popular media have begun to reflect this sentiment (e.g. Andersen and Kuhn, 2014; Monbiot, 2017).

There is also some evidence to suggest that developed societies may voluntarily move away from reliance on livestock products over time (FAOstat, 2017), and trends such as vegetarianism or veganism may occur as societies become richer and better educated (Ruby and Heine, 2012). Despite this, however, there currently exists a pressing demand for meat which is likely to continue to increase in the short-to-medium term (FAO, 2011). Managing the demand for livestock products is an important approach towards maintaining food security and minimising environmental concerns (Steinfeld and Gerber, 2010), but the currently prominent role of livestock production in the global economy (Opio et al., 2011) suggests that this will be a relatively slow process. Demand for animal protein and predicted peak population are both demonstrably linked to demographic transition (Lutz et al., 2001; Simpson, 2014; Sans and Combris, 2015), and the critical phase of action to maintain the 2°C limit falls ahead of this demand peak (Griscom et al., 2017b). This chapter will be developed in this context; namely, the requirement to address the immediate need to improve efficiency within livestock production systems, whilst acknowledging that active management of demand for animal-based protein is an important strategy for long-term mitigation of environmental impacts from this sector and for preservation of global food security at peak population.

In order to define effective strategies for the mitigation of emissions, it is necessary to understand the system in question. Farming systems, particularly those involving livestock production, are complex; Janzen et al. (2006) made the observation that interactions, feedbacks and trade-offs between on-farm practices and processes are inevitable, necessitating the use of broad-scope, holistic modelling approaches in order to avoid false economies when defining mitigation practices. At an even simpler level, modelling represents a cost-effective approach to the question of quantifying system-level emissions; whilst direct measurement is not impossible (e.g. Taylor et al., 2017), it is complex and costly and fails to account holistically for upstream emissions, which can be important in trade-offs (Janzen et al., 2006). In enabling consideration of hypothetical systems and system changes, modelling approaches offer valuable support to farm and policy-level decision makers (Kipling et al., 2016).

Models of natural systems vary by type and application, but can broadly be divided into empirical and process-based models (Gibbons et al., 2006; Del Grosso et al., 2006). Process-based models typically focus on representing underlying biological processes; by contrast, empirical models may ignore these underlying processes and instead focus on drawing conclusions based on observed data. Since most natural systems are inherently complex, accurately representing and quantifying underlying biological processes can be extremely challenging. Process-based models are therefore typically highly complex, and calibrating these models for different systems can be prohibitively data intensive (Hillier et al., 2011). Empirical models are typically much simpler and therefore less demanding of data. Whilst in some cases, these may lack the precision required for insightful analysis, counterintuitively, empirical models may be more accurate in real-world applications (Landau et al., 1999) and easier to integrate into system-level studies (Gibbons et al., 2006). Process-based models provide a means to study and understand natural processes; by design, empirical approaches are utilitarian, and eschew this role in favour of practical application.

Whilst process-based models have found considerable application in agricultural systems, use of these models has tended to be specific to academic assessments focusing on a particular emissions source (e.g. Del Grosso et al., 2005, 2006; Kröbel et al., 2016; Zhang et al., 2017). For broader, system-level assessment, simpler and more practical empirical models (which, to some extent, draw on the insights of process models) have tended to be used (e.g. Little et al., 2008; Hillier et al., 2011; MacLeod et al., 2013). The simplicity of these models permits a broader scope of assessment and generates a much lighter data demand.

1.2 The particular challenge of mitigating agricultural emissions

Under the Kyoto Protocol, most countries are required to report an inventory of their greenhouse gas emissions. The methodology provided by the Intergovernmental Panel on Climate Change (IPCC) guidelines for national greenhouse gas reporting (IPCC, 1996, 2006) serves as a basis for the calculation of these inventories. This assessment is conducted by sector at the national level and allows governments to quantify and benchmark emissions in order to demonstrate compliance with internationally agreed targets. In the case of many industries, the methodology recommended by the IPCC for national level inventories provides enough detail for governments to effect top-down approaches to mitigate greenhouse gas emissions (IPCC, 2006). Moran et al. (2011) observe that many high-emitting industries such as power generation involve relatively few centralised enterprises and mitigation practices which are largely well understood and documented. Consequently, top-down approaches to development of optimal mitigation strategies can be effectively undertaken using models which make broad, sector-wide assumptions. Agriculture, by contrast, is much more heterogeneous, with diversity on local and regional scales. As such, mitigation of emissions from agricultural practices resists a top-down approach; broad-brush mitigation strategies are difficult to tune across a diverse range of practices. Adding to this, the relative complexity of many of the biological emission pathways (e.g. soil-based nitrous oxide emissions or methane from enteric fermentation) means that, to work around data constraints, national inventory methodologies may be simplified to the point where they do not provide enough precision to respond to system changes representing mitigation strategies (e.g. IPCC Tier 1 methodology for livestock; Dong et al., 2006). Finally, mitigation within the agricultural sector represents a challenge in that measures may involve a trade-off between emissions sources (e.g. Hünerberg et al., 2014). National inventories (e.g. Salisbury et al., 2014) calculate emissions on a source basis, rather than an end user basis, and so do not account holistically for trade-offs where a mix of direct and 'upstream' emissions are involved.

Life cycle analyses (LCAs) offer a solution to this. LCA can be broadly divided into two categories namely attributional LCA, which aims to describe this environmentally relevant physical flows to and from the life cycle of a product, and consequential LCA, which aims to describe how these flows change in response to management (Ekvall et al., 2016). Most LCAs of agricultural systems are attributional and crucially differ from guidelines-based national inventories in that they account holistically for upstream emissions sources (e.g. agrochemical production). This typically renders approaches of this type more useful in the definition and assessment of mitigation options. However, the complexity of such systems typically means that LCAs focus on either one system type or at most a limited range (e.g. Beauchemin et al., 2010; Cardoso et al., 2016). As such, it is reasonable to suggest that the same heterogeneity which negates the efficacy of top-down approaches in agricultural systems also represents a barrier to the application of the results of individual LCAs to policy. Policy makers seek to legislate for a broad and diverse sector, and results garnered from a single specific system or system type are difficult to generalise. Differences in aims, methods and scope represent a barrier to review-based and meta-analytical synthesis of results derived from LCA studies.

This landscape represents the niche in which farm-level greenhouse gas modelling tools exist. In order to overcome the challenges discussed above, such tools must be:

1 Holistic in approach, in order to capture upstream emissions and associated trade-offs,
2 Precise enough to respond flexibly to system changes,
3 Simple enough to not impose prohibitively high data input or calibration burden, and,
4 Flexible enough to capture a wide range of scenarios.

In this chapter we describe a range of farm-level greenhouse gas modelling tools in line with the above constraints and critically assess their ability to function as a tool for farm-level benchmarking and mitigation assessment within these boundaries (Sykes et al., 2017). We also discuss the technical aspects of modelling farm-level emissions and environmental impacts within the defined framework, in order to provide a synthesis of the current state of the art and a basis for the consideration of specific development objectives.

1.3 Greenhouse gas emissions and the environmental footprint of agriculture

Three major greenhouse gases are emitted by the processes associated with agricultural production, with particular reference to the livestock sector, given the importance of this sectoral contribution to greenhouse gas emissions as discussed above. Almost all farm greenhouse gas calculators and life cycle assessments assess these gases in some capacity, though frequently vary in terms of the exact sources and sinks assessed (Schils et al., 2007). These gases are methane (CH_4), nitrous oxide (N_2O) and carbon dioxide (CO_2). There is also the potential for chlorofluorocarbon release (e.g. from refrigeration equipment), though this is likely to be relatively limited in the majority of agricultural systems. A graphical representation of the sources of these gases is shown in Fig. 1.

Enteric CH_4 is a by-product of the ruminant digestive system and is known to be regulated by a number of factors, including feed intake, structure and nutrient composition (Cederberg et al., 2013). It is estimated that CH_4 emission typically accounts for around 6.5% of a ruminant's gross energy intake (Dong et al., 2006). Manure storage also contributes significantly to CH_4 emissions; this is most notable in cattle and pig production systems (Cederberg et al., 2013). These are typically calculated according to IPCC (2006) Tier 2 guidelines; the IPCC provides a methane conversion factor for various systems and a range of annual temperatures. Accuracy can be improved where region-specific data are used (Cederberg et al., 2013).

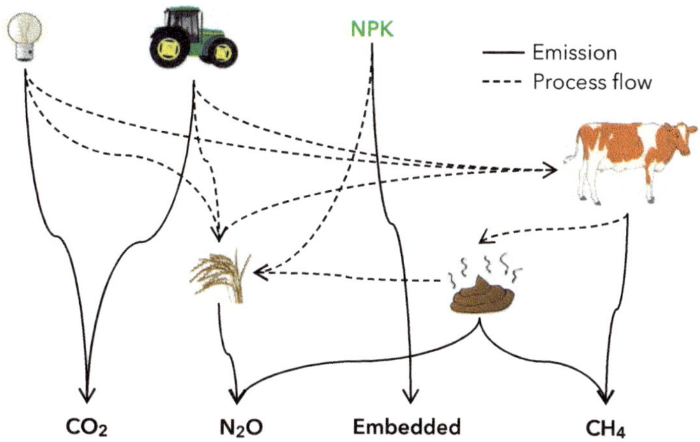

Figure 1 Simplified scope diagram for a farm-level model of a livestock production system, showing sources of carbon dioxide (CO_2), nitrous oxide (N_2O), methane (CH_4) and embedded (indirect) emissions. The dashed arrows indicate how processes on the farm interact with one another.

Nitrous oxide is emitted from a variety of nitrogen sources. A major source originates from the soil, which emits N_2O via the process of denitrification and nitrification (Cederberg et al., 2013). These processes are complex; emissions can be both direct and indirect and can be affected by local conditions such as climate, soil type, soil structure and drainage as well as nitrogen fertiliser application, manure application and crop residue management. Consequently, N_2O emissions from managed grassland can vary widely; Rees et al. (2012) showed that emissions collected from a compilation of experiments on European arable land can vary from 0.04 to 21.2 kg N_2O-N ha^{-1} yr^{-1}. This variability and array of contributing factors combine to make N_2O emissions from soil amongst the hardest to assess accurately. Nitrous oxide can also be released directly from manure (Dong et al., 2006), meaning that livestock numbers, breeds and diets, together with manure excretion to pasture and management regimes can also impact the emissions of this greenhouse gas.

Carbon dioxide emissions from agricultural production are largely dominated by land use change. When land under natural vegetation is converted to agricultural production, it is assumed that there is a change in the amount of soil organic carbon that is retained. The soil organic carbon content of cropland and grazed grasslands is often less than the carbon stocks associated with the native vegetation that they replaced. Carbon is lost from the soils as the organic matter decomposes and released to the atmosphere as CO_2. As agriculture has expanded over the past 150 years, it has led to a very significant contribution to GHG emissions. It is estimated that between 1870 and 2016, about 31% of cumulative emissions could originate from land use change (Le Quéré et al., 2016). Other more minor sources of CO_2 include farm sources primarily from burning of fossil fuels for transport or processing operations. In some regions the burning of crop residue may also contribute significantly but because of its importance in contributing to soil fertility and concerns about negative impacts on air quality, this is strongly discouraged or outlawed in regions such as Europe (Dong et al., 2006). Machinery and vehicle operation and heating of buildings are the prime reasons for on-farm fossil fuel use; consequently, intensive systems tend to produce more CO_2 from this source (Cederberg et al., 2013). Electricity supplied for use on farm will come from a range of sources, and it is likely that a portion of these will be fossil fuel powered (Sainz, 2003).

Farms are not closed systems, and farm-level assessments must take stock of greenhouse gas emissions associated with inputs and outputs. These are termed 'embedded' emissions and are frequently made up of a range of greenhouse gases; consequently, they are generally assessed at farm level in the form of CO_2 equivalents (CO_2-eq) (Gerber et al., 2013). Frequently included are emissions embedded in imported livestock feed/bedding, fertilisers, biocides and other agrochemicals (Gerber et al., 2013). It is also possible to assess embedded emissions associated with the construction of vehicles, machinery, buildings and so on; some LCAs do this, though proportional allocation of these centralised embedded emissions to a specific product or even an annual whole-farm assessment is problematic.

System boundaries must also be defined for emissions relating to farm exports, whether this is a waste product or a saleable by-product or main product. Exportation of by-products such as manure raises complex issues; a whole-farm assessment may choose to account for potential greenhouse gas emissions from this exported product or may ignore that proportion which leaves the farm gate. Likewise, the same assessment may include or ignore embedded emissions from imported manure. How these boundaries are defined will impact, perhaps significantly, upon the overall results of such an assessment, and comparisons between differently scoped assessments must be made with care to avoid misleading conclusions.

A greenhouse gas sink is something which facilitates the removal of a greenhouse gas from the atmosphere (Smith et al., 2008). These are mostly associated with land use change where land is converted from one associated with a lower soil carbon stock to a higher carbon stock. This may occur, for example, when the land used for arable cropping is converted into grassland or forestry. Other carbon sinks exist alongside many agricultural systems, in the form of cultivated and natural soils, field margins, hedgerows, woodland and wetlands (Griscom et al., 2017a). Whole-farm assessments will sometimes assess the CO_2 sequestration potential of these (e.g. Hillier et al., 2011). There are challenges associated with quantifying CO_2 removal by certain sinks, particularly where these are soil based (e.g. Campbell and Paustian, 2015), and under certain management practices, soils can become net emitters of CO_2. There are also issues associated with the permanence and fate of forestry-based sinks; whilst quantifying CO_2 removal by forestry is comparatively simple, if the resulting biomass is burned or allowed to fully decompose, the net CO_2 sequestration is returned to zero (Cannell, 1999). Land use may also change as a result of agricultural practices; a land use change assessment examines the difference in sequestration or emission potential between the two land cover types (Gerber et al., 2013). This can improve sequestration rates (e.g. if an area of grassland is planted with trees), but can be negative (e.g. deforestation to create land for cropping) (Steinfield et al., 2006). Difficulties associated with this approach largely stem from (a) defining the change in sequestration potential between land use types and (b) defining a baseline land use from which to quantify changes.

1.4 Other environmental impacts of agriculture

Aside from the well-documented effect of greenhouse gas emission, ammonia production from livestock is an important environmental concern. Around two-thirds of global ammonia release stems from the livestock sector, and this contributes heavily to acid rain, indirect greenhouse gas emissions and eutrophication and acidification of ecosystems (Steinfield et al., 2006). Sensitive ecosystems are particularly affected, with damage also caused by nitrogen deposition or acidification of soils (Demmers et al., 1999). Measuring ammonia release from UK cattle buildings, Demmers et al. (1998) estimated 6.0 kg NH_3 (500 kg live weight)$^{-1}$ (190 days)$^{-1}$ for a slurry-based dairy unit and 3.7 kg for a straw-bedded beef unit. Building on this work, Misselbrook et al. (2000) put forward an estimated UK agriculture ammonia emission factor of 226 kt NH_3-N yr^{-1}. The largest contributors to this figure on an individual basis were dairy cattle, at 16.9 g animal^{-1} day^{-1}. Beef cattle were allocated a much lower emission factor at 4.7 g animal^{-1} day^{-1}.

Water use is also under considerable pressure particularly from the livestock sector (Steinfield et al., 2006). Agriculture accounts for around 92% of humanity's global freshwater footprint (Gerbens-Leenes et al., 2013) and around a third of this is associated with livestock products; ruminant-based production systems have been shown to be particularly costly in this respect (Hoekstra and Chapagain, 2006; Eshel et al., 2014). As much of this footprint is associated with feed crop irrigation, the poor feed conversion rates of livestock is a contributing factor in this area.

Alongside water usage, of increasing concern is water pollution; nutrient runoff, often associated with application of artificial fertiliser or manure, is a leading cause of eutrophication of marine and aquatic ecosystems, causing losses of biodiversity and ecosystem services (Mason, 2002; Steinfield et al., 2006; Niemann et al., 2011). The EU Nitrates Directive (OJEC, 1991) is one example of recognition of this issue in policy. Of additional concern are other impacts to water courses, including increased runoff (from soil compaction and deforestation), reduced infiltration and degradation of watercourse banks (Steinfield et al., 2006).

Biodiversity is also impacted by agricultural production, with livestock and arable production likely to be amongst the leading causes of the current high rates of species extinction and biodiversity loss (Steinfield et al., 2006). In the context of the UK livestock sector, Sutherland et al. (2007) suggest that modifications to current livestock grazing strategies will be necessary in order to maintain biodiversity. However, livestock farming may be less damaging in this respect than equivalent arable production; the latter also requires significant management intervention (e.g. Meek et al., 2002) to prevent adverse biodiversity impacts.

2 Methodologies for accounting for agricultural emissions

2.1 The IPCC (2006) guidelines and the tier system

The IPCC (2006) guidelines for national greenhouse gas reporting provide a reference methodology and decision support tool to standardise the derivation of national-level greenhouse gas emissions inventories. The 2006 guidelines integrate and expand upon an original set of guidelines published a decade previously (IPCC, 1996). In addition to their intended use in national-level inventories, they also form the basis for many assessments of smaller scope or scale (e.g. Beauchemin et al., 2010; Hillier et al., 2011; Cardoso et al., 2016).

The IPCC (2006) guidelines utilise a three-tier system to categorise the depth of assessment of an emission source. These are published together with 'good practice' guidelines to support the decision as to the choice of tier for a particular emissions source. Moving to a higher tier generally improves the accuracy and reduces uncertainty, but requires more detailed inputs. Whilst some general methods are applicable for each tier, ascending tiers become more situation specific and hence require tailored methodologies. When conducting a calculation based on IPCC tiered methodology, various tiers can validly be employed as deemed appropriate; that is if data are available to conduct a Tier 2 assessment in one area, then this can be done alongside Tier 1 assessments in other, data-deficient areas.

Tier 1 methods are designed to be the simplest to use and contain the highest proportion of generic methodology and parameter values. Whilst country-specific data are needed, globally available, low-resolution data are generally sufficiently accurate to the level required by this methodology. Tier 2 assessments follow the same basic methodology as those employed by Tier

1, though they employ country- or region-specific data in place of generic emission factors and spatially coarse activity data. Often a mix is necessary, but Tier 2 methodology should seek to utilise higher-resolution data for the majority of livestock or land use categories.

Tier three assessments involve considerably higher order methodologies, and as such are designed to describe the underlying processes to high levels of detail. These are not standardised as in lower tiers, but should be determined according to the required application. The IPCC (2006) provides good practice guidelines to standardise development of these methods, including prescription of quality checks, audits and validations. Datasets will account for high levels of environmental detail, such as soil type, age, management activities, land use change over time and so on. If livestock are included in the assessment then there will be detailed levels of data disaggregation on the basis of factors such as species, breed, body weight, age and so on (Dong et al., 2006).

The IPCC (2006) provide detailed guidelines for choosing the methodology tier. Choice is broadly based upon data availability and requirement for accuracy. The latter is in turn based upon the significance of the LCA category or sub-category to be assessed (Dong et al., 2006). In line with the tier system, the IPCC (2006) details methodologies for calculating emissions from a range of biological, industrial and agricultural processes. These form the backbone of many whole-farm and LCA calculators, and as such, it is worth giving a brief overview of these methodologies. Sections 2.3–2.6 consider these methodologies and their application in the process of agricultural carbon footprinting.

2.1.1 Emissions from livestock and manure management

The IPCC (2006) guidelines concerning direct livestock emissions are concerned entirely with CH_4 and N_2O. Net CO_2 emissions from livestock are assumed to be zero; the CO_2 assimilated by plants is consumed by the animal and returned to the atmosphere as respired CO_2. The IPCC (2006) standardised methodology for calculation of these emissions requires, at minimum, definitions of livestock subcategories, annual populations and, for higher tier calculations, feed intake and characterisation. Direct livestock emissions are sorted into three categories (Dong et al., 2006) (a) CH_4 emissions from enteric fermentation, (b) CH_4 emissions from manure management and (c) N_2O emissions from manure management.

2.1.2 Emissions from managed soils

The IPCC (2006) methodology considers both direct and indirect N_2O emissions from managed soils. Direct emissions can be viewed as those which occur directly from the land in question; indirect emissions are those caused by management practices on the land in question, but which ultimately occur from other areas. Indirect emissions are possible largely due to leaching, runoff and volatilisation and deposition of N_2O 'precursors', processes which serve to remove the source of N_2O emission from the geographical location of the land in question.

For direct emissions, the following sources are considered:

1 Synthetically produced nitrogen fertilisers. Note that this methodology includes only N_2O emitted as a result of fertiliser application; considerable CO_2 is released during the fertiliser production process. This is accounted for separately as embedded emissions.
2 Organic nitrogen applied as fertiliser. Consideration of this emission source must take place in the context of manure management emissions; care must be taken that the methodologies do not overlap.
3 Nitrogen returned to the soil in the form of crop residues, i.e. remaining organic material following crop harvest.
4 Urine and dung deposited on managed soils by grazing animals. Likewise, this must be calculated in consideration of manure management emissions.
5 Nitrogen mineralisation associated with loss of soil organic matter (SOM). This can result from land use change (LUC) or management of mineral soils.
6 Drainage/management of organic soils.

Two pathways for indirect emissions from managed soils are identified by the IPCC (2006). Firstly, volatilisation of NH_3 and NO_x and deposition of these gases, and their products NH_4^+ and NO_3^-, onto soils and surface water can contribute to N_2O release in an analogous way to directly applied NH_3/NO_x fertiliser. A second pathway is formed by leaching and runoff of NO_3^- and NH_4^+ from synthetic and organic fertilisers, crop residues and mineralisation of soil nitrogen caused by changes in land management. These substances may bypass biological retention mechanisms where there is sufficient water flow or where they exist in the soil in excess of biological demand. Nitrification and denitrification processes then produce N_2O in the groundwater below the soil, in riparian zones receiving runoff, or in streams, rivers and estuaries which are contributed to by this runoff. The following sources of indirect N_2O emissions are considered in the IPCC (2006) methodology:

1 Synthetic nitrogen fertilisers.
2 Organic nitrogen fertilisers (manure, compost, sewage etc.).
3 Above- and below-ground crop residue nitrogen.
4 Nitrogen mineralisation associated with a loss of SOM.

2.2 Emission metrics used in greenhouse gas reporting

Global warming potential (GWP) is the current metric used to normalise the impacts of emissions of different greenhouse gases under the UNFCCC (Reisinger et al., 2011). The first GWP metrics were produced by the IPCC (1995) as part of the Second Assessment Report. Whilst others (such as global temperature potential, GTP) have since been proposed, GWP remains the popular choice of policymakers (Manning and Reisinger, 2011) and is therefore the emission metric discussed in this section. GWP is a physical metric derived from the lifetimes and radiative forcing values of greenhouse gases (IPCC, 2013) relative to those of CO_2; as such, there exist a variety of areas for uncertainty in the production of this metric. These include uncertainties in the radiative efficiency and lifetime of the greenhouse gas, together with background concentrations. Where the gas in question is not CO_2, describing GWP in terms of CO_2 equivalents (CO_2-eq) also incorporates those uncertainties associated with the absolute global warming potential (AGWP) of CO_2, the reference gas for the GWP metric.

Manning & Reisinger (2011) show that, aside from the issues of standardising to the CO_2 AGWP, the uncertainties associated with GWP for any gas come from three different factors; the radiative forcing caused by emission of the greenhouse gas, the rate at which a concentration caused by a pulse emission declines with time, and the indirect effects of that emission on other greenhouse gases within the atmosphere. The AGWP for CO_2 is a major factor in contributing to uncertainties in GWP for all greenhouse gases (Manning and Reisinger, 2011). This metric has come under considerable scrutiny as understanding of the global carbon cycle has progressed, and as such reported values for the AGWP of CO_2 have changed considerably since the IPCC Fourth Assessment Report (IPCC, 2007).

Contributing to the Fifth Assessment Report (IPCC, 2013), Joos et al. (2013) produced a comparison of carbon cycle-climate models and found that in modelling a 100 Gt-C pulse emission of CO_2 over a 1000-year time horizon, variations in results from the models suggested that 25 Gt-C ± 9% remained in the atmosphere. Together with an assumed ±10% uncertainty in radiative efficiency, this resulted in an AGWP for CO_2 with uncertainties of ±18% at a 20-year time horizon and ±26% at the commonly used 100-year time horizon.

Uncertainties in the ongoing AGWP for CO_2 have considerable implications for calculating the GWP for other gases, a metric which is normalised to the AGWP for CO_2. Incorporating this into the calculation of uncertainty in the GWPs for CH_4 and N_2O, Reisinger et al. (2011) estimated, for both gases, an uncertainty of −15/+20 for the GWP_{20} and −20 to +30% for the GWP_{100}. The calculations did not take into account uncertainties in the AGWP for CH_4 or N_2O.

Boucher (2012) estimated a ±20% uncertainty for the GWP_{100} of methane. The methods of this study differed from those of Reisinger et al. (2011) in that the result was calculated using Monte Carlo analysis of uncertainties in perturbation lifetime and radiative efficiency, Boucher (2012) assumed a constant background atmosphere following the modelled emissions pulse, as in Reisinger et al. (2011).

Assessing and combining the results of these studies, the IPCC (2013) estimates an uncertainty of ±30% for CH_4 GWP_{20} and ±40% for CH_4 GWP_{100}. This uncertainty is dominated by the uncertainties in AGWP for CO_2 and by the indirect effects of CH_4 emission. For longer lived gases, including N_2O, the IPCC (2013) provide a GWP uncertainty estimate of ±20% for the GWP_{20} and ±30% for the GWP_{100}.

3 Farm-level greenhouse gas tools

A wide range of freely available carbon calculators are available to the farming community that offer to quantify the carbon footprint of a farm enterprise. Different tools can result in substantially different estimates of greenhouse gas emissions (Sykes et al., 2017). To help explain this, this section aims firstly to review the role and approaches of existing farm-level greenhouse gas footprinting tools in the context of mitigation of agricultural emissions, in order to gain insight into their rational, scope and methods and to provide a basis for critique of the current approaches.

A farm-level greenhouse gas footprinting tool can be defined as an empirical, broad-scope tool designed for the purpose of providing better understanding of greenhouse gas emissions from a farming enterprise. Here, we review a selection of tools that are commonly used to develop carbon footprints in agricultural systems.

3.1 Profiles of farm-level greenhouse gas tools

3.1.1 The AgRE Calc tool

AgRE Calc (SRUC, 2014), standing for Agricultural Resource Efficiency Calculator, was developed by Scotland's Rural College. The calculator uses an IPCC (2006) Tier 1 and 2 methodology and is PAS2050 certified. It was developed initially in Microsoft Excel, though it is available to the public in Web-based form only. The model is used extensively as a consulting tool and provides a series of key performance indicators alongside greenhouse gas emissions estimates.

IPCC (2006) Tier 2 calculations are employed to calculate livestock and manure management emissions; this level of detail is rare amongst carbon calculators and requires detailed data input in this section. Fertiliser-embedded emissions are calculated using Carbon Trust (2010) emission factors, whilst soil emissions from fertiliser application and crop residues follow IPCC (2006) Tier 1 methodology. System boundaries include embedded emissions for material usage (e.g. plastics) and for imported feed; these are sourced from Footprint Expert v3.1 (Carbon Trust, 2010).

Electricity, renewable energy and fossil fuel emissions are calculated using emission factors from the DEFRA/DECC (2011) Conversion Factors for Company Reporting. Finally, carbon sequestration from woodland is included in the system boundaries and estimates follow IPCC (2006) methodology. Though not yet available in the online version of the model, SRUC is undertaking work to extend the system boundaries of the model to include a post-farm gate emissions estimate and to include other environmental impacts such as LUC and eutrophication in the calculations.

3.1.2 The Farm (CFF) Carbon Calculator

The Climate Friendly Food (CFF) Carbon Calculator (CFF, 2012) is an online tool funded with the support of Nesta. It places a strong emphasis on organic agriculture and horticulture. The model is based on Tier 1 methodology (IPCC, 2006) and follows a whole-farm approach. The calculator is JavaScript based and available on the Web.

The model has the capability to assess greenhouse gas emissions from fuel and electricity use, material consumption (e.g. plastics), crop production (including embedded emissions in imported feeds), fertiliser use, direct livestock emissions and manure management. There is also a strong emphasis on new-start or expanding enterprises, with the facility to assess emissions associated with building materials and capital items such as farm machinery. Lastly, there is a function to assess post-farm gate haulage costs (in terms of additional fuel use) and waste.

Aside from consideration of inputs, the model also provides an assessment of carbon sequestration. Sequestration potential is assessed for on-farm 'wild' areas such as field margins, hedges, wetlands and woodlands. The model also has the capability to assess managed land such as orchards and vineyards. The Farm Carbon Calculator does not directly assess grazed or managed grassland, but does have the capability to assess the carbon sequestration by soil, based on annual SOM change, if this data is input by the user. In the results, this sequestration potential is offset against on-farm emissions.

The model system boundaries are limited in that very little emphasis is placed upon N_2O emissions. Where these are associated with crop residues, they are considered in the model; however, the calculations take no account of N_2O emissions from fertiliser spread or from manure. Fertiliser is accounted for in terms of embedded emissions only; likewise, manure is accounted for only in terms of CH_4 emission. The authors state that this omission is due to the significant variation in these emissions caused by different climatic and soil conditions and that collecting and inputting this data would be too onerous for users (CFF Carbon Calculator, 2012).

Colomb et al. (2012) reviewed the tool and assessed it in comparison with others for skill and time requirements to complete a study. The authors used a scale of 1–4, with 1 representing the minimum skill and time requirements. The CFF tool was assessed as level 2 for time requirement and as level 3 for skill requirements. Whittaker et al. (2013) also produced a multi-criteria analysis which included the CFF tool; in this, the calculator scored highly for user-friendliness and information provision, though less well in the transparency and comprehensiveness criteria.

3.1.3 The CPLANv0 tool

CPLANv0 (SEE360, 2007) is a free-to-use carbon calculator set up by tenant farmers based in central Scotland. The development was supported by public funding provided by the South Lanarkshire 'LEADER +' grant. The model forms a key component of the agricultural consultancy business SEE360 Ltd.

The model is based on IPCC (2006) methodology and follows a whole-farm approach including carbon sequestration. It is Web based and uses JavaScript to perform calculations. Other than the statement that IPCC (2006) methodology has been observed, there is little detail given as to the specific calculations performed by the calculator.

The model methodology is therefore relatively opaque; this assessment is reflected by Whittaker et al. (2013) who awarded the model a transparency score of 17% (average = 49%). The system boundaries include CH_4 from enteric and manure sources, though exclude N_2O from manure or fertiliser spread. Nitrous oxide from crop residues is assessed. Embedded emissions from imported fertilisers, including organic fertilisers, are included, as are those associated with fossil fuel and electricity use. Renewable energy use is not considered. The sequestration potential of standing woodland is assessed, as well as impacts from forestry and LUC. The model assesses a total of 57 subcategories, with the heaviest focus on sequestration and LUC.

3.1.4 The CALM tool

The CALM Carbon Calculator was developed by the Country Land and Business Association, in partnership with Savills (CLA, 2009). Funding was supplied by the East of England Development Agency and the Crown Estates, with later updates sponsored by Natural England.

The model methodology is described as following the most recent National Inventory Report (published by the National Atmospheric Emissions

Inventory) and was recently updated in 2009. With no specific source cited for calculations, it is difficult to ascertain the provenance of model methodology.

Amani and Schiefer (2011) found that the CLA (2009) made some adjustments to the National Inventory Report methodology to allow the calculator to be applicable on a farm level. The most significant change involved the re-apportioning of manure emissions. These include only the emissions from manure used or brought onto the farm. In this respect, embedded emissions for manure brought onto the farm are assessed (these are often neglected by other models), whilst emissions for manure produced on farm but exported are neglected.

Despite this lack of transparency, the model has reasonably comprehensive system boundaries. The methodology assesses N_2O emissions from crop residues, fertiliser spread and manure management, within the boundaries described by Amani and Schiefer (2011). Enteric CH_4 emissions and those from manure management are considered, as are emissions associated with on-farm fuel and electricity use, including a variety of renewables. A variety of contracting activities are also included. The model has the capacity to assess emissions and sequestration associated with forestry, soil organic carbon and LUC. Notable system boundary omissions are embedded emissions associated with imported livestock feed and bedding. The system boundary ends at the farm gate and does not account for transport off-farm or emissions associated with product distribution.

3.1.5 The FCAT calculator

The Farm Carbon Assessment Tool (FCAT) is a free online self-assessment tool developed and provided by the Soil Association as part of the EU Low Carbon Farming Project (Soil Association, 2013). The tool focuses heavily on soil management and inputs, though includes facility for livestock data input.

Aside from an assessment of farm emissions from energy usage, the tool provides its report in the form of a series of performance indicator scores (given in arbitrary units, 1–5). This is a relatively opaque methodology, and there is little information available on how these scores are calculated. Provision of results in this form severely limits comparison with other calculators; aside from the energy usage emissions, no further comparison was made in this review.

3.1.6 The Cool Farm Tool

The Cool Farm Tool (Hillier et al., 2011) was developed at the University of Aberdeen as a mid-level calculator designed to cover a range of farming practices. It was originally engineered in Microsoft Excel and is now freely available under a Creative Commons (CC) licence as both an Excel-based and

Web-based programme. Hillier et al. (2011) state that the tool was designed to function at an intermediate level. Requirement for the high levels of data input of process-based models such as DAYCENT (Del Grosso et al., 2006) was avoided, but provision for data input beyond the standard Tier 1 inventory methods (Dong et al., 2006) were included so as to provide useful insight on a local scale.

Hillier et al. (2011) made use of a variety of emission factor databases and sub-models to provide farm-level functionality for Cool Farm Tool. Firstly, the Ecoinvent emission factor inventory (Ecoinvent Centre, 2007) was used to provide emission factors for fertiliser production and renewable electricity usage. Hillier et al. (2011) also incorporated a sub-model developed by Bouwman et al. (2002) to determine N_2O emissions relating to fertiliser usage; this model is based upon a global dataset of over 800 sites. IPCC (2006) methodology was used for livestock and manure management calculations. The model is stated to contain provision for Tier 1 or Tier 2 level calculations, as allowed by input data (Hillier et al., 2011).

Ogle et al. (2005) conducted a meta-analysis which quantified the impacts of agricultural land management on organic carbon content of a variety of soil types. The results of this study were incorporated into the Cool Farm Tool by Hillier et al. (2011), enabling assessment of the impacts of management change by the model. Defined management practices included input application levels and tillage practices; coefficients were provided for four soil types (two of which are relevant to UK agriculture) changing between three levels of each practice. A separate set of coefficients, derived from a long-term study by Smith et al. (1997), allow for separate consideration of organic manure application and the effect of this practice on soil organic carbon (SOC).

The Cool Farm Tool is intended for application worldwide, and so has provision for global variability in the inputs (Hillier et al., 2011). These include consideration of tropical soils in the incorporated dataset from Ogle et al. (2005), and inclusion of a database from greenhouse gas protocol (2003) which provides CO_2 emission factors for grid electricity by country.

3.1.7 The Carbon Calculator for New Zealand Agriculture and Horticulture

The Carbon Calculator for New Zealand Agriculture and Horticulture (AERU, 2008) was developed by the Agribusiness and Economics Research Unit (AERU) of the University of Lincoln, NZ, and is freely available online as part of the official university website.

The calculator provides the facility to assess the emissions relating to farm energy and fuel use, including that of contractors. Methane and N_2O emissions relating to livestock and manure are assessed, though no differentiation between manure storage strategies is apparent. Embedded emissions in fertiliser are considered, as are N_2O emissions related to application rate. There is no consideration of N_2O emissions relating to crop residues produced on farm; whilst the model does consider an embedded emission factor for imported feed, it is not clear where the system boundaries lie in this area. Carbon sequestration is not considered by the model.

The extent to which the model is specific to New Zealand is not clear; no indication is given as to the source of the methodology utilised in the model. The tool name suggests it is regionally specific. A search of the published literature did not reveal any case studies performed using the calculator. Colomb et al. (2012) produced a review which included the calculator, rating at the minimum level for time and skill requirements. The authors suggested that the tool is most suited to raising awareness of major agricultural greenhouse gas sources, rather than for specific footprinting, decision making or performing greenhouse gas audits.

3.1.8 The Farming Enterprise Greenhouse Gas Calculator

The Farming Enterprise Greenhouse Gas Calculator was developed at the University of Queensland to assess the emissions from a variety of agricultural enterprises in this area. Limited published information is available on the tool, though it became clear during initial assessment of the tool that the methodology was specific to the Queensland area, with no option for user-defined inputs to expand the model capabilities beyond this point.

3.1.9 The CCaLC Carbon Footprinting Tool

The CCaLC Carbon Footprinting Tool (CCaLC, 2014) was developed by the University of Manchester's School of Chemical Engineering and Analytical Science. The tool differs from the majority of reviewed calculators in that the functional unit of output (in most cases whole-farm annual emissions) is entirely user defined; the model is, in essence, a multi-purpose LCA calculator.

Thus, CCaLC can be used to footprint products and processes from a wide variety of industries. Whilst this additional functionality represents potential for a wide range of applications, it also means that the tool is considerably more complex than many others; crucially, there is no guided, step-by-step approach to utilising the tool (as with others reviewed here) and system boundaries are entirely user defined. This means gaining an accurate result is time-consuming and is likely to require training for most users. This problem is addressed with the provision of training courses by the developers of CCaLC.

The tool is Excel based, with a free download available from the Web. Functionality is provided entirely by VBA scripts, with very limited user access to database calculations. The tool incorporates an extensive Ecoinvent LCA database from which it derives a great deal of functionality. Because of the wide application potential of the CCaLC tool, and lack of specificity to the livestock sector, this tool was excluded from further review.

3.1.10 The Muntons Carbon Calculator

Muntons PLC is an international company specialising exclusively in the production and supply of malts to the food and drinks industry. Initial assessment served to identify that this model is specifically designed to provide assessments within this sector, and thus, the system boundaries are confined entirely to assessment of cereal yield, diesel consumption, seeds and pesticides (Whittaker et al., 2013), making it incompatible with wider agricultural practice, particularly in the area of livestock production.

3.2 Summary of reviewed farm-level tools

Table 1 provides a summary and breakdown of the major attributes, methodological approaches and system boundaries of the tools included in this review.

In addition to the qualitative reviews discussed in this section, Sykes et al. (2017) conducted a quantitative assessment of some publicly available and commonly cited tools. The authors found that tools could produce markedly different results from common input data (Fig. 2). Despite some commonly stated assumptions, it is apparent that tool developers, in combining and adapting methodologies, can follow quite different approaches.

The tools reviewed above are relatively heterogeneous in origin, design and approach. Nevertheless, some commonalities were present, and these are worth discussing. Almost all tools chose to include direct and indirect emissions from livestock in the scope, reflecting the magnitude and perceived importance of this source. Those which did not include this emission source were designed specifically for non-livestock enterprises. Inclusion of nitrous oxide from managed soils was relatively universal, though some tools (e.g. CFF, 2012) differed slightly in the specific sources they included. Inclusion of emission from energy use on farm was also an almost universal choice; whilst energy use in livestock production forms a relatively small proportion of the footprint (Opio et al., 2013), its inclusion here may reflect its relative simplicity of calculation and prominent inclusion in greenhouse gas accounting methodologies (e.g. GHG Protocol, 2012; DEFRA and DECC, 2015). Interestingly, though the efficacy of forestry in offsetting emissions is the subject of ongoing discussion (Cannell, 1999), the majority of tools appeared to include this as a carbon sink.

Table1 Summary of attributes for tools reviewed in section

		Emissions sources/sinks								Methodology		
Model name	Organisation	CO_2 from energy use	Enteric CH_4	CH_4 from manure	N_2O from fertiliser	N_2O from manure	N_2O from residues	Post-farm gate	Carbon sequestration	IPCC Tier 2 (livestock)	Non-GHG impacts	Uncertainty estimate
AgRE Calc	SRUC	Y	Y	Y	Y	Y	Y	N	Y	Y	N	N
CFF Calculator	Farm Carbon Calculator	Y	Y	Y	Y	Y	N	Y	Y	N	N	N
CPLANv0 Calculator	SEE360 Ltd	Y	Y	Y	Y	Y	Y	N	Y	N	N	N
CALM Calculator	Country Land and Business Association	Y	Y	Y	Y	Y	Y	N	Y	N	N	N
FCAT	The Soil Association	Y	N	N	N	N	N	N	N	N	Y	N
Cool Farm Tool	Cool Farm Alliance	Y	Y	Y	Y	Y	Y	N	Y	Y	Y	N
Calculator for NZ Agriculture	Lincoln University, NZ	Y	Y	Y	Y	Y	N	N	N	N	N	N
Farming Enterprise GHG Calc.	University of Queensland	*	Y	?	Y	?	Y	N	N	N	N	N
CCaLC	University of Manchester	Y	Y	Y	N	N	N	Y	N	N	Y	N
Muntons	Muntons PLC	*	N	N	Y	N	N	Y	N	N	N	N

* = fossil fuel emissions only.
? = data unavailable, Y= Yes, N=No.

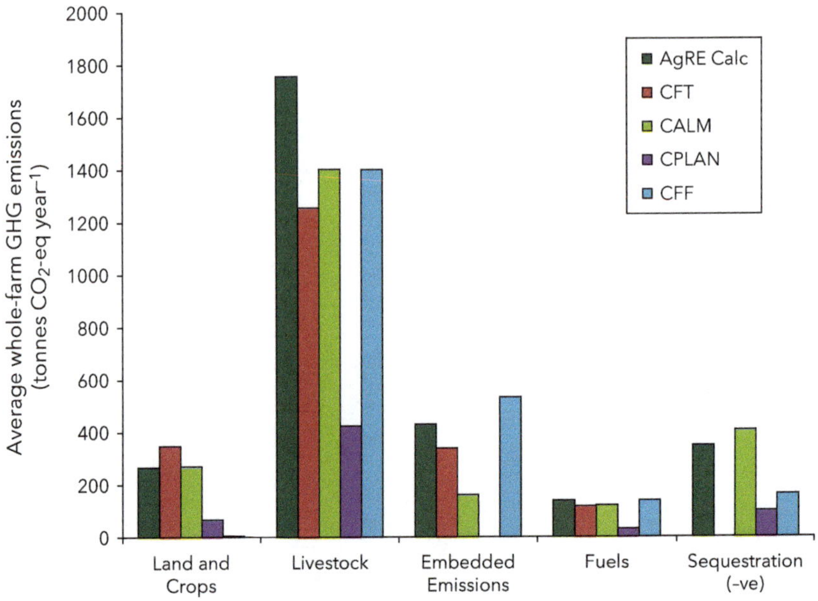

Figure 2 Average whole-farm emissions calculated across seven mixed enterprise United Kingdom farms. Estimates for disaggregated emissions sources from five different farm-level GHG tools using common input datasets are shown. Source: data from Sykes et al. (2017).

Del Grosso et al. (2006) suggested that IPCC Tier 1 level methodologies may not be sufficiently precise to give useful insight at the farm level. This conclusion appears to have been echoed by some tool developers (e.g. Hillier et al., 2011) in addition to AgRE Calc, but the majority of tools appeared to follow this simplified approach. Uncertainty assessment, identified in the literature as being of considerable potential importance in greenhouse gas footprinting (Gibbons et al., 2006; Röös and Nylinder, 2013), was not incorporated by any tools. The reason for this is not clear – some developers acknowledge its relative importance (e.g. CFF, 2012) – but may be related to the computational demand this calculation is likely to impose. Several tools incorporated estimates of non-greenhouse gas environmental impacts, though the methods by which these were calculated were not always clear.

One notable point which became apparent when researching the tools was the considerable inconsistency of methodological documentation, leading to a lack of transparency in the modelling approaches used. Some developers (e.g. Hillier et al., 2011) thoroughly document the model methodology for users, whilst others, for example, CPLANv0 (SEE360, 2007), provide little or no supporting documentation. With the notable exception of the Cool Farm Tool, the majority tool development appeared to have taken place outside of the peer-review system.

4 Developing farm-level modelling approaches

This section builds on the above review of extant tools and considers the challenges, limitations and opportunities offered by the development of the approaches discussed up to this point.

4.1 *Challenges of model sensitivity, flexibility and data input burden*

Developing a farm-level greenhouse gas tool requires methodology simple enough to keep data input burden low, but detailed enough to allow the simulation of complex scenarios and mitigation options. A fundamental difference between farm-level tools (such as AgRE Calc and the Cool Farm Tool) and scenario-specific LCA studies is that tools must be flexible enough to accommodate an undefined range of scenarios and options, whilst LCAs represent only the defined scenario. An important aspect of maintaining this flexibility is maintaining the interconnectivity between model processes. Emphasis is often put on the ability of modelling-based farm-level tools to capture the impacts of mitigation options, particularly if the tool in question is expected to influence policy decisions (e.g. Hall et al., 2010). Projects undertaken on behalf of governmental decision-making bodies (e.g. MacLeod et al., 2017) suggest that the challenges presented to policy definition by the heterogeneity of agricultural practices can be solved with the appropriate modelling approach; however, a key component of this must be the ability to respond flexibly to a variety of scenarios.

Where LCA studies model mitigation scenarios, these scenarios tend to be modelled as an empirical calibration of the study methodology or input dataset (e.g. Cardoso et al., 2016). As observed, LCA studies are in effect point estimates, meaning that the extrapolation of modelled mitigation options outside the defined scenario is problematic. This may be because the option itself is scenario specific or because the way it is captured within the methodology relies on empirical estimates which are not readily adapted outside the modelled scenario. Development of estimates for a wider range of scenarios requires considerable expertise on the part of the modeller, and a high degree of confidence that the estimates are applicable. This specificity, coupled with the heterogeneity of livestock systems, understandably represents a barrier to the incorporation of LCA-modelled mitigation strategies into policy.

The solution to this challenge is to maintain links between model processes; this means that whenever changes to the system are made, the model can adapt without the requirement for additional data input or empirical estimation by the user. As such, any mitigation options involving the interaction between these variables can be captured much more fully in the tool, avoiding the potential for accidental bias and negating the requirement for particular expertise on the part of the user. Development of farm-level greenhouse gas models following an approach which maximises connectivity between data processes in this way will greatly increase a model's worth in assessing mitigation strategies.

4.2 Consideration of spatially specific environmental data in farm-level footprints

In the context of development and improvement of farm-level models, one important issue to consider is the potential role of environmental input data. For example, the inclusion of environmental variables such as latitude, temperature, rainfall and altitude in any biological process model has the potential to considerably improve the precision and accuracy of estimates; these factors can impact animal performance, N leaching and volatilisation rates, enteric emissions, and therefore, the emission intensity of production as a whole. An existing example employing IPCC methodology is the Tier 2 approach for calculation of manure CH_4 (Dong et al., 2006); modelled emissions are highly dependent on average daily temperature. Many of the reviewed tools utilise this calculation stream, though do not require or estimate temperature as an input.

The process of making environmental variables available for use in farm-level tools would be relatively straightforward, given the availability of spatial data; from a simple input such as a postal/ZIP code, a host of spatially explicit data can be derived to a high degree of accuracy. However, the question which must be considered is the extent to which this should be allowed to influence the calculated footprint; if financial incentives exist to reduce emissions, care must be taken not to unfairly penalise certain enterprises where unfavourable environmental conditions, rather than undesirable management practices, result in a higher carbon footprint.

A counter-argument to this is that the heterogeneity of farming enterprises is in part at least due to differences in environmental factors, both directly and indirectly where environmental heterogeneity precipitates differences in management strategies. A strong argument for the usefulness of farm-level tools stems from this heterogeneity and the difficulties it engenders in defining strategies for emissions reduction, so in this sense, it is counter-intuitive to hamper the ability of tools to account for environmental factors.

There is no clear solution to this issue, but arguably there exists a 'glass ceiling' in terms of the insight that farm-level tools can give whilst assuming spatial homogeneity. As such, rendering tools spatially specific in their calculations could serve to provide much greater insight in terms of emissions mitigation and may be necessary if mitigation strategies are to be finely tuned in this respect. One potential solution to the conflict described above could be an approach which accounts for both scenarios; a non-spatially specific 'baseline' estimate for benchmarking, accompanied by a spatially specific footprint to give greater insight into farm- or area-specific mitigation practices. In either case, it is important to develop tools with an eye to their potential use by policy makers and to clearly define the purpose of a particular approach to ensure it is not applied in a way which subverts its original intention.

4.3 Challenges in modelling non-greenhouse gas environmental impacts in farming systems

Climate change, in comparison to the other impacts discussed above, has received considerable attention in the global environmental agenda (Röös et al., 2013). This means that the commonly used GWP metric, whilst not beyond critique (e.g. Smith, 2003; Shine, 2009), is generally well accepted. Greenhouse gas emissions are also relatively unique in being almost entirely non-spatially specific; emissions of climate-forcing gases diffuse quickly and cause similar impacts regardless of the precise point of origin (Röös et al., 2013). By contrast, impacts such as eutrophication and acidification are much more spatially specific (Posch et al., 2008; Röös et al., 2013), and real-world impacts can vary over small geographic scales. Where a farm-level tool or LCA modelling approach is not spatially specific in this sense, characterisation of these factors may be problematic.

Whilst there exist generally accepted metrics for impacts such as eutrophication and acidification (PO_4^{3-}-eq and SO_2-eq respectively; Williams et al., 2006), others are harder to quantify. Toxicity, for example, is an important environmental impact, but heterogeneity in accounting methods has been cited as a reason for its exclusion from previous assessments (Pant et al., 2004). Biodiversity, also, is not an inherently quantifiable concept, and selection of an appropriate metric can be a challenging problem even for specifically defined studies and assessments (Gotelli and Colwell, 2001). Penman et al. (2010) considered the adaptation of a methodology for accounting for biodiversity in LCA, but concluded that considerable research and development would have to be conducted in order to render it feasible in the foreseeable future. Key issues in this respect were those of small-scale landscape heterogeneity, meaning practices could have different impacts across small geographic scales and definition of a baseline biodiversity state in order to separate the process of interest to the LCA from previous changes or disturbances.

Providing a perspective specific to the livestock sector, Röös et al. (2013) conducted a meta-analysis which analysed a number of environmental impacts (primary energy use, land use, acidification potential, eutrophication potential, and pesticide use) alongside a measurement of GWP for livestock production systems. The authors found that in general, greenhouse gas emissions showed a strong positive correlation with other environmental impacts and concluded that the omission of additional impacts from an LCA assessment was consequently unlikely to lead to false economies based on interpretation of the results. The authors also found that a key reason for this is that many impacts (e.g. eutrophication, acidification) have similar precursors to greenhouse gas emissions (namely NH_3 and NO_x); following this, the development of methodologies to account for these impacts relies to a large extent on a common base methodology (e.g. Williams et al., 2006).

In summary, modelling the majority of non-greenhouse gas impacts represents a considerable challenge, which stems largely from the difficulties associated with quantifying these impacts. Accurately and precisely accounting for impacts which are highly variable over small geographical scales (such as eutrophication, acidification and biodiversity) is a particular challenge, and may be relatively incompatible with the concept of a non-spatially explicit model (see above). This is compounded by the small scale for which the model is designed; whilst non-spatially specific assays exist for non-greenhouse gas impacts (e.g. Williams et al., 2006; Nguyen et al., 2013), these tend to be large scale, where the effects of small-scale regional heterogeneity will largely be smoothed out. Added to this, Röös et al. (2013) demonstrated that the risk of jeopardising other impact categories by focusing on GWP is relatively slight in the case of livestock production systems.

5 Conclusion

Agricultural emissions are of considerable importance to the GHG budget, and their mitigation is a challenge made more difficult by the increasing demands placed on global agricultural production. Livestock agriculture is a sector particularly tested by this demand for increased mitigation and production. Farm-level greenhouse gas accounting tools provide a simple aid to quantifying and understanding agricultural emissions and processes and have a role in the adaptation of agriculture to global change.

It is clear, however, that the majority of farm-level tools do not effectively account for their methodological choices, either in the peer-reviewed or grey literature. Nonetheless, almost all tools draw on peer-reviewed approaches in one form or another and often share a common methodological basis, though disparity in interpretation and adaptation of available methodologies can and does lead to considerable differences in output. Adding to this complexity, farm-level greenhouse gas accounting tools are conceptual models, and there is limited possibility to validate results against observed data. These considerations highlight the importance for tool users of recognising the differences which may be present, and for developers of thoroughly documenting assumptions made in the adaptation of any approach. The requirement for more information to be made available on these tools has been implicitly acknowledged by the multitude of reviews which have been carried out for these models (e.g. Colomb et al., 2012; Whittaker et al., 2013; Keller et al., 2014; Sykes et al., 2017). Nonetheless, these tools provide valuable insight into agricultural processes and emissions and their development and are a critical aspect of mitigation efforts in this sector.

6 Acknowledgements

The authors would like to thank the Scottish Government, Innovate UK and J Sainsbury PLC for funding to support this research.

7 Where to look for further information

A good general introduction to environmental modelling is provided by Smith and Smith (2007), where they discuss the underlying principles of different modelling approaches, and consider how models can be used as research tools.

The role of agriculture in contributing to greenhouse gas emissions is described in more detail by the FAO: http://www.fao.org/resources/infographics/infographics-details/en/c/218650/

A global inventory of models has been created and maintained by the Global Research Alliance on Agricultural Greenhouse Gas Emissions: http://gramp.org.uk/

An international inter-comparison of agricultural models has been undertaken in a separate exercise by the Global Research Alliance on Agricultural Greenhouse Gas Emissions: http://www.agmip.org/

8 References

AERU. 2008. Carbon calculator for New Zealand agriculture and horticulture (Online). Available at: http://www2.lincoln.ac.nz/carboncalculator/ (accessed on 13 October 2014).

Amani, P. and Schiefer, G. 2011. Data availability for carbon calculators in measuring GHG emissions produced by the food sector. *International Journal of Food System Dynamics* 2(4), 392–407.

Andersen, K. and Kuhn, K. 2014. *Cowspiracy: The Sustainability Secret*. A.U.M. Flims & First Spark Media, United States.

Beauchemin, K. A., Henry Janzen, H., Little, S. M., McAllister, T. A. and McGinn, S. M. 2010. Life cycle assessment of greenhouse gas emissions from beef production in western Canada: a case study. *Agricultural Systems* 103(6), 371–9. doi:10.1016/j.agsy.2010.03.008.

Boucher, O. 2012. Comparison of physically- and economically-based CO_2-equivalences for methane. *Earth System Dynamics* 3(1), 49–61. Available at: http://www.earth-syst-dynam.net/3/49/2012/ (accessed on 22 January 2015).

Bouwman, A. F., Boumans, L. J. M. and Batjes, N. H. 2002. Modeling global annual N_2O and NO emissions from fertilized fields. *Global Biogeochemical Cycles* 16(4), 28–1. doi:10.1029/2001GB001812.

Campbell, E. E. and Paustian, K. 2015. Current developments in soil organic matter modeling and the expansion of model applications: a review. *Environmental Research Letters* 10(12), 123004. Available at: http://stacks.iop.org/1748-9326/10/i=12/a=123004?key=crossref.d7e96c966984dd06beac7fc8f424bba7.

Cannell, M. G. R. 1999. Growing trees to sequester carbon in the UK : answers to some common questions. *Forestry* 72(3), 237–47. doi:10.1093/forestry/72.3.237.

Carbon Trust. 2010. *Footprint Expert v3.1*. Available at: https://www.carbontrust.com.

Cardoso, A. S., Berndt, A., Leytem, A., Alves, B. J. R., de Carvalho, IdN. O., de Barros Soares, L. H., Urquiaga, S. and Boddey, R. M. 2016. Impact of the intensification of beef production in Brazil on greenhouse gas emissions and land use. *Agricultural Systems* 143, 86–96. doi:10.1016/j.agsy.2015.12.007.

Caro, D., Davis, S. J., Bastianoni, S. and Caldeira, K. 2014. Global and regional trends in greenhouse gas emissions from livestock. *Climatic Change* 126(1–2), 203–16. doi:10.1007/s10584-014-1197-x.

Caro, D., Davis, S. J., Bastianoni, S. and Caldeira, K. 2016. Greenhouse gas emissions due to meat production in the last fifty years. In: Ahmed, M. and Stockle, C. O. (Eds), *Quantification of Climate Variability, Adaptation and Mitigation for Agricultural Sustainability*. Springer, pp. 27–37. Chapter 2.

Cassidy, E. S., West, P. C., Gerber, J. S. and Foley, J. A. 2013. Redefining agricultural yields: from tonnes to people nourished per hectare. *Environmental Research Letters* 8(3), 1–8. Available at: http://stacks.iop.org/1748-9326/8/i=3/a=034015?key=crossref.531cd54b9d3cb296e11617201b50db14.

CcaLC. 2014. CCaLC: carbon calculations over the life cycle of industrial activities. Online. Available at: http://www.ccalc.org.uk/index.php (accessed on 10 October 2014).

Cederberg, C., Henriksson, M. and Berglund, M. 2013. An LCA researcher's wish list-data and emission models needed to improve LCA studies of animal production. *Animal: an International Journal of Animal Bioscience* 7(Suppl. 2), 212–9. Available at: http://www.ncbi.nlm.nih.gov/pubmed/23739464 (accessed on 26 September 2014).

CFF. 2012. Calculator resources | Farm Carbon Calculator (Online). Available at: http://www.cffcarboncalculator.org.uk/resources/calculator-resources (accessed on 22 July 2015).

CFF Carbon Calculator. 2012. Farm Carbon Calculator (Online). Available at: http://cffcarboncalculator.org.uk (accessed on 8 October 2014).

CLA. 2009. CALM carbon calculator (Online). Available at: http://www.calm.cla.org.uk (accessed on 13 October 2014).

Colomb, V., Bernoux, M., Bockel, L., Chotte, J.-L., Martin, S., Martin-Phipps, C., Mousset, J., Tinlot, M. and Touchemoulin, O. 2012. *Review of GHG Calculators in Agriculture and Forestry Sectors – A Guideline for Appropriate Choice and Use of Landscape Based Tools*. FAO.

Davis, K. F. and D'Odorico, P. 2015. Livestock intensification and the influence of dietary change: a calorie-based assessment of competition for crop production. *The Science of the Total Environment* 538, 817–23. doi:10.1016/j.scitotenv.2015.08.126.

Davis, K. F., Gephart, J. A., Emery, K. A., Leach, A. M., Galloway, J. N. and D'Odorico, P. 2016. Meeting future food demand with current agricultural resources. *Global Environmental Change* 39, 125–32. doi:10.1016/j.gloenvcha.2016.05.004.

DEFRA/DECC. 2011. *GHG Conversion Factors for Company Reporting*. Department for Environment, Food and Rural Affairs, London, UK. Available at: https://www.gov.uk/government/publications/2011-guidelines-to-defra-decc-s-greenhouse-gas-conversion-factors-for-company-reporting-methodology-paper-for-emission-factors.

DEFRA/DECC. 2015. *GHG Conversion Factors for Company Reporting*. Department for Environment, Food and Rural Affairs, London, UK. Available at: https://www.gov.uk/government/publications/greenhouse-gas-reporting-conversion-factors-2015.

Del Grosso, S. J., Mosier, A., Parton, W. and Ojima, D. 2005. DAYCENT model analysis of past and contemporary soil NO and net greenhouse gas flux for major crops in the USA. *Soil and Tillage Research* 83(1), 9–24. doi:10.1016/j.still.2005.02.007.

Del Grosso, S. J., Parton, W. J., Mosier, A. R., Walsh, M. K., Ojima, D. S.. and Thornton, P. E. 2006. DAYCENT national-scale simulations of nitrous oxide emissions from cropped soils in the United States. *Journal of Environmental Quality* 35(4), 1451–60. Available at: http://www.ncbi.nlm.nih.gov/pubmed/16825465 (accessed on 13 October 2014).

Demmers, T. G. M., Burgess, L. R., Short, J. L., Phillips, V. R., Clark, J. A. and Wathes, C. M. 1998. First experiences with methods to measure ammonia emissions from naturally ventilated cattle buildings in the U.K. *Atmospheric Environment* 32(3), 285–93. Available at: http://linkinghub.elsevier.com/retrieve/pii/S1352231097001970.

Demmers, T. G. M., Burgess, L. R., Short, J. L., Phillips, V. R., Clark, J. A. and Wathes, C. M. 1999. Ammonia emissions from two mechanically ventilated UK livestock buildings. *Atmospheric Environment* 33(2), 217–27. Available at: http://linkinghub.elsevier.com/retrieve/pii/S1352231098001502.

Dong, H., Mangino, J. and McAllister, T. A. 2006. Emissions from livestock and manure management. In: Eggleston, S., Buendia, L., Miwa, K., Ngara, T. and Tanabe, K. (Eds), *IPCC Guidelines for National Greenhouse Gas Inventories. Volume 4: Agriculture, Forestry and Other Land Use*. IPCC. Chapter 10.

Ecoinvent Centre. 2007. *Ecoinvent Data v2.0*. Ecoinvent reports No. 1e25. Swiss Centre for Life Cycle Inventories, Dübendorf.

Ekvall, T., Azapagic, A., Finnveden, G., Rydberg, T., Weidema, B. P. and Zamagni, A. 2016. Attributional and consequential LCA in the ILCD handbook. *The International Journal of Life Cycle Assessment* 21(3), 293–6. doi:10.1007/s11367-015-1026-0.

Eshel, G., Shepon, A., Makov, T. and Milo, R. 2014. Land, irrigation water, greenhouse gas, and reactive nitrogen burdens of meat, eggs, and dairy production in the United States. *Proceedings of the National Academy of Sciences of the United States of America* 111(33), 11996–2001. Available at: http://www.pnas.org/cgi/doi/10.1073/pnas.1402183111 (accessed on 22 July 2014).

FAO. 2011. *World Livestock 2011: Livestock in Food Security*. McLeod, A. (Ed.). FAO, Rome.

FAOstat. 2017. Crops and livestock products (Online). Available at: http://www.fao.org/faostat/en/#data/TP (accessed January 2018).

Gephart, J. A., Davis, K. F., Emery, K. A., Leach, A. M., Galloway, J. N. and Pace, M. L. 2016. The environmental cost of subsistence: optimizing diets to minimize footprints. *The Science of the Total Environment* 553, 120–7. doi:10.1016/j.scitotenv.2016.02.050.

Gerbens-Leenes, P. W., Mekonnen, M. M. and Hoekstra, A. Y. 2013. The water footprint of poultry, pork and beef: a comparative study in different countries and production systems. *Water Resources and Industry* 1–2, 25–36. Available at: http://linkinghub.elsevier.com/retrieve/pii/S2212371713000024 (accessed on 17 July 2014).

Gerber, P. J., Steinfeld, H., Henderson, B., Mottet, A., Opio, C., Dijkman, J., Falcucci, A. and Tempio, G. 2013. *Tackling Climate Change through Livestock – A Global Assessment of Emissions and Mitigation Opportunities*. Food and Agriculture Organization of the United Nations (FAO), Rome.

GHG Protocol. 2003. Emissions factors from cross-sector tools. Available at: https://ghgprotocol.org/calculation-tools.

GHG Protocol. 2012. GHG emissions from purchased electricity (Online). Available at: http://www.ghgprotocol.org/calculation-tools/all-tools (accessed on 3 August 2016).

Gibbons, J. M., Ramsden, S. J. and Blake, A. 2006. Modelling uncertainty in greenhouse gas emissions from UK agriculture at the farm level. *Agriculture, Ecosystems and Environment* 112(4), 347–55. Available at: http://linkinghub.elsevier.com/retrieve/pii/S016788090500424X.

Gotelli, N. J. and Colwell, R. K. 2001. Quantifying biodiversity: procedures and pitfalls in the measurement and comparison of species richness. *Ecology Letters* 4(4), 379–91. Available at: http://doi.wiley.com/10.1046/j.1461-0248.2001.00230.x.

Griscom, B. W., Adams, J., Ellis, P. W., Houghton, R. A., Lomax, G., Miteva, D. A., Schlesinger, W. H., Shoch, D., Siikamäki, J. V., Smith, P., Woodbury, P., Zganjar, C., Blackman, A., Campari, J., Conant, R. T., Delgado, C., Elias, P., Gopalakrishna, T., Hamsik, M. R., Herrero, M., Kiesecker, J., Landis, E., Laestadius, L., Leavitt, S. M., Minnemeyer, S., Polasky, S., Potapov, P., Putz, F. E., Sanderman, J., Silvius, M., Wollenberg, E. and Fargione, J. 2017a. Natural climate solutions. *Proceedings of the National Academy of Sciences of the United States of America* 114(44), 11645–50. Available at: http://www.pnas.org/lookup/doi/10.1073/pnas.1710465114.

Griscom, B. W., Adams, J., Ellis, P. W., Houghton, R. A., Lomax, G., Miteva, D. A., Schlesinger, W. H., Shoch, D., Siikamäki, J. V., Smith, P., Woodbury, P., Zganjar, C., Blackman, A., Campari, J., Conant, R. T., Delgado, C., Elias, P., Gopalakrishna, T., Hamsik, M. R., Herrero, M., Kiesecker, J., Landis, E., Laestadius, L., Leavitt, S. M., Minnemeyer, S., Polasky, S., Potapov, P., Putz, F. E., Sanderman, J., Silvius, M., Wollenberg, E. and Fargione, J. 2017b. Natural climate solutions. *Proceedings of the National Academy of Sciences of the United States of America* 114(44), 11645–50. doi:10.1073/pnas.1710465114.

Hall, P., Holmes-Ling, P., Stewart, K. and Sheane, R. 2010. *A Scottish Farm-Based Greenhouse Gas Accounting Tool for Scotland*. Scottish Government Offices, Laurence Gould Partnership Ltd. & Best Foot Forward Ltd, West Lothian, UK.

Hillier, J., Walter, C., Malin, D., Garcia-Suarez, T., Mila-i-Canals, L. and Smith, P. 2011. A farm-focused calculator for emissions from crop and livestock production. *Environmental Modelling and Software* 26(9), 1070–8. Available at: http://linkinghub.elsevier.com/retrieve/pii/S1364815211000892 (accessed on 4 September 2014). doi:10.1016/j.envsoft.2011.03.014.

Hoekstra, A. Y. and Chapagain, A. K. 2006. Water footprints of nations: water use by people as a function of their consumption pattern. *Water Resources Management* 21(1), 35–48. Available at: http://link.springer.com/10.1007/s11269-006-9039-x (accessed on 10 July 2014).

Hünerberg, M., Little, S. M., Beauchemin, K. A., McGinn, S. M., O'Connor, D., Okine, E. K., Harstad, O. M., Kröbel, R. and McAllister, T. A. 2014. Feeding high concentrations of corn dried distillers' grains decreases methane, but increases nitrous oxide emissions from beef cattle production. *Agricultural Systems* 127, 19–27. doi:10.1016/j.agsy.2014.01.005.

IPCC. 1995. *The Science of Climate Change - Contribution of Working Group I to the Second Assessment Report of the Intergovernmental Panel on Climate Change.* Cambridge University Press, Cambridge, UK.

IPCC. 1996. *Revised 1996 IPCC Guidelines for National Greenhouse Gas Inventories.* IPCC, Japan.

IPCC. 2006. *2006 IPCC Guidelines for National Greenhouse Gas Inventories.* Eggleston, H. S., Buendia, L., Miwa, K., Ngara, T. and Tanabe, K. (Eds). IGES, Japan.

IPCC. 2007. *Climate Change 2007: Synthesis Report. Contribution of Working Groups I, II and III to the Fourth Assessment Report of the Intergovernmental Panel on Climate Change.* Cambridge University Press, Cambridge, UK.

IPCC. 2013. *Climate Change 2013: The Physical Science Basis.* Stocker, T. F., Qin, D., Plattner, G.-K., Tignor, M., Allen, S. K., Boschung, J., Nauels, A., Xia, Y., Midgley, P. M. and Bex, V. (Eds). Cambridge University Press, Cambridge, UK and New York, NY.

Janzen, H. H., Angers, D. A., Boehm, M., Bolinder, M., Desjardins, R. L., Dyer, J., Ellert, B. H., Gibb, D. J., Gregorich, E. G., Helgason, B. L., Lemke, R., Massé, D., McGinn, S. M., McAllister, T. A., Newlands, N., Pattey, E., Rochette, P., Smith, W., VandenBygaart, A. J. and Wang, H. 2006. A proposed approach to estimate and reduce net greenhouse gas emissions from whole farms. *Canadian Journal of Soil Science* 86(3), 401–18. doi:10.4141/S05-101.

Joos, F., Roth, R., Fuglestvedt, J. S., Peters, G. P., Enting, I. G., von Bloh, W., Brovkin, V., Burke, E. J., Eby, M., Edwards, N. R., Friedrich, T., Frölicher, T. L., Halloran, P. R., Holden, P. B., Jones, C., Kleinen, T., Mackenzie, F. T., Matsumoto, K., Meinshausen, M., Plattner, G.-K., Reisinger, A., Segschneider, J., Shaffer, G., Steinacher, M., Strassmann, K., Tanaka, K., Timmermann, A. & Weaver, A. J. 2013. Carbon dioxide and climate impulse response functions for the computation of greenhouse gas metrics: a multi-model analysis. *Atmospheric Chemistry and Physics* 13(5), 2793-825. Available at: http://www.atmos-chem-phys.net/13/2793/2013/ (accessed on 2 January 2015).

Kastner, T., Rivas, M. J. I., Koch, W. and Nonhebel, S. 2012. Global changes in diets and the consequences for land requirements for food. *Proceedings of the National Academy of Sciences of the United States of America* 109(18), 6868–72. doi:10.1073/pnas.1117054109.

Keller, E., Chin, M., Chorkulak, V., Clift, R., Faber, Y., Lee, J. and King, H., Milà i Canals, L., Stabile, M., Stickler, C. and Viart, N. 2014. Footprinting farms: a comparison of three GHG calculators. *Greenhouse Gas Measurement and Management* 4, 90–123. Available at: http://www.tandfonline.com/doi/abs/10.1080/20430779.2014.984609#.VNx8hE1yaM8 (accessed on 4 January 2015).

Kipling, R. P., Virkajärvi, P., Breitsameter, L., Curnel, Y., De Swaef, T., Gustavsson, A. M., Hennart, S., Höglind, M., Järvenranta, K., Minet, J., Nendel, C., Persson, T., Picon-Cochard, C., Rolinski, S., Sandars, D. L., Scollan, N. D., Sebek, L., Seddaiu, G., Topp, C. F. E., Twardy, S., Van Middelkoop, J., Wu, L. and Bellocchi, G. 2016. Key challenges and priorities for modelling European grasslands under climate change. *The Science of the Total Environment* 566-567, 851–64. doi:10.1016/j.scitotenv.2016.05.144.

Kröbel, R., Bolinder, M. A., Janzen, H. H., Little, S. M., Vandenbygaart, A. J. and Kätterer, T. 2016 Canadian farm-level soil carbon change assessment by merging the greenhouse gas model Holos with the Introductory Carbon Balance Model (ICBM). *Agricultural Systems* 143, 76–85. doi:10.1016/j.agsy.2015.12.010.

Landau, S., Mitchell, R. A. C., Barnett, V., Colls, J. J., Craigon, J. and Payne, R. W. 1999. Response to "Comments on 'Testing winter wheat simulation models predictions against observed UK grain yields by Landau et al.[Agric. For. Meteorol. 89 (1998) 85–99]'by Jamieson et al.[Agric. For. Meteorol., this issue]". *Agricultural and Forest Meteorology* 96(1), 163–4.

Le Quéré, C., Andrew, R. M., Canadell, J. G., Sitch, S., Korsbakken, J. I., Peters, G. P., Manning, A. C., Boden, T. A., Tans, P. P., Houghton, R. A., Keeling, R. F., Alin, S., Andrews, O. D., Anthoni, P., Barbero, L., Bopp, L., Chevallier, F., Chini, L. P., Ciais, P., Currie, K., Delire, C., Doney, S. C., Friedlingstein, P., Gkritzalis, T., Harris, I., Hauck, J., Haverd, V., Hoppema, M., Klein Goldewijk, K., Jain, A. K., Kato, E., Körtzinger, A., Landschützer, P., Lefèvre, N., Lenton, A., Lienert, S., Lombardozzi, D., Melton, J. R., Metzl, N., Millero, F., Monteiro, P. M. S., Munro, D. R., Nabel, J. E. M. S., Nakaoka, S., O'Brien, K., Olsen, A., Omar, A. M., Ono, T., Pierrot, D., Poulter, B., Rödenbeck, C., Salisbury, J., Schuster, U., Schwinger, J., Séférian, R., Skjelvan, I., Stocker, B. D., Sutton, A. J., Takahashi, T., Tian, H., Tilbrook, B., van der Laan-Luijkx, I. T., van der Werf, G. R., Viovy, N., Walker, A. P., Wiltshire, A. J. and Zaehle, S. 2016. Global carbon budget 2016. *Earth System Science Data* 8(2), 605–49. doi:10.5194/essd-8-605-2016.

Little, S., Lindeman, J., Maclean, K. and Janzen, H. 2008. *Holos. A Tool to Estimate and Reduce Greenhouse Gases from Farms. Methodology & Algorithms for Version 1.1.x.* Agriculture and Agri-Food Canada, Ottawa, p. 158.

Lorenzoni, I. and Pidgeon, N. F. 2006. Public views on climate change: European and USA perspectives. *Climatic Change* 77(1–2), 73–95. doi:10.1007/s10584-006-9072-z.

Lutz, W., Sanderson, W. and Scherbov, S. 2001. The end of world population growth. *Nature* 412(6846), 543–5. doi:10.1038/35087589.

Machovina, B., Feeley, K. J. and Ripple, W. J. 2015. Biodiversity conservation: the key is reducing meat consumption. *The Science of the Total Environment* 536, 419-31. doi:10.1016/j.scitotenv.2015.07.022.

MacLeod, M., Gerber, P., Vellinga, T., Opio, C., Falcuccib, A., Tempio, G., Dietzeb, K., Giannib, G., Henderson, B., Makkarb, H., Mottet, A., Robinson, T., Steinfield, H., Thieme, O. and Weiler, V. 2013. GLEAM: an example of the potential contribution of livestock modelling to sustainable intensification. In: *Carbon Management Centre International Conference: Sustainable Intensification: The Pathway to Low Carbon Farming?.* SRUC, Edinburgh.

MacLeod, M., Sykes, A., Leinonen, I. and Eory, V. 2017. *Quantifying the Greenhouse Gas Emission Intensity of Scottish Agricultural Commodities: Technical Report.* CxC Project. SRUC, Edinburgh.

Manning, M. and Reisinger, A. 2011. Broader perspectives for comparing different greenhouse gases. *Philosophical Transactions. Series A, Mathematical, Physical, and Engineering Sciences* 369(1943), 1891–905. doi:10.1098/rsta.2010.0349.

Mason, C. F. 2002. *Biology of Freshwater Pollution*. Pearson Education, Harlow, UK.

Meek, B., Loxton, D., Sparks, T., Pywell, R., Pickett, H. and Nowakowski, M. 2002. The effect of arable field margin composition on invertebrate biodiversity. *Biological Conservation* 106(2), 259–71. Available at: http://linkinghub.elsevier.com/retrieve/pii/S000632070100252X.

Misselbrook, T. H., Van Der Weerden, T. J., Pain, B. F., Jarvis, S. C., Chambers, B. J., Smith, K. A., Phillips, V. R. and Dommers, T. G. M. 2000. Ammonia emission factors for UK agriculture. *Atmospheric Environment* 34(6), 871–80. Available at: http://linkinghub.elsevier.com/retrieve/pii/S1352231099003507.

Monbiot, G. 2017. Goodbye – and good riddance – to livestock farming. *The Guardian*. Available at: https://www.theguardian.com/commentisfree/2017/oct/04/livestock-farming-artificial-meat-industry-animals.

Moran, D., Macleod, M., Wall, E., Eory, V., McVittie, A., Barnes, A., Rees, R., Topp, C. F. E. and Moxey, A. 2011. Marginal abatement cost curves for UK agricultural greenhouse gas emissions. *Journal of Agricultural Economics* 62(1), 93–118. doi:10.1111/j.1477-9552.2010.00268.x.

Nguyen, T. T. H., Doreau, M., Eugène, M., Corson, M. S., Garcia-Launay, F., Chesneau, G. and van der Werf, H. M. G. 2013. Effect of farming practices for greenhouse gas mitigation and subsequent alternative land use on environmental impacts of beef cattle production systems. *Animal: an International Journal of Animal Bioscience* 7(5), 860–9. Available at: http://www.ncbi.nlm.nih.gov/pubmed/23190866.

Niemann, H., Kuhla, B. and Flachowsky, G. 2011. Perspectives for feed-efficient animal production. *Journal of Animal Science* 89(12), 4344–63. Available at: http://www.ncbi.nlm.nih.gov/pubmed/21705634 (accessed on 26 March 2014).

Official Journal of the European Community. 1991. Council Directive 91/676/EEC of 12 December 1991 concerning the protection of waters against pollution caused by nitrates from agricultural sources.

Ogle, S. M., Breidt, F. J. and Paustian, K. 2005. Agricultural management impacts on soil organic carbon storage under moist and dry climatic conditions of temperate and tropical regions. *Biogeochemistry* 72(1), 87–121. Available at: http://link.springer.com/10.1007/s10533-004-0360-2 (accessed on 6 October 2014).

Opio, C., Gerber, P. and Steinfeld, H. 2011. Livestock and the environment: addressing the consequences of livestock sector growth. *Advances in Animal Biosciences the Animal Consortium* 2, 601–7. Available at: https://www-cambridge-org.abc.cardiff.ac.uk/core/services/aop-cambridge-core/content/view/671AB21EC974B6FBDFBA1839E67DE677/S204047001100286Xa.pdf/livestock_and_the_environment_addressing_the_consequences_of_livestock_sector_growth.pdf.

Opio, C., Gerber, P., Vellinga, T., Falcucci, A., Tempio, G., Gianni, G., Henderson, B., Macleod, M., Makkar, H., Mottet, A., Robinson, T. and Weiler, V. 2013. *Greenhouse Gas Emissions from Ruminant Supply Chains – a Global Life Cycle Assessment*. FAO, Rome.

Pant, R., Van Hoof, G., Schowanek, D., Feijtel, T. C. J., de Koning, A., Hauschild, M., Olsen, S. I., Pennington, D. W. and Rosenbaum, R. 2004. OMNIITOX: LCA Case Studies Comparison between three different LCIA methods for aquatic ecotoxicity and a product environmental risk assessment. *The International Journal of Life Cycle Assessment* 9(5), 295–306. doi:10.1007/BF02979419.

Penman, T. D., Law, B. S. and Ximenes, F. 2010. A proposal for accounting for biodiversity in life cycle assessment. *Biodiversity and Conservation* 19(11), 3245–54. doi:10.1007/s10531-010-9889-7.

Peters, G. P., Andrew, R. M., Boden, T., Canadell, J. G., Ciais, P., Quéré, C. Le, Marland, G., Raupach, M. R. and Wilson, C. 2013. The challenge to keep global warming below 2 C. *Nature Climate* 3(1). Available at: http://www.nature.com/nclimate/journal/vaop/ncurrent/full/nclimate1783.html.

Posch, M., Seppälä, J., Hettelingh, J.-P., Johansson, M., Margni, M. and Jolliet, O. 2008. The role of atmospheric dispersion models and ecosystem sensitivity in the determination of characterisation factors for acidifying and eutrophying emissions in LCIA. *The International Journal of Life Cycle Assessment* 13(6), 477–86. Available at: http://link.springer.com/10.1007/s11367-008-0025-9 (accessed on 5 December 2014).

Rees, R. M., Augustin, J., Alberti, G., Ball, B. C., Boeckx, P., Cantarel, A., Castaldi, S., Chirinda, N., Chojnicki, B., Giebels, M., Gordon, H., Grosz, B., Horvath, L., Juszczak, R., Klemedtsson, Å. K., Klemedtsson, L., Medinets, S., Machon, A., Mapanda, F., Nyamangara, J., Olesen, J., Reay, D., Sanchez, L., Sanz Cobena, A., Smith, K. A., Sowerby, A., Sommer, M., Soussana, J. F., Stenberg, M., Topp, C. F. E., van Cleemput, O., Vallejo, A., Watson, C. A. and Wuta, M. 2012. Nitrous oxide emissions from European agriculture; an analysis of variability and drivers of emissions from field experiments. *Biogeosciences Discussions* 9(7), 9259–88. Available at: http://www.biogeosciences-discuss.net/9/9259/2012/ (accessed on 26 September 2014).

Reisinger, A., Meinshausen, M. and Manning, M. 2011. Future changes in global warming potentials under representative concentration pathways. *Environmental Research Letters* 6(2), 024020. Available at: http://stacks.iop.org/1748-9326/6/i=2/a=024020?key=crossref.d599e24b11bbe13bc1adc680a0e3a7ac (accessed on 22 January 2015).

Röös, E. and Nylinder, J. 2013. *Uncertainties and Variations in the Carbon Footprint of Livestock Products*. Report 063. Institutionen för energi och teknik, Sveriges lantbruksuniversitet, Uppsala, Sweden. ISSN 1654-9406.

Röös, E., Sundberg, C., Tidåker, P., Strid, I. and Hansson, P. A. 2013. Can carbon footprint serve as an indicator of the environmental impact of meat production? *Ecological Indicators* 24, 573–81. doi:10.1016/j.ecolind.2012.08.004.

Ruby, M. B. and Heine, S. J. 2012. Too close to home. Factors predicting meat avoidance. *Appetite* 59(1), 47–52. doi:10.1016/j.appet.2012.03.020.

Sainz, R. D. 2003. *Framework for Calculating Fossil Fuel Use in Livestock Systems*. FAO, Rome.

Salisbury, E., Thistlethwaite, G., Young, K. R. C. and Cardenas, L. 2014. *Greenhouse gas inventories for England, Scotland, Wales and Northern Ireland: 1990–2012*. National Atmospheric Emissions Inventory.

Sans, P. and Combris, P. 2015. World meat consumption patterns: an overview of the last fifty years (1961–2011). *Meat Science* 109, 106–11. doi:10.1016/j.meatsci.2015.05.012.

Schils, R. L. M., Olesen, J. E., del Prado, A. and Soussana, J. F. 2007 A review of farm level modelling approaches for mitigating greenhouse gas emissions from ruminant livestock systems. *Livestock Science* 112(3), 240–51. Available at: http://linkinghub.elsevier.com/retrieve/pii/S187114130700474X (accessed on 12 July 2014).

SEE360. 2007. *Cplan | The Carbon Emissions Calculator for Farms & Agriculture Industries*. Online. Available at: http://www2.cplan.org.uk/index.php?_load=page&_pageId=1 (accessed on 9 October 2014).

Shine, K. P. 2009. The global warming potential-the need for an interdisciplinary retrial. *Climatic Change* 96(4), 467–72. doi:10.1007/s10584-009-9647-6.

Simpson, E. S. 2014. *The Developing World: an Introduction*. Routledge, London.

Smith, S. J. 2003. The evaluation of greenhouse gas indices. *Climatic Change* 58(3), 261–5. doi:10.1023/A:1023960902970.

Smith, J. and Smith, P. 2007. *Environmental Modelling an Introduction*. Oxford. ISBN: 9780199272068.

Smith, P., Powlson, D. S., Glendining, M. J. and Smith, J. U. 1997. Potential for carbon sequestration in European soils: preliminary estimates for five scenarios using results from long-term experiments. *Global Change Biology* 3(1), 67–79. doi:10.1046/j.1365-2486.1997.00055.x.

Smith, P., Martino, D., Cai, Z., Gwary, D., Janzen, H., Kumar, P., McCarl, B., Ogle, S., O'Mara, F., Rice, C., Scholes, B., Sirotenko, O., Howden, M., McAllister, T., Pan, G., Romanenkov, V., Schneider, U., Towprayoon, S., Wattenbach, M. and Smith, J. 2008. Greenhouse gas mitigation in agriculture. *Philosophical Transactions of the Royal Society B: Biological Sciences* 363(1492), 789–813. Available at: http://rstb.royalsocietypublishing.org/cgi/doi/10.1098/rstb.2007.2184. doi:10.1098/rstb.2007.2184.

Soil Association. 2013. Soil Association – FCAT (Online). Available at: http://fcat.soilassociation.org/UserGuide.aspx?year=2014 (accessed on 10 October 2014).

SRUC. 2014. AgRE Calc (Online). SRUC, UK. Available at: http://www.sruc.ac.uk/info/120355/carbon_and_climate/1333/agricultural_resource_efficiency_calculator_agre_calc.

Steinfeld, H. and Gerber, P. 2010. Livestock production and the global environment: consume less or produce better? *Proceedings of the National Academy of Sciences* 107(43), 18237–8. Available at: http://www.pnas.org/cgi/doi/10.1073/pnas.1012541107. doi:10.1073/pnas.1012541107.

Steinfeld, H., Gerber, P., Wassenaar, T., Castel, V., Rosales, M. and de Haan, C. 2006. Livestock's long shadow: environmental issues and options. *Food & Agriculture Org.* Available at: http://www.fao.org/3/a0701e/a0701e00.htm.

Subak, S. 1999. Global environmental costs of beef production. *Ecological Economics* 30(1), 79–91. Available at: http://linkinghub.elsevier.com/retrieve/pii/S0921800998001001. doi:10.1016/S0921-8009(98)00100-1.

Sutherland, W. J., Bailey, M. J., Bainbridge, I. P., Brereton, T., Dick, J. T. A., Drewitt, J., Dulvy, N. K., Dusic, N. R., Freckleton, R. P., Gaston, K. J., Gilder, P. M., Green, R. E., Heathwaite, A. L., Johnson, S. M., Macdonald, D. W., Mitchell, R., Osborn, D., Owen, R. P., Pretty, J., Prior, S. V., Prosser, H., Pullin, A. S., Rose, P., Stott, A., Tew, T., Thomas, C. D., Thompson, D. B. A., Vickery, J. A., Walker, M., Walmsley, C., Warrington, S., Watkinson, A. R., Williams, R. J., Woodroffe, R. and Woodroof, H. J. 2007. PRIORITY CONTRIBUTION: Future novel threats and opportunities facing UK biodiversity identified by horizon scanning. *Journal of Applied Ecology* 45(3), 821–33. Available at: http://doi.wiley.com/10.1111/j.1365-2664.2008.01474.x (accessed on 3 December 2014).

Sykes, A. J., Topp, C. F. E., Wilson, R. M., Reid, G. and Rees, R. M. 2017. A comparison of farm-level greenhouse gas calculators in their application on beef production systems. *Journal of Cleaner Production* 164, 398–409. doi:10.1016/j.jclepro.2017.06.197.

Taylor, A. M., Amiro, B. D., Tenuta, M. and Gervais, M. 2017. Direct whole-farm greenhouse gas flux measurements for a beef cattle operation. *Agriculture, Ecosystems and Environment* 239, 65–79. doi:10.1016/j.agee.2017.01.011.

Vågen, T. G., Lal, R. and Singh, B. R. 2005. Soil carbon sequestration in sub-Saharan Africa: a review. *Land Degradation and Development* 16(1), 53–71. doi:10.1002/ldr.644.

Whittaker, C., McManus, M. C. and Smith, P. 2013. A comparison of carbon accounting tools for arable crops in the United Kingdom. *Environmental Modelling and Software* 46, 228–39. Available at: http://linkinghub.elsevier.com/retrieve/pii/S1364815213000789 (accessed on 8 September 2014).

Williams, A. G., Audsley, E. and Sandars, D. 2006. *Determining the Environmental Burdens and Resource Use in the Production of Agricultural and Horticultural Commodities.* Main Report. Defra Research Project IS0205. Cranfield University and Defra, Bedford. Available at: www.silsoe.cranfield.ac.uk, and www.defra.gov.uk.

World Economic Forum. 2019. *The Global Risks Report 2019* (14th edn.). World Economic Forum, Geneva.

Zhang, J., Hu, K., Li, K., Zheng, C. and Li, B. 2017. Simulating the effects of long-term discontinuous and continuous fertilization with straw return on crop yields and soil organic carbon dynamics using the DNDC model. *Soil and Tillage Research* 165, 302–14. doi:10.1016/j.still.2016.09.004.

Chapter 4

Assessing the environmental impact of ruminant production systems

Taro Takahashi, Rothamsted Research and University of Bristol, UK; Graham A. McAuliffe, Rothamsted Research, UK; and Michael R. F. Lee, Rothamsted Research and University of Bristol, UK

1 Introduction

As discussed elsewhere in this book, one of the most common methods to evaluate environmental footprints of farming systems is life cycle assessment (LCA). Although LCA itself is suitable for and indeed adopted by a wide range of industries far beyond agriculture, what separates agriculture, and in particular pasture-based ruminant production systems, is the high degree of uncertainties associated with physical, chemical and biological processes that underpin production (McAuliffe et al., 2018a). In the presence of uncertainties, point-estimates provided by LCA models are unlikely to be informative enough to offer robust policy implications (Chen and Corson, 2014); when this is the case, the resultant environmental burdens must be expressed in the form of probability distributions and interpreted accordingly (McAuliffe et al., 2017).

For carbon footprint (CF) analysis of ruminant systems, one significant challenge of collating a life cycle inventory is uncertainty associated with emission factors (EFs), or parameters linking nutrient inputs into the system to greenhouse gas (GHG) emissions arising from the system (Pouliot et al., 2012). On commercial livestock farms, various factors can affect their relationships; for example, weather, soil, plant/animal genetics, management practice and interactions between them. Despite these variabilities, the majority of LCA

http://dx.doi.org/10.19103/AS.2018.0044.14

studies adopt EFs derived outside the actual system boundary, most commonly as parameters defined as part of the Intergovernmental Panel on Climate Change guidelines (IPCC, 2006). As these 'generic' EFs are designed to be applicable to a wide spectrum of production environments within a single agroecological zone, a considerable level of uncertainty surrounds each of these values (Dudley et al., 2014). As a case in point, the two parameters for nitrous oxide (N_2O) emissions suggested by IPCC (2006), commonly known as EF_1 (% fertiliser N lost as N_2O) and EF_{3PRP} (% urine and dung N deposited on pasture lost as N_2O), are both deemed to have a 95% confidence interval between −67% and +300% of the respective point estimates.

To facilitate evidence-based debates about the environmental impact of ruminant production systems and, by extension, the role of ruminants in global food security, it is therefore imperative to improve reliability of EFs in a more location-specific context. This, however, requires a significant investment into field-based research, something that is not always feasible for practical reasons. Through a review of recent literature and a quantitative case study, this chapter explores how this practical trade-off between feasibility and scientific rigour should be addressed.

2 LCA applied to ruminant production systems

LCA has been applied to all major ruminant production systems (beef, dairy, lamb and wool), albeit with different degrees of scrutiny into system-wide uncertainties. Frequently cited examples of works on the sheep sector include Biswas et al. (2010), Ripoll-Bosch et al. (2013) and Wiedemann et al. (2015a). For dairy, Üçtuğ (2019) provides an extensive literature review encompassing 31 studies. In addition, Poore and Nemecek (2018) is accompanied by a global database of agri-food LCA results covering both plant-based and animal-based commodities.

For the beef sector, de Vries et al. (2015) review and compare findings from 14 studies based on contrasting farming systems from around the world. It is noteworthy, however, that the popularity of beef LCA research has substantially increased over the last 3 years, likely due to reports indicating that the sector, and in particular grazing systems, are extremely heavy contributors to global GHG emissions (Herrero et al., 2016; Springmann et al., 2016). In response to such a rapid rise in attention paid to the industry, the remainder of this section will give a summary of 14 beef LCA studies that have been published between 2015 and 2018. Papers that are not written in English and that focus on end-point modelling are excluded here.

In Brazil, Dick et al. (2015) conducted an LCA of beef cattle in two grassland systems. The first system was based on traditional grazing practices where animals can wander freely and receive little or no supplementation. The second system, termed 'improved', involved weekly rotational grazing and the introduction of winter forage species. The system boundary was from raw material extraction to farm gate and the functional unit (FU) was 1 kg liveweight gain (LWG). Data on beef production within the two systems were sourced from published literature, and GHG emissions were calculated according to IPCC guidelines. Global warming potential (GWP) for the traditional system was found to be 22.5 kg CO_2-eq per kg LWG, while GWP for the improved system was 9.2 kg CO_2-eq per kg LWG. This dramatic reduction in GWP was attributed to higher quality forage with increased digestibility in the alternative system, resulting in faster LWG.

Mogensen et al. (2015) estimated CFs of beef production systems in Denmark and Sweden. The system boundary was from cradle to farm gate and the FU was 1 kg carcase weight (CW). Five Danish and four Swedish beef farming scenarios were developed, which were categorised depending on intensive or extensive production, and dairy or beef breeds of cattle. For feed production (pasture, silage and concentrates), carbon sequestration was considered based on IPCC guidelines and published literature. Grass-clover swards were included as part of an arable rotation where the swards remained for 2–3 years in a 5-year rotation. The GHG emissions were estimated using a combination of IPCC values and published data for Nordic conditions, the latter to estimate methane (CH_4) emissions from enteric fermentation and manure management. The resultant CF ranged from 8.9 to 17.0 kg CO_2-eq per kg CW for the dairy-bull fattening systems, while the CF for cow–calf systems ranged from 23.1 to 29.7 kg CO_2-eq per kg CW. Carbon sequestration resulted in GWP mitigation across all scenarios; among them, CO_2 reduction was largest in the grass-based systems, although these systems still generated the highest CF values despite elevated carbon uptake.

Wiedemann et al. (2015b) utilised LCA to examine the environmental impact of Australian beef and lamb being exported to the United States. The system boundary was from cradle (in Australia) to the distribution warehouse in the United States, and the FU was 1 kg retail ready meat. For beef systems, the study considered beef cattle bred in rangelands and finished on pasture, and dairy steers finished on grain feedlots for either 115 days or, for specialised breeds such as Wagyu, 330 days. Farm-level data were obtained from governmental surveys and published case studies. Regionally tailored herd models were used to calculate feed intake and for predicting GHG emissions. Data on slaughtering and processing (such as cutting and chilling) were derived from an industry survey of meat processing plants in Australia. The GWP ranged from 23.4 to 27.2 kg CO_2-eq per kg beef, with the grass-finished cattle performing least favourably. Across the three scenarios, the farming phase generated the highest GWP (93%), meat processing accounted for 4%, transportation 3%, while the warehousing had negligible impacts. However, the authors also considered differences in human edible protein conversion efficiency. Under this FU, pasture-based beef production performed considerably better than grain-fed beef by converting more non-human-edible protein into human-edible protein.

In an effort to capture temporal variations in on-farm GHG emissions, Hyland et al. (2016) assessed the CF of 15 livestock enterprises over 2 time periods 3 years apart (2009/10) and 2012/13). In addition to calculating farm-level emissions intensities, the authors also used a range of sensitivity analyses to investigate potential mitigation strategies. The system boundary of the study was set as cradle to farm gate, and the FU as 1 kg liveweight (LW). Across the 15 livestock enterprises examined, five specialised in lamb, four specialised in beef, while six were mixed beef and sheep farms. Emissions intensities were calculated according to IPCC (2006) Tier 1 and 2 guidelines. In tackling the issue of mixed farming allocation, where possible Hyland et al. (2016) used system expansion; however, in certain cases, this was not possible due to a lack of differentiation and economic allocation was used instead. Between two data periods, lamb emissions were found to increase by 12%, while beef emissions decreased by 12%. However, these differences were not found to be statistically significant. Unsurprisingly, CH_4 emissions primarily resulting from enteric fermentation were the greatest GHG burdens across all enterprises. Regarding the scenario analysis aimed at reducing on-farm emissions, the authors suggested that the primary focus for farmers should be on improved resource use efficiency. The inclusion of legumes such as red (*Trifolium pratense*) and white (*Trifolium repens*) clover on suitable soils was also highlighted as an important technique to reduce fertiliser requirements.

Examining the impacts of Canadian grazing management strategies on GHG intensities from beef herds, Alemu et al. (2017) modelled a typical herd structure of 120 cows, 4 bulls and their progeny over an 8-year period. A range of different grazing strategies were considered: light continuous grazing for all cattle; heavy continuous grazing for all cattle; light continuous grazing for cow–calf pairs and moderate rotational grazing for backgrounded cattle; and heavy continuous grazing for cow–calf pairs and moderate rotational grazing for backgrounded cattle. The system boundary was set as cradle-to-farmgate, and results based on two FUs (LW and CW) were reported side-by-side. The authors used Holos (a Canadian whole-farm model) to estimate farm-level emissions, and soil carbon changes were considered using the Introductory Carbon Balance Model, while farm management data were sourced from previous management studies. Emissions intensities were found to have narrow ranges (14.5–16.0 kg CO_2-eq/kg LW and 24.1–26.6 kg CO_2-eq/kg CW) across the grazing scenarios; however, GHG emissions tended to decrease as stocking density increased. Inclusion of soil as a carbon sink reduced impacts by up to 25%. The authors highlight the complexities in crediting a grassland system as a carbon sink due to the extremely dynamic nature of carbon flows.

Berton et al. (2017) applied the LCA method to examine the environmental footprint of the integrated French–Italian beef production system. The system boundary was set as cradle-to-farmgate; however, unlike many other studies, this boundary accounted for a cow–calf operation in one country, France, with animals ready for fattening transported to another country, Italy. All inputs and outputs (including transportation) associated with each stage were accounted for, and impacts were scaled to a FU of 1 kg LW (described as bodyweight). The authors considered a range of impact categories made up of GWP, acidification potential (AP), eutrophication potential (EP), cumulative energy demand (CED) and land use (LU) reported as land occupation. Regarding allocation of burdens to coproducts of the cow–calf operation, a mass approach was adopted along with a sensitivity analysis to consider the effect of this assumption. French farms (40) and Italian farms (14) were modelled based on the best available data. The authors found that the burdens arising directly from the farms were greater than upstream processes in general; the only exception to this finding was CED, where energy demand was higher for off-farm processes for production of feed and agrochemicals. In terms of total impacts, the authors highlighted positive correlations between direct environmental burdens (GWP, AP and EP) and resource requirements (CED and LU) and pointed out that agricultural policy design needs to account for multiple indicators rather than focusing on one.

In Italy once again, Buratti et al. (2017) compared the CF of conventional and organic beef production systems. Data were collected from two case study farms in the Umbria region of Italy, both of which operated as cow–calf systems rather than specialist fattening operations. The system boundary was from cradle-to-farmgate, and the FU was 1 kg LW of heifers and bullocks ready for slaughtering. Feed production primarily occurred on each of the farms, and burdens arising from fodder were modelled based on production data provided by the farms. The few imported products were treated as background processes and sourced from *ecoinvent* V3. Fertilising strategies differed between the enterprises. For example, the 'organic' system solely used livestock manure to fertilise feed crops, while the 'conventional' system used mineral N in addition to manure. Both systems transported excess manure to nearby but external cropland. The GHG emissions were estimated using IPCC (2006) Tier 2 guidelines for all foreground sources, and, regarding enteric fermentation, the authors estimated Y_m values (CH_4 conversion factors) according to the digestible energy of the feed. The authors reported that lower GHG emissions were generated when producing organic feed due largely to lower mineral N requirements; however, interestingly, this did not translate to total CF rankings. The conventional system had a lower CF than the organic system, primarily driven by the shorter finishing times required.

De Figueiredo et al. (2017) examined the GHG balance and CF of three pasture-based beef finishing systems in Brazil. Three pasture systems all consisting of *Brachiaria spp.* were defined as: a degraded pasture receiving no external inputs; a managed pasture receiving annual fertiliser with animals receiving strategic supplementation consisting of maize (*Zea mays*) bran (82%), milled soybean (*Glycine max*) (14%), urea (3%) and mineral salt (1%) for a 6-month period during dry season at a rate of 4 g/kg bodyweight; and a crop-livestock-forest integration system, a more complex system involving afforestation and rotational crop production and the same supplementation described under managed pasture. Both values were calculated using IPCC (2006) guidelines; the GHG balance was reported in terms of land area (1 ha), whereas the CF was reported as 1 kg LW leaving the farmgate. On an area basis, degraded pasture was found to have the lowest GHG balance, due to considerably lower stocking rates and no fertiliser requirement. Nevertheless, this finding was reversed in terms of 1 kg LW and degraded pasture was found to be the least efficient system due to low animal productivity. Between the two improved pastures, managed pasture was found to have considerably lower emissions (in terms of LW) than crop-livestock-forest, with livestock productivity again being a key factor. The crop-livestock-forest system brings its own merits in terms of other impact categories not assessed, such as improved biodiversity and utilising land to produce timber and crops as coproducts from the system. Overall, the authors conclude that land designated as degraded pasture should be improved wherever feasible. This study further questions the use of area as an FU for system-level environmental evaluation.

Florindo et al. (2017) used the LCA methodology in combination with life cycle costing (LCC) to evaluate both the CF and economic performance of beef cattle in the Brazilian Midwest. The authors point out that LCA studies often recommend mitigation strategies to reduce environmental footprints while failing to account for economic viability, a trade-off they explicitly consider. Primary data for the study, including machinery costs and management activity, were collected directly from a beef farm comprising 1350 ha of grassland. The farm maintains 1830 animals consisting of breeding stock as well as growing and finishing cattle. As part of the diversification strategy, the farm is split into four different production systems, differentiated by feeding regimes, stocking densities and slaughter weights. Feeding regimes were determined as with or without strategic supplementation which varied depending on the life stage of the cattle (e.g. creep feed). The protein mineral supplement included cornmeal (36%), soybean meal (12%) and urea (11%). Creep feed was made up of 30% cornmeal and 51% soybean meal, while a 14% crude protein ration provided based on LW consisted of 72% cornmeal and 18% soybean meal. GHG emissions were calculated according to IPCC (2006) Tier 2 guidelines. Regarding LCC, the production system with the longest duration in terms of grazing was found to be the most cost-effective feed source, due to reduced supplementary feeding requirements. However, despite this positive aspect, it also resulted in the largest total financial cost due to lower stocking densities, and therefore, greater capital expenditure for land use. The same finding was true for GHG emissions; higher stocking rates and lower grazing durations generated lower CFs, despite the subsequent lower finishing weights. This demonstrates the benefits of strategic supplementation, particularly in geographical regions affected by severe weather (extremely dry seasons in this instance). Care must be taken, however, at interpretation of these results, as strategic supplementation could potentially increase the level of food-feed competition (Wilkinson and Lee, 2018).

Utilising interdisciplinary skills and expertise, Hessle et al. (2017) examined how Swedish beef and milk production systems could be environmentally and economically optimised under a range of different scenarios. Input was provided by experts in economics, LCA and supply chain management. The focus of this study was the environmental comparison of the reference situation (business-as-usual) with three hypothetical yet realistic scenarios. The three expertly designed scenarios were based around Swedish environmental objectives and set as follows: an 'ecosystem' scenario aimed at reducing impacts on biodiversity; a 'nutrient' scenario which focused on optimising plant nutrient use and supply; and a 'climate' scenario primarily concerned with reducing anthropogenic GHG impacts. The overarching goal of each alternative scenario was to maintain or improve production efficiency, while simultaneously mitigating environmental impacts. Once the study panel had agreed upon the alternative systems, LCA models were constructed using a combination of literature and expert opinion. In most instances, the improved systems demonstrated reduced negative impacts. However, there were notable trade-offs; for example, the ecosystem scenario required more land being used as grassland to improve biodiversity, which in turn caused negative impacts on eutrophication (freshwater and marine) and CED across both beef and dairy systems. Despite this, the authors concluded that a common denominator in improving these livestock systems was a more efficient use of resources such as energy and feed.

Tichenor et al. (2017) analysed differences in environmental performances between intensively managed grass-fed beef production and confinement dairy beef production systems in the Northeast United States. The system boundary was from cradle-to-farmgate and the authors considered hot CW as the FU to maximise comparative potential with previous North American studies. The impact categories considered were GWP, AP, EP, fossil fuel demand, water depletion and LU. For dairy beef, the authors adopted biophysical allocation at the ratio of 9.4%/0.4%/90.2% for beef/veal/milk, respectively. They also considered economic allocation in a sensitivity analysis, at the rate of 7.8%/0.9%/91.3%. Across GWP, EP, AP and LU, grass-fed was found to have higher burdens than dairy beef. On the other hand, dairy beef required more fossil fuel and water than grass-fed. The authors also considered impacts on a per ha basis, which resulted in lower AP and EP burdens for grass-fed. A sensitivity analysis to account for carbon sinks in grassland was also considered. While this inclusion substantially reduced the GWP of grass fed, it was not enough to offset the benefits of productivity from DB. The authors echoed the argument of Berton et al. (2017) that future research should consider multifaceted aspects of grass-fed systems that are socially important.

Wiedemann et al. (2017) examined resource use and GHG emissions associated with seven Australian feedlot beef systems. The authors adopted a gate to gate approach, with a primary focus on impacts arising from the grain-finishing stage. The FU for comparisons between the finishing stages was 1 kg LWG, while values for the entire system (including cow–calf enterprise) were reported as 1 kg LW. Three classes of cattle were considered: short-fed (55–80 days) for domestic market; mid-fed (108–164 days); and long-fed (>300 days) for alternative export markets. Similar to Hyland et al. (2016), Wiedemann et al. (2017) found that CH_4 emissions aggregated across enteric fermentation and manure management were the most significant contributors to emissions intensities. Across the three management strategies, long-fed generated more GHG emissions than mid-fed which in turn generated more emissions than short-fed, due largely to the length of production cycles. The same rankings were observed for fossil energy demand. However, the opposite rankings were noted for water consumption, an impact category with high importance in the arid regions of Australia. While the differences were not significant between short- and mid-fed, long-fed cattle had considerably lower freshwater usage due to reduced irrigated water usage. In terms of cradle-to-gate analysis, the finishing systems were found to contribute 26–44% of the total emissions intensity, with higher maximum impacts (up to 72%) recorded for total energy demand. The authors note that switches from pasture-based to grain-based systems have reduced Australia's national emissions intensity from beef cattle, but these switches have been met with a trade-off of increased national energy demand. This signifies the complexities of drawing conclusions across multiple impact categories.

Willers et al. (2017) sought to identify environmental hotspots in semi-intensive beef production systems in Brazil's Northeast. The study accounted for two farms: the cow–calf operation and a separate but nearby finishing system. Similar to most beef LCA studies, the authors adopted a cradle-to-farmgate system boundary and an FU of 1 kg LW leaving the finishing farm. Primary data were gathered from the managers of both farms, while background processes were sourced from *ecoinvent* V2. The authors considered five impact categories: GWP (reported as climate change); AP (reported as terrestrial acidification); EP (reported as freshwater eutrophication), LU (reported as agricultural land occupation) and fossil fuel depletion. Following Berton et al. (2017), Willers et al. (2017) used mass allocation to disentangle burdens arising from coproducts at the cow-calf stage. Regarding the identification of hotspots, the authors diverted from conventional approaches and considered pasture processes as separate entities to their modelled livestock. This resulted in an unusual attribution of the overall burdens, whereby 'grassland production' has higher effects on all impact categories than 'livestock burdens', making inter-study comparison of the results (de Vries et al., 2015) rather difficult.

Bragaglio et al. (2018) analysed the environmental footprints of a range of different beef production systems in Italy utilising data collected from 25 farms. The systems studied were: specialised extensive; high grain fattening; intensive cow-calf constantly kept in confinement and native breed (Podolian) maintained on pasture and finished in housing. The authors considered GWP, water depletion, LU, AP and EP within a system boundary set as cradle-to-farmgate and an FU of 1 kg LW. In terms of GWP, the intensive systems (high grain fattening and cow-calf confinement) were found to have lower impacts due largely to improved growth rates. However, the authors found that the systems with durations of pasture grazing (specialised extensive and Podolian) had lower AP than cow-calf confinement. There was no significant difference noted for water depletion, while high grain fattening and Podolian demonstrated the lowest burdens in terms of water quality (EP). Significantly higher LU was required for specialised extensive and Podolian; however, the authors also acknowledged that competition with human edible feed was lower for the grazing systems, particularly Podolian. A theme recurrent throughout grazing livestock LCA studies, namely the omission of ecosystem services and other societal benefits (e.g. improved animal welfare and meat quality) provided by grassland systems, is also highlighted by the authors. Bragaglio et al. (2018) conclude by acknowledging the importance of future LCA studies addressing these aspects of livestock systems that are more difficult to quantify.

Analytical approaches adopted by the above 14 studies are summarised in Table 1, with particular attention to the treatment of major sources of uncertainty inherent in beef production systems. Overall, it demonstrates a considerable gap in knowledge regarding uncertainty within the existing literature. For example, none of the 14 studies used individual livestock data, meaning that intra-herd distributions of animal properties and performances could not be considered. Eight out of 14 papers did use farm-level aggregated data; nonetheless, only one of them included primary information on forage quality, a parameter widely known to be affected by farm management and, in turn, contribute to the uncertainty surrounding CH_4 emissions through enteric fermentation. Only three studies conducted Monte Carlo analysis, reiterating the lack of attention bestowed upon uncertainty on the whole among LCA studies (Imbeault-Tétreault et al., 2013).

Table 1 Sources of uncertainty identified in recent LCA studies of beef production systems

Study	Animal growth over time	Animal categories	Foreground data source	Feed quality	Emission factors for carbon footprints	Monte Carlo analysis
Dick et al. (2015)	Split between first year and subsequent years	9	National statistics	Literature	IPCC Tier 2	✗
Mogensen et al. (2015)	Fixed average daily gain (ADG) per system	6	Literature	Literature	Danish specific and IPCC Tier 2	✗
Wiedemann et al. (2015b)	Fixed ADG	4	Farm	National statistics	Australian specific	✔
Hyland et al. (2016)	Monthly growth rates for growing stock	Unspecified	Farm	National statistics	UK specific and IPCC Tiers 1/2	✗
Alemu et al. (2017)	Fixed ADG per animal category	7	Literature and experimental data	Measured data	Canadian specific and IPCC Tier 2	✗
Berton et al. (2017)	Split into three growth stages on each farm	5	Farm	Literature	French specific and IPCC Tier 2	✗
Buratti et al. (2017)	Fixed ADG per animal category	7	Farm	Literature	IPCC Tier 2	✗
de Figueiredo et al. (2017)	Fixed ADG per system	Unspecified	Literature	Unspecified	Brazilian specific and IPCC Tier 1	✗
Florindo et al. (2017)	Varied ADG by age and scenario	4	Farm	Unspecified	IPCC Tier 2	✗
Hessle et al. (2017)	Unspecified	5	National statistics	Literature	IPCC (unspecified tier)	✗
Tichenor et al. (2017)[d]	Unspecified	6	Literature	Not applicable (only examines land use)	Not applicable (only examines land use)	✗
Wiedemann et al. (2017)	Varied ADG by farm and scenario	3	Farm	National statistics	Australian specific	✔
Willers et al. (2017)	Unspecified	4	Farm	Unspecified	IPCC (unspecified tier)	✔
Bragaglio et al. (2018)	Fixed ADG per system	3	Farm	Literature	IPCC Tier 2	✗

Furthermore, none of the above studies adopted site-specific EFs for calculating GHG emissions. This finding is perhaps unsurprising given the considerable effort required to measure GHG fluxes at the farm scale; conducting field trials specifically for an LCA study is unlikely to be financially justified. Should an opportunity exist, however, introduction of site-specific EFs may provide an effective and computationally straightforward means to improve accuracy of CF estimation because, as already discussed in Section 1, uncertainty associated with these parameters is one of the important practical barriers to draw useful policy implications for farm management. An important operational question, then, is as follows: under practical constraints concerning staff and budgetary availability, which EFs should we prioritise to measure on the farm? The next section outlines the procedure of a virtual experiment designed to solve this problem. To the best of our knowledge, such an attempt has not previously been made in LCA literature.

3 Case study: materials and methods

The case study was carried out using data from the permanent pasture-based beef enterprise (farmlet) of the North Wyke Farm Platform National Capability (Orr et al., 2016), an instrumented farm-scale grazing trial located in Devon, UK (50°46′10″N, 3°54′05″W). Under the attributional approach, the system boundary was defined as 'cradle-to-gate' or from the production of raw materials to the departure of live animals for slaughter, and encompassed both cow-calf and finishing operations, which are adjacent to each other but do not share pasture or other resources. The FU was set as 1 kg LW of prime beef (calves) departing the farm. Environmental burdens attributable to the sale of culled cows were portioned out from the system-wide CF using economic allocation.

Farm management practices and the data collection strategy at the study site are detailed elsewhere (Takahashi et al., 2018). Briefly, 30 Charolais × Hereford-Friesian calves and their dams constitute each year's herd, with the number of cows in the LCA model adjusted to account for extra heifers required to replace culled cows. As with the majority of beef farms in South West England, cattle graze in summer and are housed in winter, with both LWG of calves and the forage quality of pasture and silage evaluated at regular (2-4 week) intervals. All physical inputs into the system were appropriately recorded.

The present study utilised data associated with a generation of calves born in spring 2015 and slaughtered in winter 2016. On-farm GHG emissions from both livestock and pastures were calculated using the IPCC (2006) Tier 2 approach. Based on our earlier finding that ignoring the inter-animal difference in growth efficiency leads to a biased estimate of the farm-scale CF

(McAuliffe et al., 2018b), emissions for finishing animals were initially calculated for each individual animal separately and subsequently pooled together to create a whole-farm inventory. The resultant CF (kg CO_2-eq/kg LW) was expressed as a 95% confidence interval, which was derived by a Monte Carlo analysis with 1000 iterations. For each iteration, values for EFs were drawn from the distributions recommended by IPCC (2006); further details of this process are described in McAuliffe et al. (2018b).

Following the baseline CF estimation under which all EFs were assumed to be uncertain and follow predetermined probability distributions, eight groups of EFs (Table 2) were individually assumed to be certain at three levels defined by IPCC (2006): point estimate as well as lower and upper limits of the 95% confidence interval. By design, the CF derived under this setting was expected to have a narrower range than the baseline result, as one source of uncertainty had been eliminated from the model. These outputs represent hypothetical CFs when a particular group of site-specific EFs are perfectly quantified on the farm and therefore indicate the information value of pinpointing the corresponding EFs. Alternatively, as the true EFs are likely to lie somewhere between lower and upper limits of the IPCC 95% range, the difference in CF distributions derived under these two values can be seen as the level of uncertainty associated with the relevant EFs; a larger difference here suggests a higher priority for field measurements to obtain locally more accurate EF values. Overall, 25 CFs were derived, a baseline and 24 variants with distinct EFs (8 groups x 3 values).

Throughout the analysis, emissions pertaining to background processes were sourced from *Agri-footprint* v3 (Durlinger et al., 2017) and *ecoinvent* V3 (Wernet et al., 2016) databases. All CFs were calculated under the IPCC (2013) 100-year average impact assessment method on *SimaPro* v8.2.3.

Table 2 Emission factors considered in the case study

Source	IPCC (2006) notation	IPCC (2006) reference
CH_4 enteric fermentation	EF	Equation 10.21
CH_4 manure management	$EF_{(T)}$	Equation 10.23
N_2O manure management	EF_3	Table 10.21
N_2O inorganic fertiliser (ammonium nitrate)	EF_1 (for $F_{SN)}$)	Table 11.1
N_2O organic fertiliser (manure)	EF_1 (for $F_{ON)}$)	Table 11.1
N_2O grazing	EF_{3PRP}	Table 11.1
N_2O volatilisation	EF_4	Table 11.3
N_2O leaching	EF_5	Table 11.3

Note: Emissions associated with other sources (including background processes) were assumed to be uncertain throughout the case study.

4 Case study: results and discussion

The mean baseline CF was estimated to be 22.8 kg CO_2-eq/kg LW, with a confidence interval of 21.3–25.2 kg CO_2-eq/kg LW (Fig. 1). These results are at the higher end of values and ranges reported by previous studies undertaken in comparative environments (reviewed in Section 1); while the reason behind this has not been completely elucidated, it is thought to be due to a combination of the low stocking rate and the high replacement of the breeding herd to ensure safe delivery of calves for fattening experiments. Assuming subsets of EFs to take point estimate values without uncertainty did not change the CF distribution to any noticeable extent, although knowledge in CH_4 emissions from manure management and N_2O emissions from inorganic fertiliser application reduced the width of the confidence interval by 0.4 and 0.3 kg CO_2-eq/kg LW, respectively (Fig. 1). Interestingly, certainty regarding CH_4 emissions from enteric fermentation, widely considered to be the single largest source of ruminant-originated GHG emissions, contributed very little to the certainty regarding the resultant CF, likely because of the symmetric (triangular), rather than asymmetric (lognormal), nature of the probability distribution assumed under the baseline model (McAuliffe et al., 2018b).

Comparisons of CFs derived under lower and upper limits of EF values revealed, however, that on-farm measurement of enteric CH_4 may still be one

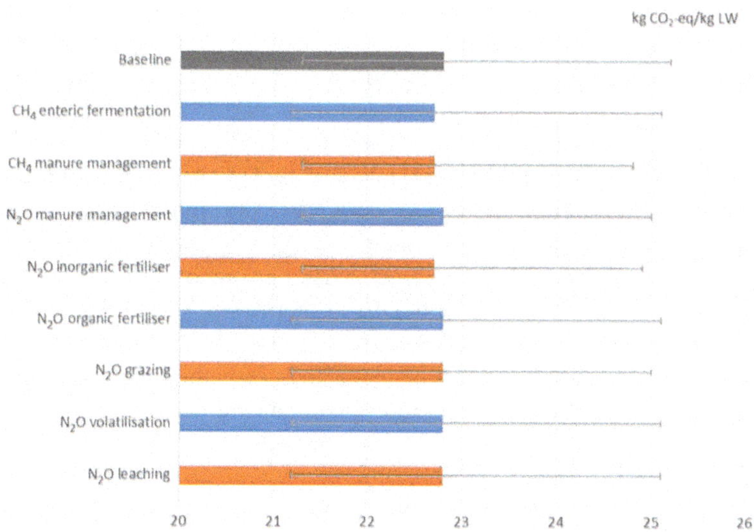

Figure 1 Carbon footprints of beef production systems estimated with one group of emission factors fixed at the IPCC point estimate. Error bars show the 95% range derived from Monte Carlo simulation, where all but one group of emission factors were assumed to follow IPCC uncertainty distributions. The baseline result accounts for all sources of uncertainty.

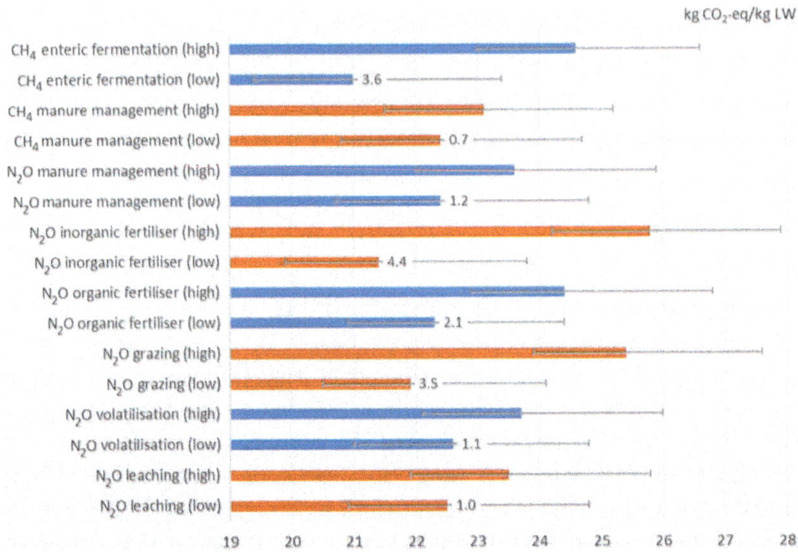

Figure 2 Carbon footprints of beef production systems estimated with one group of emission factors fixed at lower and upper limits of the IPCC 95% confidence interval. The number under each group depicts the difference in means between the two scenarios, which represents the value of knowing that particular emission factor through site-specific field trials. Error bars show the 95% range derived from Monte Carlo simulation, where all but one group of the emission factors were assumed to follow IPCC uncertainty distributions.

of the most effective approaches to reduce uncertainty (Fig. 2). The difference in mean CFs between two scenarios was estimated to be 3.6 kg CO_2-eq/kg LW, the second largest after N_2O emissions from inorganic fertiliser application (4.4 kg CO_2-eq/kg LW) and closely followed by N_2O emissions from excreta deposited during grazing (3.5 kg CO_2-eq/kg LW). The high information value of the latter two EFs is attributable to the high degree of uncertainty identified for these parameters by IPCC (2006), as discussed in Section 1 and quantitatively supported by Fig. 1. On the other hand, improved knowledge of enteric CH_4 EF reduced uncertainty surrounding the overall CF because of its large contribution to the total environmental burdens, even though its own variation is confined to a relatively small range. The remaining five groups of EFs were shown to have considerably less impact on the CF.

5 Conclusion

The above analysis revealed that uncertainty surrounding climate impacts of ruminant systems can potentially be reduced through on farm measurements of GHG fluxes, but not all measurements carry the same degree of information value.

In temperate grassland regions, the priority for measurements should be given to N_2O from inorganic fertilisers applied and manure deposited to the soil, as well as CH_4 from enteric fermentation. It is acknowledged that, strictly speaking, these site-specific EFs will still be accompanied by their own uncertainty, which stems from intra-farm variability in soil, weather, plant genetics, animal genetics, rumen microbial ecology and other confounding factors. Nevertheless, these local sources of uncertainty are likely to be considerably smaller than uncertainty about the farm location that needs to be embedded into generic EFs. In turn, CF analysis carried out under reduced uncertainty will likely offer more policy implications that are directly applicable to local production environments.

6 Acknowledgements

The case study component of this chapter was supported by the Biotechnology and Biological Sciences Research Council (BBS/E/C/000I0320 and BBS/E/C/000J0100).

7 Where to look for further information

- The foundation theory behind the above discussion is summarised in: Heijungs R and Suh S (2002) *The Computational Structure of Life Cycle Assessment* (https://doi.org/10.1007/978-94-015-9900-9), Kluwer Academic Publishers.
- Oxford University Research Archive (https://doi.org/10.5287/bodleian:0z9MYbMyZ) stores a streamlined global database of CFs associated with food and beverage production. Also see Poore and Nemecek (2018).
- Global Farm Platform network (http://www.globalfarmplatform.org) is a worldwide initiative to compare sustainability of livestock production systems across different agroecological zones. Also see Eisler et al. (2014).
- North Wyke Farm Platform data portal (https://www.rothamsted.ac.uk/north-wyke-farm-platform) provides a wealth of raw data collected from farm-scale grazing trials, including those used in the present case study. Also see Orr et al. (2016) and Takahashi et al. (2018).

8 References

Alemu, A. W., Janzen, H., Little, S., Hao, X., Thompson, D. J., Baron, V., Iwaasa, A., Beauchemin, K. A. and Kröbel, R. 2017. Assessment of grazing management on farm greenhouse gas intensity of beef production systems in the Canadian Prairies using life cycle assessment. *Agricultural Systems* 158, 1–13. doi:10.1016/j.agsy.2017.08.003.

Berton, M., Agabriel, J., Gallo, L., Lherm, M., Ramanzin, M. and Sturaro, E. 2017. Environmental footprint of the integrated France–Italy beef production system assessed through a multi-indicator approach. *Agricultural Systems* 155, 33–42. doi:10.1016/j.agsy.2017.04.005.

Biswas, W. K., Graham, J., Kelly, K. and John, M. B. 2010. Global warming contributions from wheat, sheep meat and wool production in Victoria, Australia: a life cycle assessment. *Journal of Cleaner Production* 18(14), 1386–92. doi:10.1016/j.jclepro.2010.05.003.

Bragaglio, A., Napolitano, F., Pacelli, C., Pirlo, G., Sabia, E., Serrapica, F., Serrapica, M. and Braghieri, A. 2018. Environmental impacts of Italian beef production: a comparison between different systems. *Journal of Cleaner Production* 172, 4033–43. doi:10.1016/j.jclepro.2017.03.078.

Buratti, C., Fantozzi, F., Barbanera, M., Lascaro, E., Chiorri, M. and Cecchini, L. 2017. Carbon footprint of conventional and organic beef production systems: an Italian case study. *The Science of the Total Environment* 576, 129–37. doi:10.1016/j.scitotenv.2016.10.075.

Chen, X. and Corson, M. S. 2014. Influence of emission-factor uncertainty and farm-characteristic variability in LCA estimates of environmental impacts of French dairy farms. *Journal of Cleaner Production* 81, 150–7. doi:10.1016/j.jclepro.2014.06.046.

de Figueiredo, E. B., Jayasundara, S., de Oliveira Bordonal, R., Berchielli, T. T., Reis, R. A., Wagner-Riddle, C. and La Scala Jr., N. 2017. Greenhouse gas balance and carbon footprint of beef cattle in three contrasting pasture-management systems in Brazil. *Journal of Cleaner Production* 142, 420–31. doi:10.1016/j.jclepro.2016.03.132.

de Vries, M., van Middelaar, C. E. and de Boer, I. J. M. 2015. Comparing environmental impacts of beef production systems: a review of life cycle assessments. *Livestock Science* 178, 279–88. doi:10.1016/j.livsci.2015.06.020.

Dick, M., Abreu da Silva, M. and Dewes, H. 2015. Life cycle assessment of beef cattle production in two typical grassland systems of southern Brazil. *Journal of Cleaner Production* 96, 426–34. doi:10.1016/j.jclepro.2014.01.080.

Dudley, Q. M., Liska, A. J., Watson, A. K. and Erickson, G. E. 2014. Uncertainties in life cycle greenhouse gas emissions from U.S. beef cattle. *Journal of Cleaner Production* 75, 31–9. doi:10.1016/j.jclepro.2014.03.087.

Durlinger, B., Koukouna, E., Broekema, R., van Paassen, M. and Scholten, J. 2017. *Agri-Footprint 3.0.* Blonk Consultants, Gouda.

Eisler, M. C., Lee, M. R. F., Tarlton, J. F., Martin, G. B., Beddington, J., Dungait, J. A. J., Greathead, H., Liu, J., Mathew, S., Miller, H., Misselbrook, T., Murray, P., Vinod, V. K., Saun, R. V. and Winter, M. 2014. Agriculture: step to sustainable livestock. *Nature* 507(7490), 32–4. doi:10.1038/507032a.

Florindo, T. J., de Medeiros Florindo, G. I. B., Talamini, E., da Costa, J. S. and Ruviaro, C. F. 2017. Carbon footprint and Life Cycle Costing of beef cattle in the Brazilian midwest. *Journal of Cleaner Production* 147, 119–29. doi:10.1016/j.jclepro.2017.01.021.

Heijungs, R. and Suh, S. 2002. *The Computational Structure of Life Cycle Assessment.* Kluwer Academic Publishers. doi:10.1007/978-94-015-9900-9.

Herrero, M., Henderson, B., Havlík, P., Thornton, P. K., Conant, R. T., Smith, P., Wirsenius, S., Hristov, A. N., Gerber, P., Gill, M., Butterbach-Bahl, K., Valin, H., Garnett, T. and Stehfest, E. 2016 Greenhouse gas mitigation potentials in the livestock sector. *Nature Climate Change* 6(5), 452–61. doi:10.1038/nclimate2925.

Hessle, A., Bertilsson, J., Stenberg, B., Kumm, K. I. and Sonesson, U. 2017. Combining environmentally and economically sustainable dairy and beef production in Sweden. *Agricultural Systems* 156, 105–14. doi:10.1016/j.agsy.2017.06.004.

Hyland, J. J., Styles, D., Jones, D. L. and Williams, A. P. 2016. Improving livestock production efficiencies presents a major opportunity to reduce sectoral greenhouse gas emissions. *Agricultural Systems* 147, 123–31. doi:10.1016/j.agsy.2016.06.006.

Imbeault-Tétreault, H., Jolliet, O., Deschênes, L. and Rosenbaum, R. K. 2013. Analytical propagation of uncertainty in life cycle assessment using matrix formulation. *Journal of Industrial Ecology* 17(4), 485–92. doi:10.1111/jiec.12001.

IPCC. 2006. *Guidelines for National Greenhouse Gas Inventories*. Institute for Global Environmental Strategies, Hayama, Japan.

IPCC. 2013. *Climate Change 2013: the Physical Science Basis. Contribution of Working Group I to the Fifth Assessment Report of the Intergovernmental Panel on Climate Change*. Cambridge University Press, Cambridge, UK.

McAuliffe, G. A., Takahashi, T., Mogensen, L., Hermansen, J. E., Sage, C. L., Chapman, D. V. and Lee, M. R. F. 2017. Environmental trade-offs of pig production systems under varied operational efficiencies. *Journal of Cleaner Production* 165, 1163–73. doi:10.1016/j.jclepro.2017.07.191.

McAuliffe, G. A., Takahashi, T. and Lee, M. R. F. 2018a. Framework for life cycle assessment of livestock production systems to account for the nutritional quality of final products. *Food and Energy Security* 7(3), e00143. doi:10.1002/fes3.143.

McAuliffe, G. A., Takahashi, T., Orr, R. J., Harris, P. and Lee, M. R. F. 2018b. Distributions of emissions intensity for individual beef cattle reared on pasture-based production systems. *Journal of Cleaner Production* 171, 1672–80. doi:10.1016/j.jclepro.2017.10.113.

Mogensen, L., Kristensen, T., Nielsen, N. I., Spleth, P., Henriksson, M., Swensson, C., Hessle, A. and Vestergaard, M. 2015. Greenhouse gas emissions from beef production systems in Denmark and Sweden. *Livestock Science* 174, 126–43. doi:10.1016/j.livsci.2015.01.021.

Orr, R. J., Murray, P. J., Eyles, C. J., Blackwell, M. S. A., Cardenas, L. M., Collins, A. L., Dungait, J. A. J., Goulding, K. W. T., Griffith, B. A., Gurr, S. J., Harris, P., Hawkins, J. M. B., Misselbrook, T. H., Rawlings, C., Shepherd, A., Sint, H., Takahashi, T., Tozer, K. N., Whitmore, A. P., Wu, L. and Lee, M. R. F. 2016. The North Wyke Farm Platform: effect of temperate grassland farming systems on soil moisture contents, runoff and associated water quality dynamics. *European Journal of Soil Science* 67(4), 374–85. doi:10.1111/ejss.12350.

Poore, J. and Nemecek, T. 2018. Reducing food's environmental impacts through producers and consumers. *Science* 360(6392), 987–92. doi:10.1126/science.aaq0216.

Pouliot, G., Wisner, E., Mobley, D. and Hunt Jr., W. 2012. Quantification of emission factor uncertainty. *Journal of the Air and Waste Management Association* 62(3), 287–98. doi:10.1080/10473289.2011.649155.

Ripoll-Bosch, R., de Boer, I. J. M., Bernués, A. and Vellinga, T. V. 2013. Accounting for multi-functionality of sheep farming in the carbon footprint of lamb: a comparison of three contrasting Mediterranean systems. *Agricultural Systems* 116, 60–8. doi:10.1016/j.agsy.2012.11.002.

Springmann, M., Godfray, H. C. J., Rayner, M. and Scarborough, P. 2016. Analysis and valuation of the health and climate change cobenefits of dietary change. *Proceedings of the National Academy of Sciences of the United States of America* 113(15), 4146–51. doi:10.1073/pnas.1523119113.

Takahashi, T., Harris, P., Blackwell, M. S. A., Cardenas, L. M., Collins, A. L., Dungait, J. A. J., Hawkins, J., Misselbrook, T., McAuliffe, G. A., McFadzean, J., Murray, P. J., Orr, R. J., Rivero-Viera, M. J., Wu, L. and Lee, M. R. F. 2018. Roles of instrumented farm-scale trials in trade-off assessments of pasture-based ruminant production systems. *Animal* 12, 1766–76. doi: 10.1017/S1751731118000502

Tichenor, N. E., Peters, C. J., Norris, G. A., Thoma, G. and Griffin, T. S. 2017. Life cycle environmental consequences of grass-fed and dairy beef production systems in the Northeastern United States. *Journal of Cleaner Production* 142, 1619–28. doi:10.1016/j.jclepro.2016.11.138.

Üçtuğ, F. G. 2019. The environmental life cycle assessment of dairy products. *Food Engineering Reviews* 11(2), 104–21. doi:10.1007/s12393-019-9187-4.

Wernet, G., Bauer, C., Steubing, B., Reinhard, J., Moreno-Ruiz, E. and Weidema, B. 2016. The ecoinvent database version 3 (part I): overview and methodology. *The International Journal of Life Cycle Assessment* 21(9), 1218–30. doi:10.1007/s11367-016-1087-8.

Wiedemann, S. G., Ledgard, S. F., Henry, B. K., Yan, M., Mao, N. and Russell, S. J. 2015a. Application of life cycle assessment to sheep production systems: investigating co-production of wool and meat using case studies from major global producers. *The International Journal of Life Cycle Assessment* 20(4), 463–76. doi:10.1007/s11367-015-0849-z.

Wiedemann, S. G., McGahan, E., Murphy, C., Yan, M. J., Henry, B. K., Thoma, G. and Ledgard, S. 2015b. Environmental impacts and resource use of Australian beef and lamb exported to the USA determined using life cycle assessment. *Journal of Cleaner Production* 94, 67–75. doi:10.1016/j.jclepro.2015.01.073.

Wiedemann, S. G. , Davis, R., McGahan, E., Murphy, C. and Redding, M. 2017. Resource use and greenhouse gas emissions from grain-finishing beef cattle in seven Australian feedlots: a life cycle assessment. *Animal Production Science* 57(6), 1149–62. doi:10.1071/AN15454.

Wilkinson, J. M. and Lee, M. R. F. 2018. Use of human-edible animal feeds by ruminant livestock. *Animal* 12(8), 1735–43. doi:10.1017/S175173111700218X.

Willers, C. D., Maranduba, H. L., de Almeida Neto, J. A. and Rodrigues, L. B. 2017. Environmental Impact assessment of a semi-intensive beef cattle production in Brazil's Northeast. *The International Journal of Life Cycle Assessment* 22(4), 516–24. doi:10.1007/s11367-016-1062-4.

Chapter 5

Using life cycle assessment to compare intensive versus extensive agricultural systems

Matthias Meier, Research Institute of Organic Agriculture (FiBL), Switzerland

1 Introduction

The number of life cycle assessment (LCA) and LCA-based footprint studies comparing food products coming from different agricultural production systems, reflecting different production intensities, is rapidly growing providing an increasing amount of data for further analyses (Poore and Nemecek, 2018). In the same course, the number of studies synthesizing these individual LCA studies to gain an overall picture on the environmental sustainability of different farming systems or the food system as a whole is increasing too (Mondelaers et al., 2009; Tuomisto et al., 2012; Meier et al., 2015; Clark and Tilman, 2017; Meemken and Qaim, 2018; Poore and Nemecek, 2018).

LCAs of agricultural systems usually compare conventional (intensive) farming systems to certified organic (extensive) systems, based mainly on farming practice in Europe and North America. Even though production intensity (in terms of yields and inputs) varies widely in both systems (Seufert and Ramankutty, 2017), organic agriculture is generally less intensive than conventional farming, since fertilizer inputs are restricted whilst chemical fertilizers and synthetic pesticides are prohibited under organic standards (Seufert et al., 2017).

http://dx.doi.org/10.19103/AS.2018.0044.03

Whether organic agriculture, as an extensive system, is more environmentally-friendly than conventional agriculture is much debated within the LCA community and beyond. It is not the aim of this chapter to contribute to this debate. Nevertheless, comparative LCA studies on organic and conventional farming are well suited to discuss advantages and limitations of using LCA to assess the environmental sustainability of farming systems of different production intensities. Further, these existing comparative studies are an important basis to critically reflect on whether their results fuel the discussion on sustainable agriculture in the right direction and how LCA can be further improved in order to provide decision support for science and policy on the sustainability of agricultural production and food systems as a whole.

Aims and conclusions of individual comparative LCA studies and in particular of the studies synthesizing LCA results of many other individual studies are often similar. The aims are to assess agriculture's environmental impacts as affected by farming system and agricultural input efficiency in order to draw conclusions on how to make food and agricultural systems more sustainable. The outcome of the LCA is often consistent across studies: most products from organic agriculture have a lower eco-efficiency, that is organic farming results in higher environmental impacts per unit of product for many of the impact categories analyzed (Tuomisto et al., 2012; Meier et al., 2015; Meemken and Qaim, 2018). But organic farming usually also shows lower impacts per unit of land, even though the latter is not recognized as a functional unit. The recent meta-analysis of Clark and Tilman (2017) showed that across a large number of individual studies covering plant and animal products, organic food products require significantly more land, cause significantly more eutrophication, and have a slightly higher, though not significant acidification potential. However, organic food products use significantly less energy and emit similar amounts of greenhouse gas (GHG) emissions than conventional products.

Based on the product-related assessment results from LCAs, some studies draw further conclusions on the sustainability of extensive production, i.e. organic farming in the case of widespread upscaling. It is argued that due to the higher land use in organic farming as a result of the generally lower yields in organic compared to conventional agriculture, this would lead to additional loss of natural habitats, which, in turn, would lead to losses of biodiversity and the offset of carbon stored in soil and biomass within these habitats (Clark and Tilman, 2017; Meemken and Qaim, 2018). This conclusion assumes that equivalent amounts of food or in the future even higher amounts should be produced with organic farming given the facts that the human population is growing and global consumption of animal-based products is increasing.

Obviously, the focus in the discussion on sustainable agriculture based on LCA results is mainly on agriculture's function as provider of food and fiber. However, this one-sided view is hard to sustain since agriculture depends on a range of potentially scarce ecosystem services (DeClerck et al., 2016). The discussion of sustainable agriculture based on LCA studies does not take account of the scarcity of these ecosystem services. Sustainable agricultural production should be seen as the product of multuple interacting ecosystem services (DeClerck et al., 2016). Based on this awareness sustainability, the assessment of agriculture needs to be refined and assessment tools improved, i.e. LCA in this case.

Similarly, it is hardly ever questioned in comparative LCAs on products coming from different farming systems whether the obtained results are valid. Normally, validity of the results is taken for granted despite the high level of uncertainty at inventory and impact assessment levels. Therefore, in the following sections, I critically discuss the validity of LCAs analyzing agricultural products from different farming systems.

2 Validity of LCA studies comparing intensive and extensive agricultural systems

Life cycle assessment was originally developed to assess the environmental impacts of industrial processes and of products that result from these processes. Industrial processes usually take place within closed systems, and the energy and mass flows in and out of these systems are relatively well defined. In addition, land use of industrial products is relatively low compared to agriculture. Because of its many advantages, LCA has been more and more adapted to assess the environmental impacts of agricultural processes and products. However, except from soil-less vegetable and fruit production in greenhouses, agricultural production takes place in an open system, requires large areas of land, and is subject to complex interactions with the environment under spatially highly variable conditions.

Even if energy and mass flows on the input side of agricultural processes can be determined with relatively high accuracy, it is a challenge to estimate emissions from these processes with reliable precision due to complex interactions with the abiotic and biotic environments. Simplifications in emissions models at the inventory level are inevitable to come to an assessment result. In addition, for biodiversity and soil quality aspects, both highly relevant impact categories in an agricultural context, the development of suitable impact assessment methods is still in its infancy. Things become even more difficult if LCA is used to compare agricultural products originating from different production systems especially if production methods differ only slightly (Meier et al., 2015). Even though it is likely that the resolution of emission models at the inventory level is too low to capture the real differences between the two production systems and even though it is obvious that an environmental assessment of agriculture without considering its impacts on biodiversity and soil quality aspects will never be conclusive, potential biases in LCA results of agricultural products are hardly ever questioned. Therefore, there is an urgent need to critically discuss the validity of LCA results of agricultural products especially if LCA is used to compare intensive versus extensive agricultural systems and to find answers on the overall environmental sustainability of agricultural production systems.

3 Assessing validity at the inventory level: reactive nitrogen emission modeling

Comparative LCAs between organic and conventional products often result in higher global warming potential (GWP) as well as higher acidification and eutrophication potential (Tuomisto et al., 2012; Meier et al., 2015; Clark and Tilman, 2017). However, according to the meta-analysis by Clark and Tilman (2017), only for the latter, there seems to be a statistically significant difference between organic and conventional agriculture. Nevertheless, in all three impact categories, reactive nitrogen (N) emissions such as ammonia, nitrate, nitrogen oxides, and nitrous oxide are a dominant factor in causing the environmental impacts. The emissions primarily result from surplus nitrogen in the agricultural production system necessary to achieve stable yields. Therefore, the nitrogen balance describes approximately the overall nitrogen-loss potential without indicating as to through what reactive nitrogen compound the nitrogen is lost. Higher impacts in the case of organic products would mean higher nitrogen losses per product unit by reactive nitrogen emissions. This suggests lower nitrogen use efficiency in organic farming representing an extensive production system.

Looking at individual empirical studies measuring and comparing nitrogen emissions between organic and conventional crop rotations and relating them to yields, the picture regarding nitrogen use efficiency is ambiguous. There are studies showing higher N-leaching losses in organic systems per unit of yield (Aronsson et al., 2007; Korsaeth, 2008), as well as studies measuring several reactive N-emissions showing higher N use efficiency in organic systems (Benoit et al., 2015). Further, results from two farm balance assessments suggest overall a higher N-use efficiency in organic systems when accounting for changes in the soil C–N pool (Knudsen et al., 2006; Lin et al., 2016). Although there is no clear answer from empirical studies measuring nitrogen emissions in different farming systems, the evidence is in contrast to the clear results obtained from LCA studies. An explanation for this contrast in results could be due to the modeling approach often implemented in LCA inventories of plant products.

Already in Meier et al. (2014, 2015), it was pointed out that nitrogen emissions modeling at the inventory level is often too simple to differentiate between farming systems. In particular, it is problematic to use mutually independent N-emission models for the different reactive nitrogen species (ammonia, nitrate, nitrogen oxides, and nitrous oxide) within plant production inventories as it is, for example, the case of the respective inventories in the ecoinvent 3 database (Nemecek and Schnetzer, 2011). The different emissions are estimated primarily based on the total nitrogen input from fertilizer use multiplied with an emission factor. In this simple emission modeling, fertilizer types are differentiated to a limited extent only and changes to the soil C-N pool due to mineralization resulting from agricultural management are not considered. However, there are fundamental differences between organic and chemical fertilizers regarding their mode of action. In the latter, the total N is 100% plant available, whereas in the former, depending on the type of organic fertilizer, only a fraction of the total N is plant available. The rest is organically bound and needs to be mineralized by microorganisms in the soil in order to feed the plant. In organic farming, only organic fertilizers are used with the aim of increasing soil fertility. However, in conventional agriculture, most of the N in plant production is supplied by chemical fertilizers.

Due to the use of independent N emission models within plant production inventories, the amount of N estimated as losses to the environment hardly ever corresponds to the actual N surplus in the system. In other words, the N flows into an agricultural system are not equal to the N flows out of the system meaning that the law of mass conservation is not met, which is a fundamental violation of an underling principle in LCA. In addition, meeting the law of mass conservation is also the main requirement for calculating N use efficiencies (Godinot et al., 2014).

As N emissions from plant production arise from microbial processes that are linked with each other, it is obvious that losses of one species of reactive nitrogen to the environment are not independent from the losses of another species of reactive nitrogen. From empirical measuring studies, we know that nitrogen emissions are the result of complex interactions among soil microorganisms and abiotic factors (Chirinda et al., 2010a). Further, the soil microbial activity is also influenced by management practices showing differences between organic and conventional agriculture (Chirinda et al., 2010b).

Accurate nitrogen emission modeling is challenging and more reliable mechanistic models may be too data demanding. In particular, accurate nitrate leaching in plant production is difficult to simulate as complex soil hydrological and biogeochemical processes are involved (Li et al., 2006; Pedersen et al., 2007). Further, our understanding of the long-term fate of fertilizer-derived nitrogen in the plant-soil-water system is still limited (Sebilo et al., 2013). Despite these challenges, the validity of comparative assessments of intensive and extensive agriculture, in general, and plant production systems, in particular, could be substantially improved by at least aligning the N emissions with the N surplus existing in the system by taking into account the nitrogen balance at the plant level as outlined in Meier et al. (2012, 2014). The nitrogen balance describes approximately the overall nitrogen-loss potential without indicating through what reactive nitrogen compound the nitrogen is lost. With regard to calculating the N balance within a plant production system, it is important to account for all N inputs such as fertilizer input, symbiotic N fixation, atmospheric N deposition, management-induced changes in the soil C–N pool as well as for the N output by the harvested products.

In particular, inclusion of the management-induced changes in the soil C–N pool helps to account for the different modes of action of organic fertilizers including green manure, which are typical for extensive agricultural production systems. But inclusion of the management-induced changes in the soil C–N pool also improves N emission modeling of chemical fertilizer applications. This is because plant production systems using chemical fertilizers always involve some nitrogen mineralized from the C–N pool that is taken up by the crops. A formula to calculate management-induced changes in the soil C–N pool is, for example, provided in Brock et al. (2012), which was included in the modeling of Meier et al. (2014). As nitrate leaching is the most difficult emission to model, it may be calculated from the N surplus in the system minus the N-losses of the other reactive nitrogen species (ammonia, nitrogen oxides, and nitrous oxide). This would represent the maximum amount of N that is potentially lost as nitrate in plant production systems irrespective of the site-specific leaching loss. The proposed procedure to calculate reactive nitrogen emissions is still a considerable over-simplification of real emission patterns. However, the simplification represents a systematic mistake independent of the farming system assessed, and this way it is guaranteed that the law of mass conservation is met for nitrogen flows throughout plant production inventories, which will eventually improve the comparability of farming systems.

4 Assessing validity at the inventory level: heavy metal emissions in different farming systems

At the inventory level, heavy metal emissions from plant production processes are another issue for which the validity of results between intensive and extensive farming systems needs to be reviewed. In comparative LCA studies on arable crops, heavy metal flows to soil are often reported as considerably higher per product unit in organic farming (Meier et al., 2015). Although the yield gap between organic and conventional arable crops typically ranges from 5% to 34%, with lower yields in organic systems (Seufert et al., 2012), comparative LCA studies on different arable crops report heavy metal flows to the soil that are more than 100% to over 1000% higher per product unit in organic systems (Nemecek et al., 2007; Meier et al., 2015). Usually, in the cases where differences are within the upper boundary, copper was used as a fungicide in the organic crop (usually potatoes). Thus, this certainly explains some of the differences, but also in cases where no copper fungicides were used in organic systems, the differences are still much higher than could be expected from the yield gap.

The considerably higher heavy metal flows in inventories of organic crops compared to conventional systems seem remarkable when taking into account where heavy metals in cropping systems originate from and how they cycle through farming systems. Typically, heavy metals enter plant production systems via fertilizers and pesticides, the former representing the major source. Among the mineral fertilizers, it is especially the phosphate fertilizers that contribute most to heavy metal inputs in cropping systems, which also includes rock phosphate (Chien et al., 2011). Rock phosphate may be also used in organic farming. However, it is used to a limited extent only as usually enough phosphate is supplied via organic fertilizers. Some of the heavy metals are taken up by the crops and end up in the harvested product and are, therefore, transferred to animals if used as feed (Makridis et al., 2012). Finally, a fraction of the heavy metals ingested by animals via feed ends up in the manure (Zhang et al., 2012) and further accumulation may take place if the manure is composted (Vukobratović et al., 2014).

From the above-mentioned heavy metal cycle between cropping and animal systems, it becomes evident that organic fertilizers from animal production may also contribute considerably to heavy metal loads in agricultural soils. Therefore, models quantifying heavy metal loads in agricultural soils within crop inventories also take into account the heavy metal inputs from organic fertilizers. Most often, the heavy metal model of Freiermuth (2006) (see also Nemecek and Schnetzer, 2011) is used in LCA inventories, which is for example implemented in the plant production processes of the ecoinvent database (Wernet et al., 2016). The model is an input–output balance taking the cropped area as the system boundary and considering heavy metal inputs from seed, fertilizers, and pesticides. Outputs are taken into account via harvested products, erosion, and leaching.

For the heavy metal inputs from organic fertilizers, standard values are assumed in the model of Freiermuth (2006) for the different heavy metals, which were derived from three empirical studies carried out in Switzerland. Measurements in cattle, pig, and poultry manure were carried out in the 1980s and 1990s (Desaules and Studer, 1993; Menzi and Kessler, 1998). Even though it is not explicitly stated in the studies, it can be assumed that the manure analyzed mainly came from conventional farms since in the beginning of the 1990 less than 1% of the agricultural used area in Switzerland was cultivated under organic agriculture; by 1999, the share grew up to 7.4%

(Willer and Lernoud, 2018). Heavy metals were analyzed for composts made from green and food waste only (Candinas et al., 1999). Therefore, these values do not necessarily correspond to values in composts using manure as substrate.

Heavy metal contents per kg of fertilizer are within the same range for mineral and organic fertilizers (Desaules and Studer, 1993; Menzi and Kessler, 1998; Freiermuth, 2006). However, as organic fertilizers are brought out in much higher masses, high overall heavy metal loads may occur per hectare. In organic farming, synthetic fertilizers are banned and only organic fertilizers are used in plant production besides rock phosphate. That is the main reason why heavy metal loads in organic cropping systems turn out to be usually considerably higher than in conventional cropping systems as reported by the models at the inventory level. However, what is overlooked is that the heavy metal content in organic fertilizers used on organic farms might be lower in certain cases than on conventional farms.

Assuming a mixed organic farm with forage-based cattle production and arable cropping where nutrient flows are nearly closed as manure from livestock is used to fertilize the cropland and external feed inputs to livestock are zero to minimal, there is no or only little external heavy metal input related to agricultural management on the farm. In fact, in this case, heavy metals are recycled on farm via manure. Therefore, in the case of an extensive mixed farm, heavy metal content in cattle manure related to agricultural management can be assumed to be lower than on a more intensive conventional farm relying on external inputs.

However, in another setting, the situation could look quite different. As there is also a tendency toward specialized farms in organic production, in particular, in vegetable, pig, and poultry production, soils could indeed end up with higher heavy metal loads. These farms import most of their organic fertilizers and their feeds, they might import high loads of heavy metals; for example, in vegetable production where according to some organic standards, 100% of the manure is allowed to originate from conventional farms. Further, in organic pig and poultry production, there is usually no difference from the conventional production in terms of copper and zinc additives in the feed, which finally end up as some part of the manure.

From the above examples, it becomes clear that in order to properly differentiate between heavy metal loads to soil in extensive and intensive farming systems, system-specific average heavy metal concentrations in feed and manure are necessary. Therefore, in the case of heavy metal emissions, it is less the modeling approach that might bias the assessment result but rather the lack of specific empirical data from extensive farming systems. In order to improve the validity of assessed heavy metal loads to soil in different farming systems, these data need to be generated first.

5 References

Aronsson, H., Torstensson, G. and Bergström, L. 2007. Leaching and crop uptake of N, P and K from organic and conventional cropping systems on a clay soil. *Soil Use and Management* 23(1), 71-81. doi:10.1111/j.1475-2743.2006.00067.x.

Benoit, M., Garnier, J., Billen, G., Tournebize, J., Gréhan, E. and Mary, B. 2015. Nitrous oxide emissions and nitrate leaching in an organic and a conventional cropping system (Seine Basin, France). *Agriculture, Ecosystems and Environment* 213, 131-41. doi:10.1016/j.agee.2015.07.030.

Brock, C., Hoyer, U., Leithold, G. and Hülsbergen, K.-J. 2012. The humus balance model (HU-MOD): a simple tool for the assessment of management change impact on soil organic matter levels in arable soils. *Nutrient Cycling in Agroecosystems* 92(3), 239-54. doi:10.1007/s10705-012-9487-z.

Candinas, T., Golder, E., Kupper, T. and Besson, J. M. 1999. Nähr- und Schadstoffe im Kompost. *Agrarforschung* 6, 421-4.

Chien, S. H., Prochnow, L. I., Tu, S. and Snyder, C. S. 2011. Agronomic and environmental aspects of phosphate fertilizers varying in source and solubility: an update review. *Nutrient Cycling in Agroecosystems* 89(2), 229-55. doi:10.1007/s10705-010-9390-4.

Chirinda, N., Carter, M. S., Albert, K. R., Ambus, P., Olesen, J. E., Porter, J. R. and Petersen, S. O. 2010a. Emissions of nitrous oxide from arable organic and conventional cropping systems on two soil types. *Agriculture, Ecosystems and Environment* 136(3-4), 199-208. doi:10.1016/j.agee.2009.11.012.

Chirinda, N., Olesen, J. E., Porter, J. R. and Schjonning, P. 2010b. Soil properties, crop production and greenhouse gas emissions from organic and inorganic fertilizer-based arable cropping systems. *Agriculture, Ecosystems and Environment* 139(4), 584-94. doi:10.1016/j.agee.2010.10.001.

Clark, M. and Tilman, D. 2017. Comparative analysis of environmental impacts of agricultural production systems, agricultural input efficiency, and food choice. *Environmental Research Letters* 12(6). doi:10.1088/1748-9326/aa6cd5.

DeClerck, F. A. J., Jones, S. K., Attwood, S., Bossio, D., Girvetz, E., Chaplin-Kramer, B., Enfors, E., Fremier, A. K., Gordon, L. J., Kizito, F., Lopez Noriega, I., Matthews, N., McCartney, M., Meacham, M., Noble, A., Quintero, M., Remans, R., Soppe, R., Willemen, L., Wood, S. L. R. and Zhang, W. 2016. Agricultural ecosystems and their services: the vanguard of sustainability? *Current Opinion in Environmental Sustainability* 23, 92-9. doi:10.1016/j.cosust.2016.11.016.

Desaules, A. and Studer, C. 1993. NABO - Nationales Beobachtungsnetz, Messresultate 1985-1991. Schriftenreihe Umwelt Nr. 200. Bundesamt für Umwelt, Wald und Landschaft BUWAL, Bern.

Freiermuth, R. 2006. Modell zur Berechnung der Schwermetallflüsse in der Landwirtschaftlichen Ökobilanz - SALCA-Schwermetall. Forschungsanstalt Agroscope Reckenholz-Tänikon, Zürich, p. 28.

Godinot, O., Carof, M., Vertès, F. and Leterme, P. 2014. SyNE: an improved indicator to assess nitrogen efficiency of farming systems. *Agricultural Systems* 127, 41-52. doi:10.1016/j.agsy.2014.01.003.

Knudsen, M. T., Kristensen, I. S., Berntsen, J., Petersen, B. M. and Kristensen, E. S. 2006. Estimated N leaching losses for organic and conventional farming in Denmark. *The Journal of Agricultural Science* 144(2), 135-49. doi:10.1017/S0021859605005812.

Korsaeth, A. 2008. Relations between nitrogen leaching and food productivity in organic and conventional cropping systems in a long-term field study. *Agriculture, Ecosystems and Environment* 127(3–4), 177–88. doi:10.1016/j.agee.2008.03.014.

Li, C., Farahbakhshazad, N., Jaynes, D. B., Dinnes, D. L., Salas, W. and McLaughlin, D. 2006. Modeling nitrate leaching with a biogeochemical model modified based on observations in a row-crop field in Iowa. *Ecological Modelling* 196(1–2), 116–30. doi:10.1016/j.ecolmodel.2006.02.007.

Lin, H.-C., Huber, J. A., Gerl, G. and Hülsbergen, K.-J. 2016. Nitrogen balances and nitrogen-use efficiency of different organic and conventional farming systems. *Nutrient Cycling in Agroecosystems* 105(1), 1–23. doi:10.1007/s10705-016-9770-5.

Makridis, C., Svarnas, C., Rigas, N., Gougoulias, N., Roka, L. and Leontopoulos, S. 2012. Transfer of heavy metal contaminants from animal feed to animal products. *Journal of Agricultural Science and Technology A* 2, 149–54.

Meemken, E.-M. and Qaim, M. 2018. Organic agriculture, food security, and the environment. *Annual Review of Resource Economics* 10(1), 39–63. doi:10.1146/annurev-resource-100517-023252.

Meier, M. S., Schader, C., Berner, A. and Gattinger, A. 2012. Modelling N_2O emissions from organic fertilisers for LCA inventories. In: Corson, M. S. and van der Werf, H. M. G. (Eds), *8th International Conference on Life Cycle Assessment in the Agri-Food Sector (LCA Food 2012)*, 1–4 October 2012. INRA, Rennes and Saint-Malo, France, pp. 177–82.

Meier, M. S., Jungbluth, N., Stoessel, F., Schader, C. and Stolze, M. 2014. Higher accuracy in N modeling makes a difference. In: *9th International Conference LCA of Food*, 8–10 October 2014, San Francisco, USA.

Meier, M. S., Stoessel, F., Jungbluth, N., Juraske, R., Schader, C. and Stolze, M. 2015. Environmental impacts of organic and conventional agricultural products – are the differences captured by life cycle assessment? *Journal of Environmental Management* 149, 193–208. doi:10.1016/j.jenvman.2014.10.006.

Menzi, H. and Kessler, J. 1998. Heavy metal content of manures in Switzerland. In: Martinez, J. and Maudet, M. N. (Eds), *Proceedings of the 8th International Conference on the FAO ESCORENA Network on Recycling of Agricultural, Municipal and Industrial Residues in Agriculture (RAMIRAN 98)*, Rennes, pp. 495–506.

Mondelaers, K., Aertsens, J. and Van Huylenbroeck, G. 2009. A meta-analysis of the differences in environmental impacts between organic and conventional farming. *British Food Journal* 111(10), 1098–119. doi:10.1108/00070700910992925.

Nemecek, T. and Schnetzer, J. 2011. Methods of assessment of direct field emissions for LCIs of agricultural production systems. Data v3.0 (2012). Agroscope Reckenholz-Tänikon Research Station ART, Zurich, Switzerland.

Nemecek, T., Heil, A., Huguenin, O., Meier, S., Erzinger, S., Blaser, S., Dux, D. and Zimmermann, A. 2007. *Life Cycle Inventories of Agricultural Production Systems*. Agroscope FAL Reckenholz and FAT Taenikon, Swiss Centre for Life Cycle Inventories, Dübendorf, Switzerland.

Pedersen, A., Petersen, B. M., Eriksen, J., Hansen, S. and Jensen, L. S. 2007. A model simulation analysis of soil nitrate concentrations–does soil organic matter pool structure or catch crop growth parameters matter most? *Ecological Modelling* 205(1–2), 209–20. doi:10.1016/j.ecolmodel.2007.02.016.

Poore, J. and Nemecek, T. 2018. Reducing food's environmental impacts through producers and consumers. *Science* 360(6392), 987–92. doi:10.1126/science.aaq0216.

Sebilo, M., Mayer, B., Nicolardot, B., Pinay, G. and Mariotti, A. 2013. Long-term fate of nitrate fertilizer in agricultural soils. *Proceedings of the National Academy of Sciences* 110(45), 18185–9.

Seufert, V. and Ramankutty, N. 2017. Many shades of gray–the context-dependent performance of organic agriculture. *Science Advances* 3(3), e1602638. doi:10.1126/sciadv.1602638.

Seufert, V., Ramankutty, N. and Foely, J. A. 2012. Comparing the yields of organic and conventional agriculture. *Nature* 485(7397), 229–32. doi:10.1038/nature11069.

Seufert, V., Ramankutty, N. and Mayerhofer, T. 2017. What is this thing called organic? - how organic farming is codified in regulations. *Food Policy* 68, 10–20. doi:10.1016/j.foodpol.2016.12.009.

Tuomisto, H. L., Hodge, I. D., Riordan, P. and Macdonald, D. W. 2012. Does organic farming reduce environmental impacts? A meta-analysis of European research. *Journal of Environmental Management* 112, 309–20. doi:10.1016/j.jenvman.2012.08.018.

Vukobratović, M., Vukobratović, Z., Lončarić, Z. and Kerovac, D. 2014. *Heavy Metals in Animal Manure and Effects of Composting on It*. International Society for Horticultural Science (ISHS), Leuven, Belgium, pp. 591–7.

Wernet, G., Bauer, C., Steubing, B., Reinhard, J., Moreno-Ruiz, E. and Weidema, B. 2016. The ecoinvent database version 3 (part I): overview and methodology. *The International Journal of Life Cycle Assessment* 21(9), 1218–30. doi:10.1007/s11367-016-1087-8.

Willer, H. and Lernoud, J. (Eds). 2018. *The World of Organic Agriculture Statistics and Emerging Trends 2018*. Research Institute of Organic Agriculture (FiBL), Frick and IFOAM - Organics International, Bonn.

Zhang, F., Li, Y., Yang, M. and Li, W. 2012. Content of heavy metals in animal feeds and manures from farms of different scales in Northeast China. *International Journal of Environmental Research and Public Health* 9(8), 2658–68. doi:10.3390/ijerph9082658.

Part 2

Modelling particular impacts

Chapter 6

Modelling impacts of agriculture on freshwater

Stephan Pfister, ETH Zurich, Switzerland; and Sandra Payen, CIRAD, France and AgResearch, New Zealand

1 Introduction

Agriculture is the main human activity affecting water resources, due to the high volumes of irrigation water consumed for crop production. 80% to 90% of human freshwater consumption is occurring in this sector. Additionally, agricultural activities affect the hydrological cycles through land-use changes, soil modifications and lead to pollution of water bodies. On a local level, these impacts and specific improvement options can be assessed by risk assessment and integrated water resource management.

Modelling the impacts of agriculture on freshwater comprehensively is a difficult task. As illustrated by Fig. 1, freshwater can be impacted by water consumption and also by water pollution. These are influenced by direct agricultural activities (e.g. irrigation) as well as by indirect activities for the

http://dx.doi.org/10.19103/AS.2018.0044.06

production of agricultural inputs (e.g. fertilizers, agrochemicals). Assessing multiple environmental impacts and multiple contributing processes in the supply chain across the planet is a core strength of life cycle assessment (LCA), which aims to address this complexity in a comprehensive way.

In the past, water use and consumption has not been the focus of LCAs due to various reasons. The first missing part was a clear concept for water flows between the technosphere (man-made environment) and the natural environment. Additionally, the understanding of water 'consumption' differed widely: water use (such as on the bill of water suppliers) is not water consumption from a water resource perspective if it is released back to the environment (Fig. 1). Therefore, a large share of industrial water use is not water consumption. Water consumption mainly occurs through evaporation (in agriculture, from reservoirs and for cooling purposes) and to a lesser

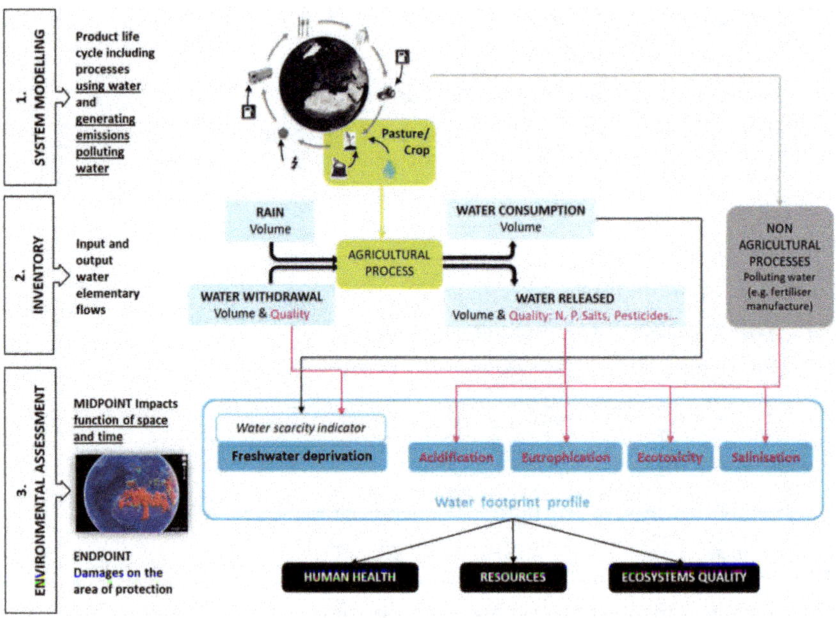

Figure 1 Water-use impact modelling framework in life cycle assessment (LCA): (1) Product life cycle modelling including many processes using and/or polluting water, (2) inventory of water flows for each process in terms of volume and quality (inventory flows requirement depends on the method), (3) environmental impact assessment (function of time and space): water footprint profile includes water deprivation impacts (based on scarcity indicator), but also acidification, eutrophication, ecotoxicity and other impacts related to water degradation. Source: based on Payen (2015).

extent on freshwater transfer in products to different watersheds. Life cycle databases such as Ecoinvent version 3 and Gabi implemented consistent modelling of water flows to allow for assessment of water consumption effects in LCA (Pfister, 2015). However, water consumption estimates have high uncertainties for several reasons: (1) water is usually cheap and exact measurements of inflow and outflow are not of core interest, (2) it is often a question of water-use efficiency, which can be highly variable and (3) climatic effects on water availability and water losses are difficult to predict. Global databases mainly provide simplified results of water consumption based on global models that do not account for all hydrological parameters. These databases usually differentiate irrigation and industrial water ('blue water') from natural precipitation water supply ('green water'). Additionally, agricultural production is more complex than technical systems, as agriculture is at the interface of natural and technical systems. Irrigation and precipitation (natural flow) are both directly applied to and stored in the soil, which leads to the general issue in agricultural LCA of the technosphere - ecosphere boundary, both in space and in time (further discussed in Pfister, 2015). The duration of the impacts is not limited to the use phase of the soil, but also after, as soil moisture and nutrients might be depleted, soils compacted or eroded. This has been addressed for the livestock sector in a recent FAO activity.

Since water (precipitation and soil moisture) has a large impact on soils, linkages between water consumption and use to other environmental impacts including pollution should be addressed. A framework to address freshwater as a natural resource highlighted the importance for quality aspects beyond quantity (Pradinaud et al., 2019). Pollution of freshwater is treated in other chapters of this book. Of special importance in agriculture are toxic and eutrophying emissions.

Land-use effects on the albedo and thus GHG/energy balance from the timing and type of crop plantations or pastures need to be considered in addition to the effects of land-use change from nature to agriculture on the hydrological cycle and thus water availability. Irrigation also influences albedo and related climate change effects can be considered as impacts related to freshwater use (Munoz et al., 2010).

In the last decade, the UNEP-SETAC working group (WULCA) established a framework to assess water consumption in LCA, which was largely adopted by the ISO 14046 standard on water footprinting and the upcoming guidelines on water footprinting in the livestock sector by the technical advisory group on water of FAO (FAO, 2018a). Internationally agreed terms of importance for this chapter are presented in Box 1.

Box 1 Terms and definitions

Blue water	Freshwater flows originating from run-off or percolation, contributing to freshwater lakes, dams, rivers and aquifers. Soil moisture is blue water if it originates from blue water added through irrigation or owing to hydrological events, such as flooding, from springs or capillary rise.
Green water	Precipitation that is stored as soil moisture and eventually transpired or evaporated from the unsaturated zone.
Precipitation	Liquid or solid products of the condensation of water vapour falling from clouds or deposited from air.
Run-off	Part of the precipitation, which flows towards a water body on the ground surface (surface run-off) or within the soil.
Transpiration	Process by which water from vegetation is transferred into the atmosphere in the form of vapour. Together with evaporation it forms evapotranspiration.
Water availability	Renewable water available in a catchment (blue water).
Water body	Entity of water with definite hydrological, hydro-geomorphological, physical, chemical and biological characteristics in a given geographical area, such as aquifers, lakes, rivers and glaciers.
Water consumption (consumptive water use)	Water consumption is a form of water use (consumptive water use). The term 'water consumption' is used to describe water removed from, but not returned to, the same drainage basin. Water consumption can be because of evaporation, transpiration, integration into a product or release into a different drainage basin or the sea.
Water scarcity footprint (WSF)	Metric that quantifies the potential environmental impacts of water consumption related to water scarcity (based on ISO, 2014).
Water use	Use of water by human activity (ISO, 2014), including consumptive and degradative use (Pfister et al., 2009).
Water withdrawal	Anthropogenic removal of water from any water body or drainage basin, either permanently or temporarily (ISO, 2014).
Water Productivity (WP)	Ratio of the benefit to the amount of green and blue water consumed to produce those benefits in a production process (product units: e.g. mass, energy or nutrition per m^3 water). The WP can be reported with fractions of green and blue water consumed or as total WP.
Water scarcity	Extent to which demand for water compares to the replenishment of water in an area (ISO, 2014).
Water stress	Water scarcity is sometimes referred as water stress and thus used as synonym. However, some authors also include water degradation in water stress.

Source: adapted from FAO (2018a).

2 Modelling impacts of water consumption: water scarcity footprints

Water scarcity impacts or footprints are generally evaluated as a consequence of freshwater consumption (blue water consumption). The ISO 14046 standard on water footprinting (ISO14046, 2014) defines that water scarcity should be addressed spatially and temporally explicitly, in order to reflect environmental relevance of water consumption. Water scarcity is also addressed in the UN Sustainable Development Goals (SDG) 6 'Ensure availability and sustainable management of water and sanitation for all'. The SDG 6.4.2 indicator defines water scarcity as 'freshwater withdrawal as a proportion of available freshwater resources is the ratio between total freshwater withdrawn by major economic sectors and total renewable freshwater resources, after taking into account environmental water requirements'. It therefore refers to withdrawal rather than consumption, which is based on early water scarcity indicators. More recent indicators are based on water consumption instead of withdrawal and it is important to make clear what is accounted for in each indicator. Water withdrawals add pressure to the water resources, especially locally, but neglect water consumption from water storage such as dams for hydropower and irrigation, which have been identified to be a major water consumer, even if often mitigating water scarcity in the driest periods (Scherer and Pfister, 2016). On the other hand, water withdrawal released back to the environment and thus also for further human use is not comparable to water consumption, which is a loss to the freshwater resources in the watershed. For instance, drip irrigation that reduces water withdrawal can lead to increased water consumption (through greater evapotranspiration), and thus increase water scarcity, even if the SDG 6.4.2 indicator is reduced. Therefore, the SDG indicator should not be used standalone to assess water availability and sustainable water management.

2.1 Water consumption and inventory modelling

Various global models assess water consumption of crop production but with a limited level of detail. Some quantify total agricultural water consumption, while others assess crop groups or vegetation classes only (detailed model comparison in Sood and Smakhtin, 2015). These models can be used to calculate total water scarcity of a region, but not for product-level assessments, since different crop characteristics are not accounted for.

Modelling impact on water scarcity at a product or process level requires an inventory of water flows. There is a variety of tools available to do so, ranging from databases to agro-hydrological models (for a detailed review, refer to Payen et al., 2017). The approach adopted should depend on the

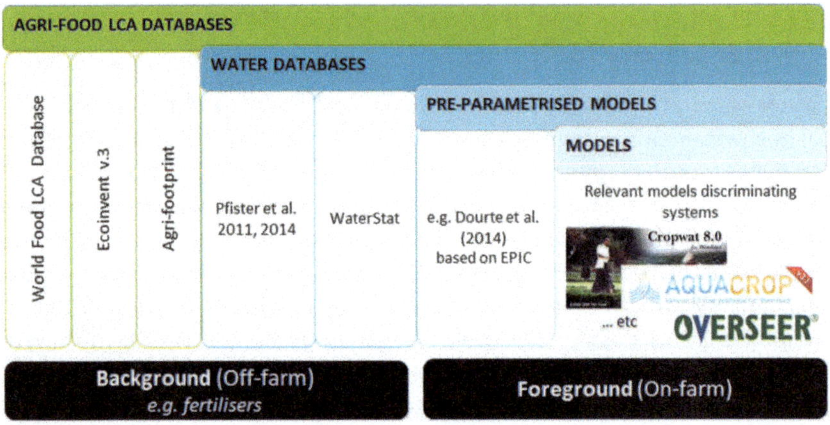

Figure 2 Gradient of possible approaches (models and databases) for the inventory of field water flows and associated type of LCA study.

scope and the objective of the study. Figure 2 illustrates the gradient of possible approaches for the inventory of field water flows and associated type of LCA study.

Water inventory and agri-food LCA databases, such as Ecoinvent (Wernet et al., 2016; Pfister et al., 2016), GABI (Thinkstep, 2019), Agri-footprint (Blonk Agri-footprint, 2014) and World Food LCA database (Nemecek et al., 2015), provide the inventory of water flows for many crops and inputs used on farm (e.g. fertilizer, electricity, agrochemical). Ecoinvent, the largest and most widely used LCI database with transparent system modelling, addressed this issue as a main improvement for their version 3: they integrated a water balance into their entire database (Pfister, 2015). All water inputs and outputs are recorded and the potential mismatch is considered for data improvement efforts. In principle, the water consumption for each process is the sum of water withdrawals and input product water content minus output product water content and water released to freshwater bodies. A process can thus also have negative consumption, for example tap water sourced from a different watershed and used to irrigate a garden: the water is integrated in the product 'tap water' and the excess water percolating to groundwater is a release to the environment. Assessing inputs and outputs separately is important in order to allow regionalized assessment as the production might be in different places, such as for the case of the South-North water transfer project in China (Lin et al., 2012).

Detailed water databases provide estimates of water consumed by a crop based on the theoretical crop water consumption (partially with adjustments as in the case of Pfister and Bayer, 2014), but not on the water actually withdrawn and consumed. In addition, databases rely on data and modelling approaches which include limitations regarding the implications of farming practices and the complex mechanisms of field water flows (Payen et al., 2017). Other database, such as Aquastat, are based on statistics but generally lack level of spatial detail (FAO, 2016). Thus, databases are relevant and can be used when the cropping system is at the background level. However, they are too generic to distinguish precise causes of impacts due to the cropping system itself as needed in most agricultural studies where it is at the foreground level, especially where an adaptation of practices is sought (eco-design). In such cases, a detailed modelling approach is required, which is in line with the FAO recommendations on water footprinting (FAO, 2018a).

For the eco-design of cropping systems, the estimation of water flows should be based on a model simulating evapotranspiration, percolation and run-off, which accounts for soil, climate, crop specificities and agricultural management. Note that better estimation of evapotranspiration relies on a separate assessment of evaporation and transpiration. In addition, distinguishing productive water (transpiration related to yield) from non-productive water (evaporation) is valuable information for water management and is in accordance with the ISO standard on water footprinting (ISO 14046, 2014). However, separation of evaporation and transpiration is quite challenging and mainly important for water productivity/efficiency and not for quantifying the impacts (FAO, 2018a).

There is a wide spectrum of models representing the soil-plant-atmosphere ranging from transient state and physically based models to steady-state and bucket models, but there is a gradient in terms of their complexity and data requirements. The agro-hydrological models with a good trade-off between accuracy, simplicity and robustness for LCA eco-design are the FAO functional models: 'CropWat' and 'AquaCrop' (for a description of the models, see Allen et al., 1998; Steduto et al., 2012). Their minimum data requirement makes them suitable for use in LCA. The CropWat model has already been applied in agricultural LCA to estimate crop water requirements; however, it is important to highlight that it may not allow sufficient discrimination of agricultural systems because salinity and nutrient stresses are not accounted for, and run-off and drainage are not differentiated. AquaCrop is an improved version of CropWat and accounts for the effect of salinity, soil fertility and water stress. The main issue with AquaCrop is that it is currently limited to herbaceous crops. However, AquaCrop is supported by a large scientific community which will improve its performance and coverage (both in terms of crop parameters and flexibility for use with GIS and remote sensing data).

Note that this section focussed on the estimate of the volume of water flows, but estimating the quality of the released water flow is also of high relevance for modelling impacts on water quality. This remains a difficult task since it often requires the combined use of databases and modelling. For example, AquaCrop provides an estimate of water salinity, whereas other water quality parameters can be obtained from a nutrient budget.

2.2 Impact assessment and water scarcity footprints

Impact assessment is a key step to identify environmental pressures due to water consumption. Various methods have been developed in the last decade. In general, the methods can be grouped as addressing (1) water scarcity, (2) impacts on human health, (3) impacts on ecosystem quality and (4) impacts on resource depletion. Water scarcity is considered a midpoint indicator, while the others are addressing endpoints (damage oriented). However, it has to be noted that midpoints do not necessarily lead to endpoints, since different mechanisms are addressed for different endpoints.

Water footprinting studies often evaluate midpoint impact on water scarcity, that is providing water scarcity footprint results. Pfister et al. (2009) and Pfister and Bayer (2014) provide the most applied indicator to date, and is based on a water stress index (WSI) representing the water withdrawal-to-availability ratio in an area. The original most-used WSI from 2009 is an annual factor for >11000 watersheds based on WaterGAP2 (Alcamo et al., 2003), while the 2014 indicator is provided on a monthly level to allow for temporally explicit assessments, as required by the ISO standard. Since 2009, various other midpoint indicators have been developed, differing in concepts but generally agreeing that the main function is use-to-availability or consumption-to-availability ratios. In order to overcome the issue that different characterization factors (CFs - refer to Fig. 3 for an illustration) result in different outcomes for comparisons (e.g. Pfister and Lutter, 2016), a consensus method was developed by the UNEP-SETAC working group 'Water Use in LCA (WULCA)' based on the same spatial resolution (using WaterGAP 2) and on monthly levels. Its goal was to provide a generic midpoint for water scarcity, equally relevant for human and ecosystem impacts. It is based on the quantification of the Available WAter Remaining (AWaRe) in an area once the demand for human and ecosystems has been met (Boulay et al., 2017). Thus, it depends not only on the use-to-availability ratio, but also on the absolute water scarcity (i.e. the availability per area). AWaRe represents the potential to deprive another user (human or ecosystem) when consuming water in a given area in relation to the global reference situation of water used. AWaRe was the approach recommended in a Pellston workshop by the United Nation Environmental Program/Society of Environmental Toxicology and Chemistry Life Cycle Initiative (UNEP/ SETAC Life Cycle Initiative) for assessing a water scarcity footprint, in combination with a conceptually different CF for sensitivity assessment (UNEP, 2016). Both indicators rely

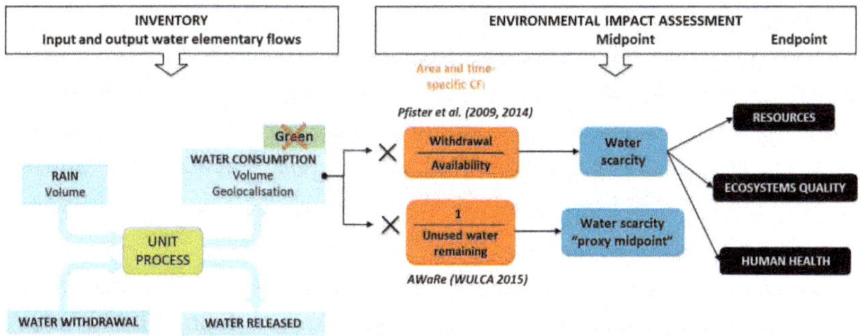

Figure 3 Water footprint impact assessment: from the inventory of water flows to midpoint (water scarcity) and endpoint damages on human health, ecosystem quality and resources. Two widely used midpoint indicators and their corresponding characterization factors (CFs) are represented, namely Pfister et al. (2009), Pfister and Bayer (2014) and AWaRe (Boulay et al., 2017). Note that midpoints do not necessarily lead to endpoints, since different mechanisms are addressed. AWaRe is considered a 'proxy-midpoint' for the endpoints' human health and ecosystem quality. Note that the consumption of green water is not considered as having an impact on water scarcity.

on spatially and temporally explicit CFs used to relate water consumption (inventory) with local pressure on the water resource (Fig. 3). Application of both indicators can be used to test the robustness of the results and compare with previous studies.

LCA methods can also account for so-called endpoint impacts on human health ecosystem quality and resources (Fig. 3). A detailed review of early methods is available in Kounina et al. (2013).

- *Impacts on human health*
  Impacts on human health by water consumption are addressed by most methods through lack of water for nutrition (e.g. Pfister et al., 2009; Motoshita et al., 2018) and potentially for lack of drinking water (Boulay et al., 2011b). These methods have also been the basis for the consensus method on water-use impacts on human health, which includes water scarcity, economic development and trade effects (Motoshita et al., 2017). Since malnutrition is related to poverty, these models consider socio-economic factors and attribute only a share of the total malnutrition impacts to water scarcity. These models generally have very high uncertainties (Pfister and Hellweg, 2011).

- *Impacts on ecosystem quality*
  Ecosystems are affected by water use in various ways. The AWaRe method accounts for environmental water requirements (i.e. in-stream flows), but groundwater may also be affected with impacts spread to terrestrial

ecosystems. The recent method of Verones et al. (2017) is consistent with the land-use consensus method by UNEP-SETAC. It combines the biodiversity assessment based on surface and groundwater use on wetlands developed by Verones et al. (2013a,b) with the generic conceptual approach assessing water stress impacts on terrestrial vegetation from Pfister et al. (2009).

- *Impacts on natural resources*
  Depletion of natural resources is mainly used for fossil and mineral resources. Additionally, stocks of water can be depleted. This mainly concerns groundwater stocks, more specifically fossil groundwater, but can also affect other large water bodies such as lakes, where future impacts are expected due to overuse of water resources. Approaches for assessing these impacts are available in Milà I Canals et al. (2009) and Pfister et al. (2009). Competitive use of renewable water resources is assessed through their consequences on human health and ecosystem quality (see above).

3 Modelling impacts on water quality

The ISO water footprint standard 14046 requires a water footprint to include all impacts on the water resources (i.e. freshwater resources) due to both water consumption and water pollution (from emissions). Water pollution is also sometimes referred to as grey water (Hoekstra et al., 2011), but generally LCA methods assess pollution through different environmental impact indicators such as eutrophication (e.g. from fertilizer emissions) and toxicity (from agrochemicals). These effects are covered in the respective chapters and thus are not further explained here. A guide for assessing eutrophication impacts in livestock systems was published by FAO (2018b), and recommendations from UNEP/SETAC for assessing eutrophication and (eco)toxicity impacts are about to be released (UNEP, 2019).

4 Modelling salinization impacts

Although salinization is a worldwide issue affecting both land and water resources, it has been only partially included in LCA. So far, four indicators have been developed to model salinization impacts in LCA, each one addressing different salinization pathways (Payen et al., 2016). The midpoint soil salinization potential developed by Feitz and Lundie (2002) addresses salinization associated with irrigation water and characterizes impacts on soil structure and the accumulation of sodium in the soil. Amores et al. (2013) evaluated the damages on biodiversity associated with a salinity increase in a Spanish coastal wetland caused by seawater infiltration in the wetland due to groundwater overexploitation for irrigation. Zhou et al. (2013) proposed a method for assessing aquatic ecotoxicity of brine disposal from seawater desalination plants. Leske and Buckley (2003, 2004a,b) developed an LCA salinity impact category for South Africa. They provide salinity potential CFs for salts release in the atmosphere, surface water, natural surfaces and agricultural surface compartments. The main limitations of the existing indicators are their restricted scope in terms of salinization pathways covered (Feitz and Lundie, 2002), their intensive inventory data requirement (Feitz and Lundie, 2002) or their restricted geographical validity (Amores et al., 2013; Feitz and Lundie, 2002; Leske and Buckley, 2003, 2004a,b).

Comprehensively modelling of salinization impacts is complex since it can be due to various types of human interventions (land-use change, irrigation, brine disposal and overuse of a water body), involve both water and soil compartments and can affect human health, ecosystem quality and resources. Payen et al. (2016) proposed a framework for modelling salinization impacts. A bottom-up approach describing the environmental mechanisms (fate, exposure and effect) is recommended because (i) salts and water are mobile and their effects are interconnected, (ii) this approach allows the evaluation of both on- and off-site impacts and (iii) it is the best way to discriminate systems and support a reliable eco-design approach.

5 Soil quality and land-use implications

The availability and quality of freshwater is closely related to soil quality. Indicators of soil quality include soil organic carbon, biotic production, erosion, mechanical filtration, groundwater regeneration and water infiltration capacity. Soil organic carbon has often been considered as a good indicator for overall soil quality (UNEP, 2019).

The land-use indicator model LANCA (Bos et al., 2016) estimates the infiltration capacity and groundwater regeneration due to different types of land use as an important indicator. Land use affects soil sealing and thus soil permeability, groundwater recharge and surface water flows. Soil quality and land use also affect the eutrophying impacts of soil erosion (Scherer and Pfister, 2015).

6 Accounting for geographical and temporal variation

Water is mainly a flow resource and the flows vary over time. Additionally, water is not a tradable good in general (although it is the most extracted resource) and thus humans rely on local water availability (as is obviously the case for natural ecosystems). This special case of freshwater resources requires geographically and temporally explicit inventory and impacts assessment modelling for agricultural systems, as described in Section 2.

AWaRe and other recent water scarcity methods thus account for monthly and watershed characteristics (refer to Section 7.1 for an illustrative case study). Since water scarcity is modelling the impact of a resource flow, temporal fluctuations are important, especially since in general scarcity is the highest during peak demand. This also applies to variability due to extreme weather periods or events. A prominent example was the 2018 summer in Europe, where even in generally water abundant locations such as Switzerland, water scarcity led to bans on water use for agriculture, when agriculture needed irrigation due to the extremely dry and hot spring/summer (which is not common). While climate change trends can be observed, the variability of weather and especially precipitation events contribute to modelling uncertainty. This applies especially for forecasting, which is evident from the low consistency of precipitation forecasts for the 2050s by different climate models (IPCC, 2007).

Another important aspect that contributes to variability in agricultural production and thus high uncertainty in generic assessment are crop yields. Yields are affected by pests and management failures, which vary over time and space.

7 Case studies

7.1 Spatio-temporal resolution matters in water scarcity footprinting: case study of New Zealand milk

A case study on milk production in New Zealand illustrates the importance of geographical and temporal variation when modelling impact on water scarcity (Payen et al., 2018). This case study estimated the water scarcity footprint of milk produced in two contrasting regions in New Zealand: 'non-irrigated moderate rainfall' (Waikato) and 'irrigated low rainfall' (Canterbury). An extensive inventory of all the consumed water flows was carried out from the production of dairy farm inputs to the milk and meat leaving the dairy farm, including the water uses on farm (e.g. irrigation water, cow drinking water). Impacts were calculated with the widely used Pfister et al. (2009) indicator, as well as methods recommended by the UNEP/SETAC Life Cycle Initiative (AWaRe and Motoshita et al., 2017) for water scarcity and human health impacts. Different spatial (country vs. region) and temporal (annual vs. monthly) resolutions were tested (for the inventory flows and CFs). Results showed that the water scarcity footprint decreased by 74% (Waikato) and 33% (Canterbury) when regional and monthly CFs were used instead of country and annual CFs, clearly demonstrating that using aggregated CFs may overestimate impacts. However, in other cases, aggregated CFs might underestimate the impacts (Pfister and Bayer, 2014). In particular, it is interesting to note that spatial resolution had a major effect on the water scarcity footprint of Waikato milk from non-irrigated pasture (i.e. localizing the farm in a specific catchment had more effect than accounting for the timing of water consumption). In contrast, temporal resolution had a major effect on the water scarcity footprint of Canterbury milk from irrigated pasture (i.e. accounting for the timing of water consumption for irrigation had more effect than specific farm location). The water scarcity footprint calculated at the higher resolution was 22 L_{world} eq/kg FPCM for Waikato milk and 1118 L_{world} eq/kg FPCM for Canterbury milk. Regarding the contribution analysis, background processes such as fertilizer manufacturing dominated for non-irrigated pasture, but was negligible for irrigated pasture where irrigation dominated the impacts.

7.2 Life cycle assessment of a perennial crop including an in-depth assessment of water-use impacts: the case of mandarin in Morocco

Mandarin production in the water-scarce Bahira plain in Morocco is an interesting case study illustrating (i) the challenge of perennial crop modelling in LCA, (ii) high valuable crop production in a water-scarce area and (iii) cultivation of a crop sensitive to salinity in an area prone to high soil salinity. Indeed, in this area the pressure on the water resource is worsened by salinity of the aquifer. The 25-year perennial crop cycle of a large commercial mandarin orchard (8-year old) was modelled from nursery to end-of-life, based on primary farm data and a projection scenario. The life cycle of the mandarin was modelled up to the French market, accounting for post-farm production stages and transportation from Morocco to France. To satisfy the water inventory requirements for a perennial crop, we had to develop and use a model of field water flows: estimating the consumed and released water flows according to actual water supply and system specificities (soil-climate-practices). Both water and salt elementary flows were estimated and used for modelling impacts on water scarcity. Impacts were calculated using Pfister et al. (2009) methodology and impacts on water scarcity stress/availability (associated with both consumptive and degradative water use) were calculated using the Boulay et al. (2011a,b) method. Other impact indicators were assessed, including climate change, eutrophication, (eco)toxicity and acidification. Results for the contribution analysis showed the dominant contribution of energy required for pumping water from the aquifer, revealing a water-energy nexus (Fig. 4). Indeed, mandarin cultivation requires a lot of water that has to be withdrawn from a deep aquifer, thus requiring the use of electricity-intensive pumps. In addition, over 50% of the Moroccan electricity mix is from fossil energy. This water-energy nexus is frequent in water-scarce

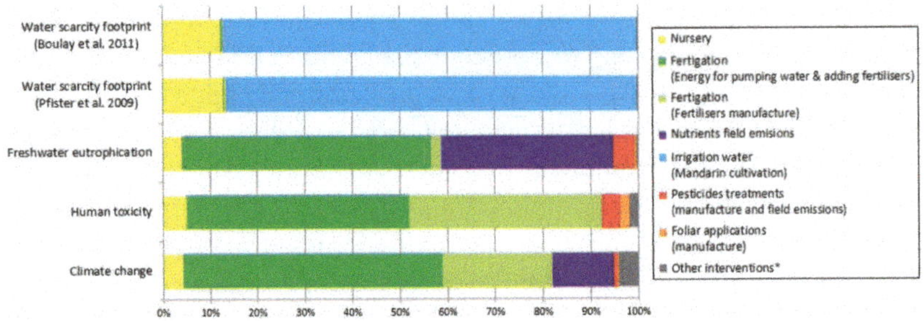

Figure 4 Contribution analysis of a mandarin crop from Morocco at the farm gate: Environmental impacts were calculated with ReCiPe (H), Boulay et al. (2011b) and Pfister et al. (2009). *Other interventions include orchard establishment, fertigation (materials), mechanical weeding and wood grinding, pesticide treatments (water, materials, energy), foliar applications (energy) and fruit harvesting (energy and materials).

countries, and was also observed for clementine (Basset-Mens et al., 2016) and tomato (Payen et al., 2015). Without surprise, irrigation was the dominant contributor to water scarcity and availability impacts. Water availability impacts calculated with Boulay et al. (2011a,b) were close to water scarcity impacts estimated with Pfister et al. (2009), respectively 181 $m^3.ton^{-1}$ and 189 $m^3.ton^{-1}$ mandarin at the farm gate. Water salinity inventory was accounted for in water stress/availability impacts, but in the absence of an operational indicator for salinization, impacts associated with soil and water salinization could not be addressed (Payen, 2015).

7.3 Water conservation and increased irrigation efficiency can add to water scarcity: the case study of drip irrigation

Drip irrigation is an important technology to save unproductive evaporation and increase water-use efficiency. As described in the study of Ward and Paulido-Velazquez (2008), increased water-use efficiency can lead to additional production and due to reduced return flows through infiltration, overall water scarcity can be increased. Ward and Paulido-Velazquez (2008) performed an integrated basin-scale analysis by assessing biophysical, hydrologic, agronomic, economic, policy, as well as institutional aspects of the Upper Rio Grande Basin of North America. They concluded that water use is unlikely to be reduced by water conservation subsidies and that more efficient irrigation technologies reduce water return flows and aquifer recharge. Additionally, reduced return flows can lead to soil salinization (e.g. Contreras et al., 2017). Therefore, it is important that water efficiency policies account not only for water-use reductions, for example through drip irrigation, but also monitor and reward reduced water depletion. Conservation programmes should also account for salinization risks.

8 Summary

8.1 Optimizing freshwater efficiency

Effects of water on productivity is a well-known field in agricultural research and various tools to improve it exist. In an LCA or water footprint assessment, it is important to first localize the hotspots of water use, since often it is in the supply chain. Further research needs to provide better tools to assess hotspots of impacts accounting for uncertainty, since the supply chain data is often not detailed in standard assessments. In the longer term, benchmarks should be established to identify the improvement potential in combination with relevance among the thousands of processes generally involved in an LCA or water footprint study to most effectively improve the production system.

8.2 Optimizing effective use of water resources

Effective use of water (such as irrigating high-value crops) is more important than water-use efficiency. This is especially important in agriculture, since biomass products can often be replaced. The major purpose - food supply - can be achieved by cultivating a variety of crops and livestock. Diet changes have been identified as having a high potential to reduce water scarcity. For example, the replacement of animal-based with plant-based protein is often considered as an alternative for reducing global blue water consumption (Jalava et al., 2014). However, when analysing such dietary changes, it is crucial to account for differences in the nutritional value and for balanced diets (which may be difficult to assess comprehensively). There are also alternative options for producing much less water-intensive fibre and biofuel products, although often at the expense of other environmental impacts. Systems analysis is therefore important to optimize agricultural production systems. While effective water use and choice of products can be strategies to reduce all impacts, it can have a greater effect on water scarcity assessments, since large-scale reduction in water consumption can also lower water scarcity indicators.

Food losses have a large impact on water scarcity, not only as part of the agricultural production, but even more importantly through the processing and consumer phases of agricultural products. A recent study showed the high potential of combining the different mechanisms of avoiding food waste with increased agricultural efficiency, but also revealed that more detailed research is required in the future (Kummu et al., 2017).

8.3 Minimizing environmental impacts of water consumption

Water consumption in different locations can lead to highly varying environmental impacts. Therefore, the efficient and effective use of water should also be linked with reducing environmental impacts. This includes accounting for the hydrology of the system as well as the vulnerability of humans and ecosystems. Although indicators are still being improved, all the recent developments in water-use impact modelling are a major contribution towards accounting for catchment specificities and seasonality. Thus, current trends in research acknowledge that the same quantity of water used may have very little or huge consequences depending on where and when it was consumed.

Reducing impacts of water consumption may affect other impact categories such as soil quality, soil and water salinization and toxicity. These impacts, which might also come with a trade-off in water scarcity impacts, need to be captured (i.e. potential burden shifts identified) by doing a multi-criteria analysis in LCA. However, some impact indicators are not yet operational (e.g. salinization) and might need to be assessed in a more qualitative way. Operationalizing CFs for these related impacts to quantify the impacts along supply chains is needed to assess water impacts in a comprehensive way.

8.4 Global dimension of water scarcity

While water scarcity is a local problem, the cause is often international due to the globalized economy. Food products are heavily traded and impacts occur often far away from consumption of the products, and thus national and international policy actions are required (IRP, 2019). While trade might alleviate water scarcity, it can also increase water stress, since trade mainly reflects affluence, rather than water scarcity (Weinzettel and Pfister, 2019). Thus, policies have to account for the economic aspects related to water scarcity in regions with low economic development.

Companies have a large influence on trade; thus, it is important to involve them in coordinating actions to prevent water scarcity. Recently, the concept of science-based targets has emerged. The goal is that companies set reduction goals based on scientific knowledge, which then get evaluated and eventually approved. Several large companies are making progress in this direction (SBT, 2019). So far, most focus has been on climate targets, but water scarcity has gained interest too, together with land use, pollution of oceans and biodiversity. Global policies such as the Paris Agreement to combat climate change might be necessary to address water scarcity on a global level.

9 Future trends

9.1 Impact assessment

Assessing the environmental impact of water consumption is still a relatively young research field and several gaps have been identified. The main methods do not distinguish groundwater and surface water, which have distinct hydrological behaviour and different values for humans and ecosystems. The main reason is that global hydrological models are very weak on groundwater and data on the state of groundwater are rather scarce. Therefore, recommendations of an international working group on water-use impacts in LCA are to create a more detailed model to better assess impacts on ecosystems (Nunez et al., 2018).

There is a methodological gap in the interaction of water use with other environmental impacts, such as assessment of salinization impacts (see Section 4 for more details). Efforts should focus on the development of operational indicators addressing impacts on soil and water. Modelling salinization impacts comprehensively is a challenge since it is strongly dependent on water availability, water and soil quality/composition.

9.2 Inventory

Agriculture consists of hundreds of millions of individual producers and thus inventory assessment requires special attention to variability and if possible to local assessment of production. We often make the distinction between conventional vs. organic production systems, but there are hundreds of agricultural management alternatives in a given pedo-climatic environment. In a context where the Global Agronomy concept is emerging (e.g. Makowski et al., 2014), we need to capture all system specificities with a global coverage. Remote sensing data are a promising basis in that regard and enable presentation of results at a high temporal frequency across large spatial areas. But using such data raises the challenge of their management and interfacing with analysis tools or LCA software. With the increasing use of spatially explicit data (both for the inventory and impact assessment), there is future requirement for models and LCA software to supporting a coupling with GIS. This coupling with GIS is already possible with the agro-hydrological model AquaCrop (inventory of water flows) and the LCA software Open LCA (life cycle modelling and impact assessment) as well as the scientific open-source LCA software Brightway, as summarized by Frischknecht et al. (2019). Additional information on inventory modelling is provided by the guidelines for assessment of water use of livestock production systems and supply chains by FAO (2018a).

9.3 Technical solutions

From hunters to rainfed and irrigated agriculture, humankind invented many technical solutions to enhance agricultural output. Beyond drip irrigation, greenhouse production can increase yield and thus water productivity significantly. Further research on where production can reduce pressure on water resources directly or through replacing other production locations is important to provide better knowledge on how to improve sustainability of the food production system. Other trends in research include urban gardening, which involves technical innovation and symbiosis with industrial systems.

10 Acknowledgements

We thank Stewart Ledgard for commenting and proofreading the final version of this chapter.

11 Where to look for further information

In recent years, many research papers have been published and an extensive review is beyond the scope of this chapter, as it would also be outdated soon. However, beyond the references in the sections above, there is a list of papers, reports and book chapters that address the topic of water use in LCA and water footprinting in higher detail.

The FAO LEAP report gives the most recent summary of water-use assessment in agriculture, including water productivity, water footprint and LCA research. The draft report is publicly available (FAO, 2018a). There has also been exchange on general discussion about the water footprint concepts and different applications (Pfister et al., 2017; Hoekstra, 2016).

Three book chapters on water use in LCA and water footprinting are useful resources to get more details on LCA methodologies and applications, even if latest scientific developments are missing (Pfister, 2015; Berger et al., 2016; Verones et al., 2016).

In terms of impact assessment, the international harmonization efforts towards a consensus method for water scarcity (midpoint, Boulay et al., 2017) and human health impacts (endpoint, Motoshita et al., 2017) were reported by UNEP in 2017. Recommendations for assessing impacts related to water quality (eutrophication and ecotoxicity) are about to be released (UNEP, 2019).

UNEP's international resource panel (IRP) published two reports dedicated to water use related to resource and material production (UNEP, 2012, 2015).

12 References

Agri-footprint, B. 2014. *Agri-Footprint–Description of Data*. V1.0. Blonk Agri-footprint BV, the Netherlands.

Alcamo, J., Doll, P., Henrichs, T., Kaspar, F., Lehner, B., Rosch, T. and Siebert, S. 2003. Global estimates of water withdrawals and availability under current and future 'business-as-usual' conditions. *Hydrolog. Sci. J.* 48(3), 339–48. doi:10.1623/hysj.48.3.339.45278.

Allen, R. G., Pereira, L. R., Raes, D. and Smith, M. 1998. *Crop Evapotranspiration: Guidelines for Computing Crop Water Requirements*. FAO Irrigation and Drainage Paper No. 56. FAO, Rome, Italy, 300pp.

Amores, M. J., Verones, F., Raptis, C., Juraske, R., Pfister, S., Stoessel, F., Antón, A., Castells, F. and Hellweg, S. 2013. Biodiversity impacts from salinity increase in a coastal wetland. *Environ. Sci Technol.* 47(12), 6384–92. doi:10.1021/es3045423.

Basset-Mens, C., Vannière, H., Grasselly, D., Heitz, H., Braun, A., Payen, S., Koch, P. and Biard, Y. 2016. Environmental impacts of imported and locally-grown fruits for the French market. *Fruits* 71(2), 93–104 doi:10.1051/fruits/2015050.

Berger, M., Pfister, S. and Motoshita, M. 2016. Water footprinting in Life Cycle Assessment - how to count the drops and assess the impacts? In: Finkbeiner, M. (Ed.), *LCA Compendium - The Complete World of Life Cycle Assessment.* Volume: Special Types of Life Cycle Assessment. Edited by Klöpffer, W. and Curran, M. A. Springer, Dordrecht. Chapter 3.

Bos, U., Horn, R., Beck, T., Lindner, J. P. and Fischer, M. 2016. *LANCA® - Characterisation Factors for Life Cycle Impact Assessment*, Version 2.0. Fraunhofer Verlag, Stuttgart. ISBN: 978-3-8396-0953-8.

Boulay, A.-M., Bouchard, C., Bulle, C., Deschênes, L. and Margni, M. 2011a. Categorizing water for LCA inventory. *Int. J. Life Cycle Assess.* 16(7), 639–51. doi:10.1007/s11367-011-0300-z.

Boulay, A.-M., Bulle, C., Bayart, J. B., Deschênes, L. and Margni, M. 2011b. Regional characterization of freshwater use in LCA: modeling direct impacts on human health. *Environ. Sci Technol.* 45(20), 8948–57. doi:10.1021/es1030883.

Boulay, A.-M., Bare, J., Benini, L., Boulay, A. M., Bare, J., Benini, L., Berger, M., Lathuillière, M. J., Manzardo, A., Margni, M., Motoshita, M., Núñez, M., Pastor, A. V., Ridoutt, B., Oki, T., Worbe, S. and Pfister, S. 2018. The WULCA consensus characterization model for water scarcity footprints: assessing impacts of water consumption based on available water remaining (AWARE). *Int. J. Life Cycle Assess.* 23(2), 368–78. doi:10.1007/s11367-017-1333-8.

Contreras, J. I., Alonso, F., Cánovas, G. and Baeza, R. 2017. Irrigation management of greenhouse zucchini with different soil matric potential level. Agronomic and environmental effects. *Agric. Water Manag.* 183, 26–34. doi:10.1016/J.AGWAT.2016.09.025.

FAO. 2016. *AQUASTAT Website.* Food and Agriculture Organization of the United Nations (FAO). Available at: http://www.fao.org/nr/water/aquastat/data/query/index.html?lang=en (accessed on 14 March 2019).

FAO. 2018a. *Water Use of Livestock Production Systems and Supply Chains - Guidelines for Assessment (Draft for Public Review). Livestock Environmental Assessment and Performance (LEAP) Partnership.* FAO, Rome, Italy.

FAO. 2018b. *Nutrient Flows and Associated Environmental Impacts in Livestock Supply Chains: Guidelines for Assessment (Version 1). Livestock Environmental Assessment and Performance (LEAP) Partnership.* FAO, Rome, Italy, 196pp.

Feitz, A. J. and Lundie, S. 2002. Soil salinisation: a local life cycle assessment impact category. *Int. J. Life Cycle Assess.* 7, 244–9. doi:10.1065/lca2002.06.084.

Frischknecht, R., Pfister, S., Bunsen, J., Haas, A., Känzig, J., Kilga, M., Lansche, J., Margni, M., Mutel, C., Reinhard, J., Stolz, P., van Zelm, R., Vieira, M. and Wernet, G. 2019. Regionalization in LCA: current status in concepts, software and databases–69th LCA forum, Swiss Federal Institute of Technology, Zurich, 13 September, 2018. *Int. J. Life Cycle Assess.* 24, 364–9. doi:10.1007/s11367-018-1559-0.

Hoekstra, A. Y. 2016. A critique on the water-scarcity weighted water footprint in LCA. *Ecol. Indic.* 66, 564–73. doi:10.1016/j.ecolind.2016.02.026.

Hoekstra, A. Y., Chapagain, A. K., Aldaya, M. M. and Mekonnen, M. M. 2011. *The Water Footprint Assessment Manual: Setting the Global Standard*. Earthscan, London.

IPCC. 2007. Contribution of working group I to the fourth assessment report of the intergovernmental panel on climate change. In: Solomon, S. D. Q., Manning, M., Chen, Z., Marquis, M., Averyt, K. B., Tignor, M. and Miller, H. L. (Eds). Cambridge University Press, Cambridge and New York.

IRP. 2019. *Global Resources Outlook 2019: Natural Resources for the Future We Want*. Oberle, B., Bringezu, S., Hatfield-Dodds, S., Hellweg, S., Schandl, H., Clement, J., Cabernard, L., Che, N., Chen, D., Droz-Georget , H., Ekins, P., Fischer-Kowalski, M., Flörke, M., Frank, S., Froemelt , A., Geschke, A., Haupt , M., Havlik, P., Hüfner, R., Lenzen, M., Lieber, M., Liu, B., Lu, Y., Lutter, S., Mehr , J., Miatto, A., Newth, D., Oberschelp , C., Obersteiner, M., Pfister, S., Piccoli, E., Schaldach, R., Schüngel, J., Sonderegger, T., Sudheshwar, A., Tanikawa, H., van der Voet, E., Walker, C., West, J., Wang, Z. and Zhu, B. A Report of the International Resource Panel. United Nations Environment Programme, Nairobi, Kenya.

ISO. 2014. *ISO 14046: Environmental Management–Water Footprint: Principles, Requirements and Guidelines*. ISO, Geneva.

Jalava, M., Kummu, M., Porkka, M., Siebert, S. and Varis, O. 2014. Diet change–a solution to reduce water use? *Environ. Res. Lett.* 9(7). doi:10.1088/1748-9326/9/7/074016.

Kounina, A., Margni, M., Bayart, J.-B., Boulay, A.-M., Berger, M., Bulle, C., Frischknecht, R., Koehler, A., Milà i Canals, L., Motoshita, M., Nuñez, M., Peters, G., Pfister, S., Ridoutt, B., Zelm, R., Verones, F. and Humbert, S. 2013. Review of methods addressing freshwater use in life cycle inventory and impact assessment. *Int. J. Life Cycle Assess.* 18. doi:10.1007/s11367-012-0519-3.

Kummu, M., Fader, M., Gerten, D., Guillaume, J. H., Jalava, M., Jägermeyr, J., Pfister, S., Porkka, M., Siebert, S. and Varis, O. 2017. Bringing it all together: linking measures to secure nations' food supply. *Curr. Opin. Environ. Sustain.* 29, 98–117. doi:10.1016/J.COSUST.2018.01.006.

Leske, T. and Buckley, C. 2003. Towards the development of a salinity impact category for South African environmental life-cycle assessments : part 1 - a new impact category. *Water SA* 29(3), 289–96. doi:10.4314/wsa.v29i3.4930.

Leske, T. and Buckley, C. 2004a. Towards the development of a salinity impact category for South African life cycle assessments : part 2 - a conceptual multimedia environmental fate and effect model. *Water SA* 30(2), 241–52. doi:10.4314/wsa.v30i2.5070.

Leske, T. and Buckley, C. 2004b. Towards the development of a salinity impact category for South African life cycle assessments : part 3 - salinity potentials. *Water SA* 30(2), 253–65. doi:10.4314/wsa.v30i2.5071.

Lin, C., Suh, S. and Pfister, S. 2012. Does south-to-north water transfer reduce the environmental impact of water consumption in China? *J. Ind. Ecol.* 16(4), 647–54. doi:10.1111/j.1530-9290.2012.00500.x.

Makowski, D., Nesme, T., Papy, F. and Doré, T. 2014. Global agronomy, a new field of research. A review. *Agron. Sustain. Dev.* 34(2), 293–307. doi:10.1007/s13593-013-0179-0.

Milà i Canals, L., Chenoweth, J., Chapagain, A., Orr, S., Antón, A. and Clift, R. 2009. Assessing freshwater use impacts in LCA: part I–inventory modelling and characterisation factors for the main impact pathways. *Int. J. Life Cycle Assess.* 14(1), 28–42. doi::doi:10.1007/s11367-008-0030-z.

Motoshita, M., Boulay, A.-M., Pfister, S., Benini, L., Brito de Figueiredo, M. C., Gheewala, S. H., Harding, K. and Schenker, U. 2017. Water use related impacts: water scarcity and human health effects - part 2: human health effects. In: Frischknecht, R. and Jolliet, O. (Eds), *Global Guidance for Life Cycle Impact Assessment Indicators - Volume 1*. United Nations Environment Programme (UNEP), Paris.

Motoshita, M., Ono, Y., Pfister, S., Boulay, A.-M., Berger, M., Nansai, K., Tahara, K., Itsubo, N. and Inaba, A. 2018. Consistent characterisation factors at midpoint and endpoint relevant to agricultural water scarcity arising from freshwater consumption. *Int. J. Life Cycle Assess.* 23(12), 2276–87 doi:10.1007/s11367-014-0811-5.

Muñoz, I., Campra, P. and Fernández-Alba, A. R. 2010. Including CO_2-emission equivalence of changes in land surface albedo in life cycle assessment. Methodology and case study on greenhouse agriculture. *Int. J. Life Cycle Assess.* 15(7), 672–81. doi:10.1007/s11367-010-0202-5.

Nemecek, T., Bengoa, X., Lansche, J., Mourin, P., Riedener, E., Rossi, V. and Humbert, S. 2015. *Methodological Guidelines for the Life Cycle Inventory of Agricultural Products*. World Food LCA Database (WFLDB). Version 3.0, July 2015. Quantis and Agroscope, Lausanne.

Núnez, M., Rosenbaum, R. K., Karimpour, S., Boulay, A. M., Lathuillière, M. J., Margni, M., Scherer, L., Verones, F. and Pfister, S. 2018. A multimedia hydrological fate modeling framework to assess water consumption impacts in life cycle assessment. *Environ. Sci Technol.* 52(8), 4658–67. doi:10.1021/acs.est.7b05207.

Payen, S. 2015. Toward a consistent accounting of water as a resource and a vector of pollution in the LCA of agricultural products: methodological development and application to a perennial cropping system. (Doctoral dissertation). Montpellier University.

Payen, S., Basset-Mens, C. and Perret, S. 2015. LCA of local and imported tomato: an energy and water trade-off. *J. Cleaner Prod.* 87, 139–48. doi:10.1016/j.jclepro.2014.10.007.

Payen, S., Basset-Mens, C., Núñez, M., Follain, S., Grünberger, O., Marlet, S., Perret, S. and Roux, P. 2016. Salinisation impacts in Life Cycle Assessment: a review of challenges and options towards their consistent integration. *Int. J. Life Cycle Assess.* 21(4), 577–94 doi:10.1007/s11367-016-1040-x.

Payen, S., Basset-Mens, C., Colin, F. and Roignant, P. 2017. Inventory of field water flows for agri-food LCA: critical review and recommendations of modelling options. *Int. J. Life Cycle Assess.* 23(6), 1331–50. doi:10.1007/s11367-017-1353-4.

Payen, S., Falconer, S. and Ledgard, S. F. 2018. Water scarcity footprint of dairy milk production in New Zealand - a comparison of methods and spatio-temporal resolution. *Sci. Total Environ.* 639, 504–15. doi:10.1016/j.scitotenv.2018.05.125.

Pfister, S. 2015. Water use. In: Hauschild, M. and Huijbregts, M. (Eds), *LCA Compendium - The Complete World of Life Cycle Assessment*. Volume: Life Cycle Impact Assessment. Edited by Klöpffer, W. and Curran, M. A. Springer, Dordrecht. Chapter 12.

Pfister, S. and Bayer, P. 2014. Monthly water stress: spatially and temporally explicit consumptive water footprint of global crop production. *J. Clean. Prod.* 73, 52–62. doi:10.1016/j.jclepro.2013.11.031.

Pfister, S. and Hellweg, S. 2011. Surface water use - human health impacts. Report of the LC-IMPACT project (EC: FP7). doi:10.13140/RG.2.2.12858.44482.

Pfister, S. and Lutter, S. F. 2016. How EU27 is outsourcing the vast majority of its land and water footprint. In: *LCA Food 2016*, pp. 838–41. Available at: http://www.wulca-waterlca.org/pdf/case_study/LCA_FOOD_Pfister_lutter.pdf.

Pfister, S., Koehler, A. and Hellweg, S. 2009. Assessing the environmental impacts of freshwater consumption in LCA. *Environ. Sci. Technol.* 43(11), 4098–104. doi:10.1021/es802423e.

Pfister, S., Vionnet, S., Levova, T. and Humbert, S. 2016. Ecoinvent 3: assessing water use in LCA and facilitating water footprinting. *Int. J. Life Cycle Assess.* 21(9), 1349–60: doi:10.1007/s11367-015-0937-0.

Pfister, S., Boulay, A. M., Berger, M., Hadjikakou, M., Motoshita, M., Hess, T., Ridoutt, B., Weinzettel, J., Scherer, L., Döll, P., Manzardo, A., Núñez, M., Verones, F., Humbert, S., Buxmann, K., Harding, K., Benini, L., Oki, T., Finkbeiner, M. and Henderson, A. 2017. Understanding the LCA and ISO water footprint: a response to Hoekstra (2016) 'A critique on the water-scarcity weighted water footprint in LCA'. *Ecol. Indic.* 72, 352–9. doi:10.1016/j.ecolind.2016.07.051.

Pradinaud, C., Northey, S., Amor, B., Bare, J., Benini, L., Berger, M., Boulay, A., Junqua, G., Lathuillière, M. J., Margni, M., Motoshita, M., Niblick, B., Payen, S., Pfister, S., Quinteiro, P., Sonderegger, T. and Rosenbaum, R. K. 2019. Defining freshwater as a natural resource: a framework linking water use to the area of protection natural resources. *Int. J. Life Cycle Assess.* 24(5), 960–74. doi:10.1007/s11367-018-1543-8.

SBT. 2019. Science based targets – companies taking action. Shaw, B. J., Gladek, E., Unneland, S., Angelucci, D., Lelouvier, R. Commissioned by WWF, the Netherlands, Produced by Metabolic. Available at: https://sciencebasedtargets.org/companies-taking-action/.

Scherer, L. and Pfister, S. 2015. Modelling spatially explicit impacts from phosphorus emissions in agriculture. *Int. J. Life Cycle Assess.* 20(6), 785–95. doi:10.1007/s11367-015-0880-0.

Scherer, L. and Pfister, S. 2016. Dealing with uncertainty in water scarcity footprints. *Environ. Res. Lett.* 11(5), 54008. doi:10.1088/1748-9326/11/5/054008.

Sood, A. and Smakhtin, V. 2015. Global hydrological models: a review. *Hydrol. Sci. J.* 60(4), 549–65. doi:10.1080/02626667.2014.950580.

Steduto, P., Hsiao, T. C., Fereres, E. and Raes, D. 2012. *Crop Yield Response to Water. FAO* Irrigation and Drainage Paper No. 66. FAO, Rome, Italy, 505pp.

Thinkstep. 2019. Water footprint: GaBi software. Available at: http://www.gabi-software.com/international/solutions/water-footprint/ (accessed on 26 January 2019).

UNEP. 2012. *Measuring Water Use in a Green Economy, A Report of the Working Group on Water Efficiency to the International Resource Panel.* McGlade, J., Werner, B., Young, M., Matlock, M., Jefferies, D., Sonnemann, G., Aldaya, M., Pfister, S., Berger, M., Farell, C., Hyde, K., Wackernagel, M., Hoekstra, A., Mathews, R., Liu, J., Ercin, E., Weber, J. L., Alfieri, A., Martinez-Lagunes, R., Edens, B., Schulte, P., von Wirén-Lehr, S. and Gee, D. (Eds). United Nations Environment Programme. ISBN: 978-92-807-3220-7.

UNEP. 2015. *Options for Decoupling Economic Growth from Water Use and Water Pollution*.

UNEP. 2016. *Global Guidance for Life Cycle Impact Assessment Indicators (Volume 1). UNEP/SETAC Life Cycle Initiative*. Edited by Frischknecht, R. and Jolliet, O. UNEP, Paris.

UNEP. 2019. *Global Guidance for Life Cycle Impact Assessment Indicators (Volume 2). UNEP/SETAC Life Cycle Initiative*. Edited by Frischknecht, R. and Jolliet, O.

Verones, F., Pfister, S. and Hellweg, S. 2013a. Quantifying area changes of internationally important wetlands due to water consumption in LCA. *Environ. Sci. Technol.* 47(17), 9799–807. doi:10.1021/es400266v.

Verones, F., Saner, D., Pfister, S., Baisero, D., Rondinini, C. and Hellweg, S. 2013b. Effects of consumptive water use on biodiversity in wetlands of international importance. *Environ. Sci. Technol.* 47(21), 12248–57. doi:10.1021/es403635j.

Verones, F., Pfister, S. and Berger, M. 2016. Water use analysis. In: Dewulf, J., De Meester, S. and Alvarenga, R. A. F. (Eds), *Sustainability Assessment of Renewables-Based Products: Methods and Case Studies*. John Wiley & Sons, Chichester UK. Chapter 7.

Verones, F., Pfister, S., Zelm, R. Van and Hellweg, S. 2017. Biodiversity impacts from water consumption on a global scale for use in life cycle assessment. *Int. J. Life Cycle Assess.* 22(8), 1247–56. doi:10.1007/s11367-016-1236-0.

Ward, F. A. and Pulido-Velazquez, M. 2008. Water conservation in irrigation can increase water use. *Proc. Natl. Acad. Sci. U. S. A.* 105(47), 18215–20. doi:10.1073/pnas.0805554105.

Weinzettel, J. and Pfister, S. 2019. International trade of global scarce water use in agriculture: modeling on watershed level with monthly resolution. *Ecol. Econ.* 159, 301–11. doi:10.1016/J.ECOLECON.2019.01.032.

Wernet, G., Bauer, C., Steubing, B., Reinhard, J., Moreno-Ruiz, E. and Weidema, B. 2016. The ecoinvent database version 3 (part I): Overview and methodology. *Int. J. Life Cycle Assess.* 21(9), 1218–30. doi:10.1007/s11367-016-1087-8.

Zhou, J., Chang, V. W. C. and Fane, A. G. 2013. An improved life cycle impact assessment (LCIA) approach for assessing aquatic eco-toxic impact of brine disposal from seawater desalination plants. *Desalination* 308, 233–41. doi:10.1016/j.desal.2012.07.039.

Chapter 7

Modelling the environmental impacts of pesticides in agriculture

Peter Fantke, Technical University of Denmark, Denmark

1 Introduction

Pesticides are defined as 'any substance, or mixture of substances of chemical or biological ingredients, intended for repelling, destroying or controlling pests, or regulating plant growth' (FAO, 2014). In this chapter, the term 'pesticides' refers to the actual active ingredients in plant protection formulations. Pesticides are widely applied around the world with an estimated use of more than 3 million tonnes in 2018, of which 80–90% are used in agriculture (Markets and Markets, 2013). Pesticides can be classified according to a wide range of criteria. This includes target organisms (e.g. herbicides applied against unwanted weeds), chemical structure (e.g. carbamates derived from carbamic acid containing alcohol and methyl functional groups), toxic mode of action (e.g. mitochondrial adenosine triphosphate (ATP) synthase inhibitors preventing the synthesis of the energy carrier coenzyme ATP), activity spectrum (e.g. selective pesticides only targeting specific organisms) and formulation mode (e.g. wettable powders). In Table 1, the most relevant classification systems are shown and applied to a set of example pesticides.

http://dx.doi.org/10.19103/AS.2018.0044.08

Table 1 Different classification systems according to target organisms, chemical structure, toxic mode of action and basic chemical type, applied to selected example pesticides

Substance name	Target class	Chemical class	Mode of action	Basic chemical type
Glyphosate (CAS: 1071-83-6)	Herbicide	Phosphonoglycine	Systemic EPSPS[a] enzyme inhibitor	Synthetic organic pesticide
Malathion (CAS: 121-75-5)	Insecticide	Organophosphate	Non-systemic AChE[b] enzyme inhibitor	Synthetic organic pesticide
Copper oxychloride (CAS: 1332-40-7)	Fungicide	Copper compound	Multi-site enzyme system disruptor	Inorganic metal-based pesticide
D-limonene (CAS: 5989-27-5)	Insect repellent	Cyclic monoterpene	Nervous system interference agent	Natural organic bio-pesticide
Calcium phosphide (CAS: 1305-99-3)	Rodenticide	Phosphide	Mitochondrial complex IV electron transport inhibitor	Inorganic non-metal pesticide

[a] 5-enolpyruvylshikimate-3-phosphate (EPSP) synthase.
[b] Acetylcholinesterase.

Agricultural pesticides are pesticides that are primarily applied in the production of agricultural crops, where pesticide amount, time and frequency differ as a function of how the crop is finally used. The main uses are on food crops for human consumption (e.g. rice, wheat, apples), fodder crops to feed domesticated animals (e.g. clover, grass, alfalfa), fibre crops to manufacture textiles and other fibres (e.g. cotton, flax, sisal). Also energy crops are used as a source for biofuels (e.g. maize, rapeseed, oil palms), crops used as sugar and starch feedstock for producing biochemicals (e.g. sugarcane, sugar beet) and other industrial crops used for example as cut flowers (e.g. roses, tulips, lilies) or medicinal plants (e.g. aloe vera, borage, eucalyptus). Across these different uses, pesticide application is usually most restrictively regulated for food crops grown for human consumption. Worldwide, there are several thousand pesticides on the market. The EU Pesticide Database contains more than 1350 pesticides, of which almost 500 are approved under regulation Reg. (EC) 1107/2009 (EC, 2009). More than 800 are not approved, and the remaining substances either are pending approval or are additives not falling under this regulation. The U.S. Environmental Protection Agency's Office of Pesticide Programs has listed more than 4600 pesticides; however, this list also contains substances other than active ingredients.

Pesticides are mainly used to increase crop yields by reducing pest pressure on marketed crops, and increase food safety by fighting pathogen residues (e.g. Cooper and Dobson, 2007). However, this comes at the cost of human and ecosystem exposure and the contamination of the environment including remote areas where pesticides are rarely used (Malaj et al., 2014; Landrigan et al., 2018). In an environmental assessment context, pesticides applied in agriculture contribute to ecotoxicity and human toxicity impacts, with varying toxicity potentials across pesticides. This means that related impacts on humans and ecosystems cannot be estimated based on application amounts, but instead require the quantification of related emissions, toxicity and ecotoxicity (Räsänen et al., 2015; Yang, 2015). Additional impacts from pesticides can be associated with their production and disposal, in fact with their entire life cycle. A generic life cycle of a pesticide is shown in Fig. 1. Along such a life cycle, inputs from the environment (e.g. water and energy, but also materials and a wide range of chemical substances from solvents to additives) are used in the different life cycle stages, and outputs from these life cycle stages (e.g. emissions of chemicals into air) are released back into the environment. Emissions and impacts of pesticides can be modelled either by using a risk assessment perspective (based on conservative assumptions) or by using a comparative impact assessment perspective (based on best estimates). The focus of the present chapter is on assessing the emissions and impacts associated with agricultural pesticides in comparative assessments. This includes environmental impact assessment, life cycle inventory, and impact assessment as part of evaluating the environmental sustainability performance of any agrifood product or system in life cycle assessment (LCA), and alternative assessment for substitution of harmful chemicals.

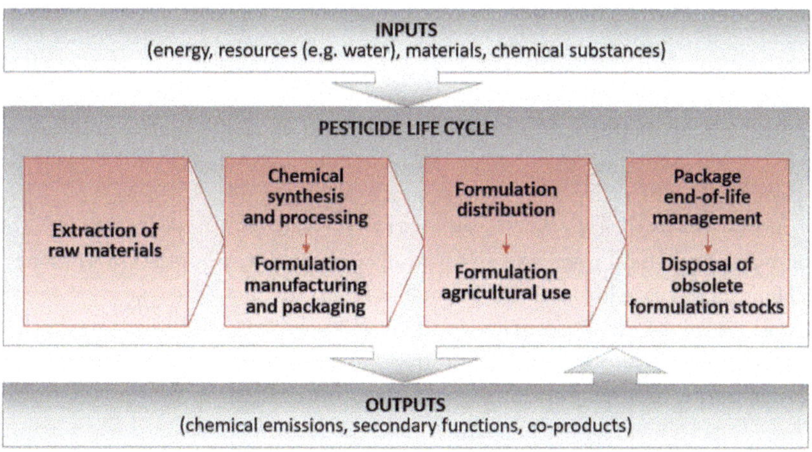

Figure 1 Generic life cycle stages of a pesticide used as an active ingredient in a plant protection formulation that is applied to agricultural crops with inputs from the environment and outputs back into the environment.

1.1 Key themes in modelling pesticides in environmental assessments

Agricultural pesticides generally impact the environment along their life cycle including during chemical manufacturing (Geisler et al., 2005; Holt et al., 2008; Natalia and Robert, 2016), use (Margni et al., 2002; Fantke et al., 2012a; Antón et al., 2004) and end of life treatment (Götz et al., 2012; Weber et al., 2013).

Nowadays, there is increasing focus on regulatory risk assessments of pesticides. However, emissions and related impacts on ecosystems and humans associated with the application of pesticides in agriculture, have long been ignored or addressed via crude assumptions in assessment frameworks like LCA. Only recently, have quantitative and mass-balance-based approaches become available in this context to consistently quantify pesticide field emissions and associated exposures, and impacts on ecosystem quality and human health (Rosenbaum et al., 2015; Fantke et al., 2017a). However, despite these approaches, there are still several gaps and unresolved questions in the environmental assessment practice, specifically in the life cycle inventory (LCI) and life cycle impact assessment (LCIA) phases of LCA for agricultural pesticides. Due to these gaps and questions, the field of addressing pesticides in LCA is continuously being advanced by researchers, modellers and practitioners. The present chapter will provide an overview of the challenges in emission and impact modelling of agricultural pesticides, with the aim of helping the reader understand how pesticides can currently be addressed in environmental assessments, and to find practical information and sources for related data and tools.

1.2 Scope of the chapter

The present chapter gives an overview of existing data and modelling approaches relevant for addressing pesticides in environmental assessments. Focus is thereby on pesticide application and emission quantification (Section 2), modelling ecotoxicity and human toxicity impacts from pesticide exposure (Sections 3 and 4), the relevance of spatial and temporal variability for modelling pesticides (Section 5), and the role of substitution scenarios to optimize agricultural pesticide use from a sustainability point of view (Section 6). Finally, pesticide impacts are quantified in an illustrative case study (Section 7), and future trends in pesticide modelling research (Section 8) along with a list of further reading and useful information around addressing pesticides in environmental assessments (Section 10) are provided.

2 Applying pesticides and quantifying emissions

When evaluating environmental sustainability, impacts of agricultural pesticides on humans or ecosystems, per functional unit (FU) of a particular product or service system, are expressed in terms of impact score, IS^i [impact/FU], for impact category i with index i = e for ecotoxicity impacts and i = h for human toxicity impacts. The impact score is thereby derived from the sum of pesticide mass, $m_{p,c}$ [$kg_{emitted}$/FU], emitted from any life cycle stage into a particular environmental compartment c, multiplied by the respective impact characterization factor, $CF^i_{p,c}$ [impact/$kg_{emitted}$], for all involved pesticides p:

$$IS^i = \sum_{p,c} \left(m_{p,c} \times CF^i_{p,c} \right) \tag{1}$$

where $m_{p,c}$ is the sum of all emissions of the same pesticide into the same environmental compartment. Since emission amounts are usually not known, they have to be estimated or modelled, which is discussed in more detail in subsequent sections. Pesticide emissions are usually estimated based on applied pesticide mass, and various statistics exist to derive applied pesticide mass. However, while summary statistics across pesticides per target class (e.g. herbicides) or chemical class (e.g. carbamates) are widely available across countries (see Fig. 2), detailed information on the actual applied amount of specific pesticides to a particular crop usually requires access to regulatory reports. Summary statistics across target or chemical classes at the global scale are provided by the Food and Agricultural Organization Corporate Statistical Database (FAOSTAT).

Global pesticide use trends from FAOSTAT (http://fao.org/faostat) between 1990 and 2015 are summarized in Fig. 2, showing that only a handful of countries account for more than 90% of the global pesticide use. High usage is seen in China, followed by the USA and Brazil, of which on average 50% are herbicides, 25% are fungicides and 20% are insecticides. However, this data should be interpreted with great care because these usage statistics are not comprehensive, for example, there is nothing reported for the USA in 2013–15 or for major African countries, both of which have a relevant influence on global pesticide use in current years. The decreasing overall trend of pesticide use according to FAOSTAT in the last 10 years is in contrast to global pesticide market price statistics. Such statistics show an increase in pesticide market value from 45 billion USD in 2009 to 59 billion USD in 2016 (Fig. 2), demonstrating that market value and use are not necessarily following the same trend.

Another challenge is the correct interpretation of 'use' data in terms of applying this correctly in environmental assessments. Often, data on production, import, export, storage (stockpiles), sales, consumption and actual use are not available in a suitable format for deriving estimates of pesticide application. We can take China as an example, where in 2013 a total of 3.2 million tonnes were produced, 76 000 tonnes imported, 1.1 million tonnes

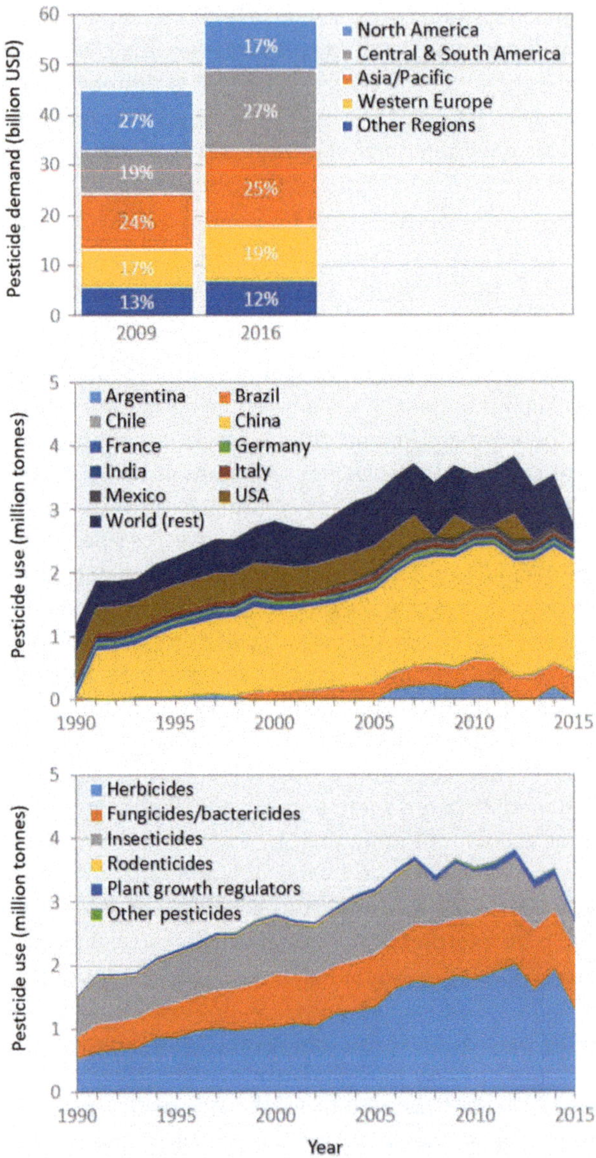

Figure 2 Global pesticide demand in 2009 and 2016 (based on market data from http://freedoniagroup.com) and global pesticide use between 1990 and 2015 (based on statistics from http://fao.org/faostat).

exported and 1.8 million tonnes used (Zhang, 2016). This, however, leaves 370 000 tonnes for which it is unclear if they have been used or not. The same caution is necessary for data of individual pesticides applied to specific crops, where treated area, harvested area, application count and other terms can easily lead to problems for accurately determining application data. In these cases, it is always recommended to check unit of input data. Finally, application dose, count and formulation content differ greatly between pesticides, and hence they cannot be averaged or totaled across pesticides per crop or even across crops. To obtain such information, it is recommended to always consult product labels or guidelines provided by the respective pesticide regulatory authorities.

2.1 Pesticide application data

For modelling pesticides in environmental assessments, one of the most relevant items of information is the amount of pesticides used related to the given FU. As for agricultural pesticides, this amount typically relates to the overall pesticide mass applied on a crop, either as field or greenhouse application. This overall mass applied per pesticide, m_{app} [$\text{kg}_{\text{applied}}$/FU], can be derived from the application dosage, $D_{\text{app},j}$ [$\text{kg}_{\text{applied}}$/ha], and the treated field or greenhouse area, A_j [ha], summed over j treatments (among which dosage and treated area might vary):

$$m_{\text{app}} = \sum\nolimits_j \left(D_{\text{app},j} \times A_j \right) \tag{2}$$

The mass of a pesticide applied is often the only information that is known to modellers and practitioners (Rosenbaum et al., 2015). However, if this information is missing, it can be derived from regulatory guidelines or pesticide formulation information, where the recommended treatment dose and count per pesticide according to good agricultural practice are indicated. However, in many countries worldwide, pesticides are not applied as officially recommended, but rather overused or misused (Ecobichon, 2001; Zhang et al., 2015a), which should be considered to reflect actual agricultural practice. In addition, pesticide application differs widely between pesticide type and crops. As an example, the insecticide imidacloprid is usually applied as foliar treatment in doses ranging from 110 g/ha on, for example banana and other tropical fruits, to 500 g/ha when applied as soil chemigation. In contrast, the widely used herbicide glyphosate is usually applied in doses of more than 4000 g/ha on, for example cereal crops, banana, olives and other fruits and vegetable crops (see reports from the Joint WHO/FAO Meetings on Pesticide Residues). A pesticide application database containing data on application dose and method, crop growth stage, pre-harvest treatment interval, and treatment count was recently made available for more than 4000 pesticide crop scenarios focusing on European conditions (Steingrimsdottir et al., 2018). This database can be used as a data source to build pesticide application scenarios in Europe, while additional data sources should be consulted for scenarios outside Europe.

2.2 Quantifying pesticide emissions

On a theoretical basis, the mass of any pesticide *p* emitted into a particular environmental compartment *c* during application, m_c [$kg_{emitted}$/FU] (see Eq. 1), is derived from the pesticide mass applied, m_{app} [$kg_{applied}$/FU] (see Eq. 2), and the emission fraction, f_c [$kg_{emitted}$/$kg_{applied}$], into that environmental compartment (Hauschild, 2000; Birkved and Hauschild, 2006):

$$m_c = m_{app} \times f_c \tag{3}$$

When degradation is not considered during the (usually very short) time between pesticide application and emission into the environment, the sum of all emission fractions into *j* environmental compartments should sum up to 100%, that is $\sum_j \left(f_{c_j}\right) = 1$. Usually, not all emission fractions are available at the same time. From the above condition, an emission fraction into a particular environmental compartment *i*, f_{c_i} (e.g. the fraction reaching the soil surface in an agricultural field) could be derived as $f_{c_i} = 1 - \sum_j \left(f_{c_j}\right)$, if all other emission fractions f_{c_j} into compartments $j \neq i$ (e.g. the total fraction leaving the agricultural field via initial wind drift and the fraction reaching the surface of the agricultural crop) are known (Juraske et al., 2007a; Fantke et al., 2011a).

For quantifying emissions from agricultural pesticides applied to field or greenhouse crops, different approaches are available and have been followed by practitioners. These approaches range from assuming fixed emission fractions to particular environmental compartments across all pesticides and crops, to mass-balance-based models distinguishing emission fractions as a function of pesticide properties, crop characteristics, environmental conditions and application scenarios. A review of various pesticide emission models is provided in Mottes et al. (2014). In LCA, one of the most widely applied assumptions for pesticide emission fractions is that 100% of the applied mass is emitted to soil (within the agricultural field), thereby considering the field soil as part of the ecosphere. This approach is for example, implemented in the Ecoinvent database (Nemecek and Kägi, 2007; Nemecek and Schnetzer, 2011), which is widely used in LCA. However, depending on how the environmental fate is subsequently modelled for pesticides, this can lead to important underestimations of ecotoxicity impacts including in freshwater (by neglecting initial deposition fractions to freshwater bodies outside the treated field) or to an overestimation of human toxicity impacts via crop residues, such as root and tuber crops. Other approaches are available differentiating emissions into more than a single environmental compartment, such as 75% emitted to field soil and 25% to air for viticulture crops (Neto et al., 2013) or 96% emitted to air and 4% to surface water outside the field (USDE, 2012). To avoid underestimating emissions (and related impacts) in any assessment, a certain fraction of applied pesticide should always be assumed to reach surface freshwater, based on the average surface area distribution between soil and freshwater in a given region.

An alternative to assuming fixed emission fractions is proposed by modelling the initial distribution of pesticides on the agricultural field, and subsequently estimating emissions into different environmental compartments. A model that follows this approach is PestLCI (Birkved and Hauschild, 2006; Hauschild, 2000), currently available as version 2.0 (Dijkman et al., 2012). PestLCI is applied in several environmental assessment studies (e.g. Renaud-Gentié et al., 2015). This was further adapted in a consensus building process to address challenges related to allocating parts of the agricultural field to either technosphere (i.e. the studied anthropogenic system) or ecosphere (i.e. the environment we want to protect), in emission quantification and impact pathway modelling of pesticides (Rosenbaum et al., 2015). Outcomes of this process include an improved emission model that now provides time- and area-integrated emission estimates, comes with a suite of improved drift functions, and a flexible approach to addressing field soil as part of the environment (Fantke et al., 2017a). In Table 2, an overview is provided of the different assumed fixed or calculated varying pesticide emission fractions into the environment that are applied by different emission inventory approaches. Varying emission fractions are thereby referring to modelled output that differs between combinations of pesticides, crops, and application scenarios (e.g. climate, soil type, application method). Other approaches exist that might be applicable in an environmental assessment context (e.g. Audsley et al., 1997; Berthoud et al., 2011; Yang and Suh, 2015), and studies applying different emission estimation approaches to pesticides are discussed by Schmidt Rivera et al. (2017).

2.3 Interface between emission quantification and impact assessment for agricultural pesticides

In environmental assessments, several emission quantification approaches for agricultural pesticides exist. There are also several impact assessment methods for modelling related ecotoxicity and human toxicity impacts (see Sections 3 and 4). However, depending on which approaches are followed for both emission inventory and impact assessment, there are potential overlaps between what is considered on the emission inventory side, and what is included in the impact assessment in terms of pesticides distributing among different environmental compartments after field application. These overlaps might occur

Table 2 Fixed or variable (var.), that is, pesticide crop application scenario-specific pesticide fractions emitted into the environment according to different life cycle emission inventory approaches

Emission fraction to	Ecoinvent[a]	Neto et al.[b]	US LCI[c]	USDA Ag-LCI[d]	JALCA[e]	WFLDB[f]	PestLCI[g]	Emission consensus model[h]
Field								
Agricultural soil	100%	75%	-	100%	100%	100%	var.[l]	var.[n]
Crops	-	-	-	-	-	-	var.[l]	var.
Air	-	25%	~95%[j]	-	-	-	var.	var.
Off-field								
Agricultural soil	-	-	-	-	-	-	-	var.[n]
Natural soil	-	-	-	-	-	-	-	var.
Surface water	-	-	~5%[k]	-	-	-	var.	var.
Groundwater	-	-	-	-	-	-	var.[m]	var.[m]
Other	-	-	-	-	-	-	var.[l]	var.[l]

[a] Ecoinvent database until version 3.4 (Nemecek and Kägi, 2007; Nemecek and Schnetzer, 2011).
[b] Fixed emission fractions specifically estimated for viticulture (Neto et al., 2013).
[c] U.S. Life Cycle Inventory database (USDE, 2012).
[d] U.S. Department of Agriculture Commons Life Cycle Inventory datasets (Cooper, 2015).
[e] National Agriculture and Food Research Organization, Japan LCA database (Hayashi et al., 2012).
[f] World Food LCA Database until version 3.3 (Nemecek et al., 2015).
[g] PestLCI model version 2.0; point emissions over distance range to field edge (Dijkman et al., 2012).
[h] Adapted and further developed from PestLCI 2.0; area-integrated emissions over distance range to field edge (Rosenbaum et al., 2015; Fantke et al., 2017a).
[i] Processes leading to a reduction of emission fractions to the environment, for example degradation and plant uptake.
[j] Emissions modelled as function of pesticide vapour pressure, but typically little deviation from values ~95% (US-EPA, 1995).
[k] Emissions modelled as function of pesticide and soil properties, but typically little deviation from values ~5% (Kellogg et al., 2002).
[l] Initial distribution fraction not ultimately reported as emission to the environment.
[m] Currently, no matching characterization factors for groundwater emissions in any available LCIA method.
[n] User can define if emissions to the agricultural field are included in emission fractions to the environment or not.

whenever it is unclear what part of the agricultural field (e.g. the treated soil), is considered part of the studied system (technosphere) or part of the environment (ecosphere). This issue was discussed in more detail by van Zelm et al. (2014), where an aligned inventory-impact assessment system was proposed for pesticides to assess the differences between efficient and inefficient agricultural practices.

To address this potential overlap and define a clear interface between emission quantification and impact assessment, a consensus process was initiated (Rosenbaum et al., 2015). This process led to preliminary recommendations on how to delineate technosphere from ecosphere, and what to report in emission inventory databases and impact assessment models in order to facilitate a consistent assessment of pesticides. A possible way of consistently linking emission estimates for agricultural pesticides to impact factors for ecotoxicity and human toxicity is shown in Fig. 3. In this approach, emissions to groundwater are allocated to freshwater in the impact assessment model in case impact factors specifically for groundwater emissions are not available.

3 Ecotoxicity impacts of pesticides

Ecotoxicity is one of the most relevant impact categories in environmental impact assessments, which may be dominated by impacts associated with the use of pesticides in the production of food, bioplastics, cotton-based textiles

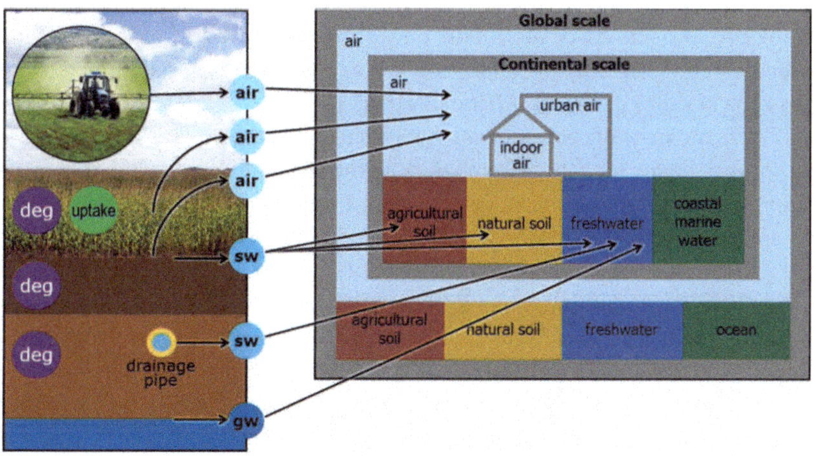

Figure 3 Recommended allocation of emission flows of agricultural pesticides to environmental compartments as defined in the USEtox impact characterization model. Abbreviations: deg - degradation, uptake - leaf and root uptake into field crops, sw - surface water, gw - groundwater.

and other agriculture-based products (e.g. Berthoud et al., 2011; Bessou et al., 2016; Tsiropoulos et al., 2015; Nordborg et al., 2014, 2017; Zhang et al., 2015b). However, ecotoxicity is also a widely discussed and controversial impact category. This is mainly related to the large number of chemical substances including agricultural pesticides that are considered in this category, the complexity of the related impact pathway, and the difficulty of translating measured or estimated ecotoxicity test results into impacts and damage to ecosystems. Ecotoxicity impact factors show large uncertainties, mainly associated with the complexity of the related impact pathways and with limitations in the underlying data, while at the same time important ecosystems are currently still missing or insufficiently represented in existing impact characterization frameworks. Assessing ecotoxicity impacts is particularly relevant for agricultural pesticides, since they are directly released into the environment in much higher doses than most other industrial chemicals that are usually emitted unintentionally. After application to agricultural fields, pesticides can easily reach not only organisms in air (via wind drift and volatilization) and agricultural soil (via deposition and wash-off from field crops), but also in adjacent water bodies including groundwater, rivers and lakes, marine water downstream, and in natural soil (via re-deposition after emission or volatilization). Whilst information on the environmental fate and transport is easily available for most agricultural pesticides in use today, data on ecotoxicological effects are not consistently available across pesticides, which is a current challenge for characterizing related ecotoxicity impacts. Several of these aspects are discussed in the present section.

3.1 A framework for characterizing ecotoxicity impacts

Ecotoxicity impacts associated with environmental emissions of agricultural pesticides can generally be characterized in the same way as impacts from emissions of other chemical substances (Fig. 4).

The ecotoxicity impact pathway for agricultural pesticides follows the general impact characterization framework combining environmental fate, ecosystem exposure and ecotoxicity effects, as implemented in different impact characterization models including USEtox (Rosenbaum et al., 2008; Henderson et al., 2011), Impact2002 (Pennington et al., 2005), CalTOX (McKone and Enoch, 2002) and USES-LCA (van Zelm et al., 2009b). Related ecosystem quality impact characterization factors at damage level are expressed as potentially disappeared fraction (PDF) of ecosystem species integrated over the exposed environmental compartment, and time per kg pesticide emitted to any environmental compartment (e.g. air, water, soil), CF^e [PDF m^3 d/kg$_{emitted}$]. CF^e is derived from fate factors from emission compartment c_1 (e.g. agricultural soil) to exposure compartment c_2 (e.g. freshwater), $FF_{c_2 \leftarrow c_1}$ [kg$_{in\ compartment}$ per kg$_{emitted}$/d], ecosystem exposure factors, XF^e [kg$_{dissolved}$/kg$_{in\ compartment}$], ecotoxicological

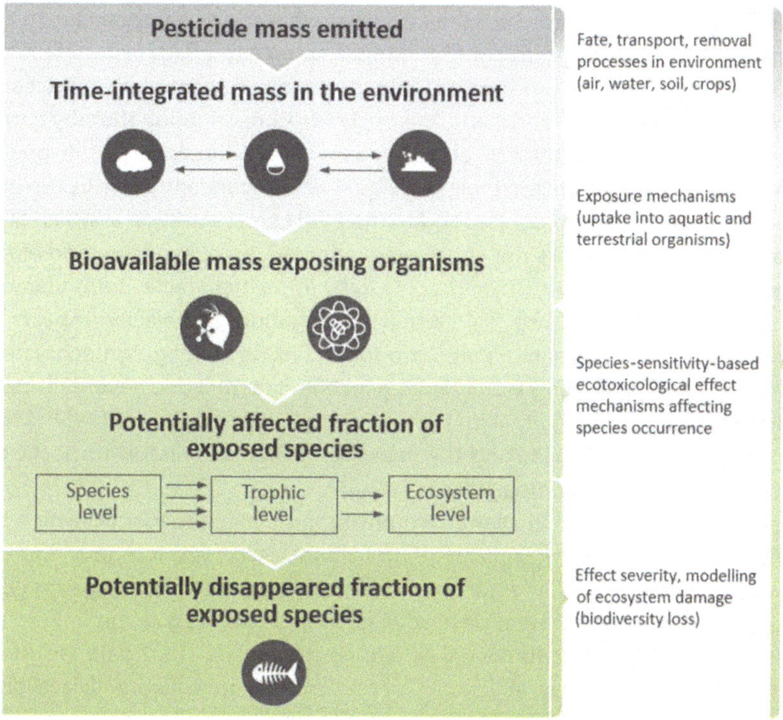

Figure 4 Generalized impact pathway for ecotoxicity from exposure to agricultural pesticides in environmental impact assessment (adapted from Fantke et al. 2018a).

effect factors for determining the potentially affected fraction (PAF) of species, EF^e [PAF $m^3/kg_{dissolved}$] and an ecosystem effect severity factor, SF^e [PDF/PAF] (Rosenbaum, 2015):

$$CF^e = FF_{c2 \leftarrow c_1} \times XF^e \times \underbrace{EF^e \times SF^e}_{DF^e} \tag{4}$$

Fate factors link the increase of pesticide mass in a given environmental compartment that is attributable to an emission of the same pesticide in the same or any other environmental compartment, after undergoing physical and chemical transport and degradation processes. More specifically, $FF_{c_2 \leftarrow c_1}$ can be interpreted as the product of the residing time of a pesticide in the receiving exposure compartment, FF_{c_2} [d], and the overall time-integrated pesticide mass fraction transferred from any given emission compartment c_1 to the exposure compartment c_2, $f_{c_2 \leftarrow c_1}$ [$kg_{in\ compartment}/kg_{emitted}$], that is $FF_{c_2 \leftarrow c_1} = f_{c_2 \leftarrow c_1} \times FF_{c_2}$. Ecosystem exposure factors link the pesticide mass in any environmental exposure compartment that is truly dissolved or otherwise effectively bioavailable to cross the cellular membrane of an organism, to the total pesticide mass in that compartment. Ecotoxicity effect factors link the PAF of exposed ecosystem species to the pesticide concentration in their surrounding exposure compartment. The pesticide concentration thereby corresponds to the point on the PAF curve where a certain fraction of the exposed species are affected above their individual effect concentration EC. In the case of EC50, this would correspond to the effect concentration at which 50% of all exposed individuals per species are affected. Fate, exposure and effect factors can be combined to yield ecotoxicity impact characterization factors expressed as PAF m^3 d/$kg_{emitted}$. When aggregating ecotoxicity impacts at damage level, severity factors are introduced linking damage on ecosystem quality to PAF. Combining ecotoxicological effect factors and severity factors yields damage-level effect factors for ecosystem quality, $DF^e = EF^e \times SF^e$ [PDF $m^3/kg_{dissolved}$] (see Eq. 4), linking the increase of PDF per dissolved pesticide mass in the exposure compartment.

Among the biggest challenges in environmental assessment practice for characterizing ecotoxicity impacts of agricultural pesticides, are deriving ecotoxicological effect data and translating affected species into damage on ecosystems. Despite the fact that pesticides have their own regulation in Europe, for the USA and other countries that require the collection of data including information on ecotoxicity, effect data for pesticides is not consistently available across substances, ecosystems, trophic levels and species (Panizzi et al., 2017). In addition, as existing effect data is primarily reported for use in regulatory risk assessment, it is not readily applicable in environmental assessment without further scrutinizing and adapting the raw data (Müller et al., 2017). Ideally, ecotoxicological effect factors are derived from repeated tests on several species across trophic levels as relevant for a particular ecosystem. However, validated data suitable for ready use in line with current impact characterization frameworks is often not available, and then needs to be extrapolated from acute tests, which increases uncertainty towards impact characterization results. Data sources for ecotoxicity effect information include, the European Food Safety Authority (EFSA) Pesticide Ecotoxicology Tier 1 Endpoints database (EFSA, 2012; Barmaz et al., 2011) and the U.S. Environmental Protection Agency (EPA) ECOTOX database (http://cfpub.epa.gov/ecotox). To arrive at ecosystem damage, any considered effect (e.g. mortality or inhibition) needs to be translated into species disappearance. Currently however, none of the approaches to translate PAF into PDF is sufficiently accepted as being recommended by the scientific community, and further research and consensus building are required to address this aspect. Meanwhile, pragmatic approaches should be followed to convert PAF into PDF using a default conversion factor of 10 when PAF is based on no-observed effect concentrations (NOEC), or a factor of 2 if PAF is based on 50% effect concentrations (EC50), as discussed in Larsen and Hauschild (2007).

3.2 Aquatic ecosystems including sediment

Currently, only the modelling of ecotoxicological effects on freshwater aquatic organisms is considered mature enough to be included in all existing impact assessment methods (Fantke et al., 2018a). Hence, ecotoxicity impact characterization factors of pesticides are usually limited to impacts on freshwater ecosystems in current environmental assessment studies (Margni et al., 2002; Juraske and Sanjuán, 2011). Relevant studies include those by van Zelm et al. (2009a) on related effect factors for pesticides based on species sensitivity distribution modelling, by Golsteijn et al. (2012) addressing warm-blooded freshwater species and by Dong et al. (2014) for metal-based substances. Where aquatic ecotoxicity beyond freshwater ecosystems is addressed in existing impact assessment methods, it is usually extrapolated from freshwater species data. Specific methods for targeting impacts of agricultural pesticides on groundwater or sediment-dwelling organisms are currently missing. To account for potential differences in sensitivities of organisms to pesticide exposure, related effect data would need to be specific to groundwater or sediment-dwelling organisms. Such data is currently rare for groundwater (Verweij et al., 2015), while data for sediment-benthic species is becoming available and could be included as contributing to freshwater aquatic impact scores when considering sediment as part of the overall water environment (Fantke et al., 2018a). For marine aquatic environments, initial ecotoxicity modelling approaches are becoming available which focus on metal-based substances and their impacts on coastal marine aquatic ecosystems (Dong et al., 2016, 2017). However, impacts from non-metal pesticides still need to be estimated from freshwater ecotoxicity, where factors might or might not include marine species already, depending on the data availability per substance.

3.3 Terrestrial ecosystems

Ecotoxicity of terrestrial soil organisms plays a key role for agricultural pesticides, where emissions after field applications to a large fraction reach and expose species living in and on soil. Wherever terrestrial soil ecotoxicity is included in current impact assessment methods, it is usually based on equilibrium partitioning approaches, and - in the case of missing soil ecotoxicity effect data - extrapolated from effect data of aquatic species. Specific approaches for metal-based substances have recently been developed considering speciation and bioavailability in soil (Owsianiak et al., 2013, 2015; Plouffe et al., 2016), showing that geographic variability is an important aspect especially for very persistent substances including metal-based pesticides (Peña et al., 2017).

3.4 Additional target organisms

In addition to the relevance for aquatic species and terrestrial soil organisms, ecotoxicity impacts from exposure to agricultural pesticides have also been shown to be important for other organisms. This includes secondary poisoning in predatory birds and mammals (Nendza et al., 1997; Elliott et al., 1997) and exposure of pollinating insects particularly to agricultural pesticides (Woodcock et al., 2017). Within environmental assessments, these target organisms are currently not included, but preliminary research shows that especially for pollinators, related impact pathways can be defined (Crenna et al., 2017). However, this research currently remains in progress. Overall, aquatic and terrestrial predatory species could be included into existing ecosystem-based factors. In contrast, modelling pesticide exposure of pollinators (more specifically domestic bees) will require a separate approach using dose-response data specifically available for bees. For example, in EFSA's OpenFoodTox database (Dorne et al., 2017), the magnitude of exposure of a single bee species towards different pesticides will have to be modelled as a function of bee nesting and (nectar and pollen) foraging activities (Arena and Sgolastra, 2014).

3.5 The global ecological burden from pesticides

Agricultural pesticides have by design a high biological activity, and both acute and chronic toxicity potential, which makes them by definition important contributors with respect to negatively affecting ecological species and contaminating the environment. It is only recently that ecotoxicity has been sufficiently studied to be applicable in environmental assessments, and many practitioners still have reservations (Rosenbaum, 2015). However, the predominant role of agricultural pesticides as contributors to environmental pollution, and loss of biodiversity and ecosystem services on a local to regional scale is well recognized. However, while pollution associated with pesticides is considered a large-scale environmental problem requiring holistic actions to restore and preserve ecosystem health, its overall relevance regarding larger spatial scales remains mostly unknown, with existing studies likely underestimating the impacts due to usually only studying or monitoring a set of known, specific substances (Malaj et al., 2014). The exceptions are persistent and mobile pesticides that can be found in the environment and ecosystems around the world, thereby constituting a potential threat on a global scale (Persson et al., 2013). Overall, environmental pollution from pesticides and other chemical substances is among the five key contributors to global biodiversity loss (SCBD, 2010). Hence, modelling ecotoxicity of agricultural pesticides on all relevant ecosystems (e.g. freshwater ecosystems in streams around agricultural fields) and target species (e.g. pollinating insects foraging on agricultural field crops) is a core component in the environmental performance evaluation of agricultural practices, as well as for designing more sustainable and less environmentally damaging pesticides. Practitioners should therefore always pay special attention to include all relevant pesticides in their assessment to avoid underestimation of related impacts on the environment.

4 Human toxicity impacts of pesticides

Human toxicity is one of the most controversially discussed impact categories in environmental impact assessments. On the one hand, human toxicity impacts are often contributing largely to overall environmental performance profiles of products and services. On the other hand, the diversity of chemical substances included in this impact category and the complexity of the human toxicity impact pathways, yield questions over data reliability. Moreover, several important exposure pathways are already assessed by other frameworks (e.g. an occupational exposure assessment), which leads to perceived overlaps with environmental assessments. All this is particularly relevant for agricultural pesticides, which are among the most widely regulated chemical substances worldwide. Modelling human toxicity impacts for agricultural pesticides has various specialist aspects compared to modelling impacts of other chemical substances in an impact assessment context. First, unlike most substances emitted from continuous industrial processes, pesticides are applied in pulses. This means that a large chemical mass is released over a very short period, such as minutes or hours, which requires a special way of quantifying emissions (see Section 2 for details). Secondly, several exposure pathways are especially important for agricultural pesticides, for example exposure of workers applying pesticides in open fields, and exposure to pesticide residues in harvested food crops. Finally, information on human dose-response is available at least for some well-studied agricultural pesticides due to the relatively high exposure and toxicity potential of this substance group. Several of these aspects are discussed in the present section.

4.1 A framework for characterizing human toxicity impacts

Toxicity impacts of the general human population associated with environmental emissions of agricultural pesticides can be characterized in the same way as impacts from emissions of other chemical substances (Fig. 5).

The emission-based impact pathway follows the general impact characterization framework, combining environmental fate, human exposure and human toxicity effects as implemented in different impact characterization models including USEtox (Rosenbaum et al., 2008, 2011), Impact2002 (Pennington et al., 2005), CalTOX (McKone and Enoch, 2002) and USES-LCA (van Zelm et al., 2009b). Related human health impact characterization

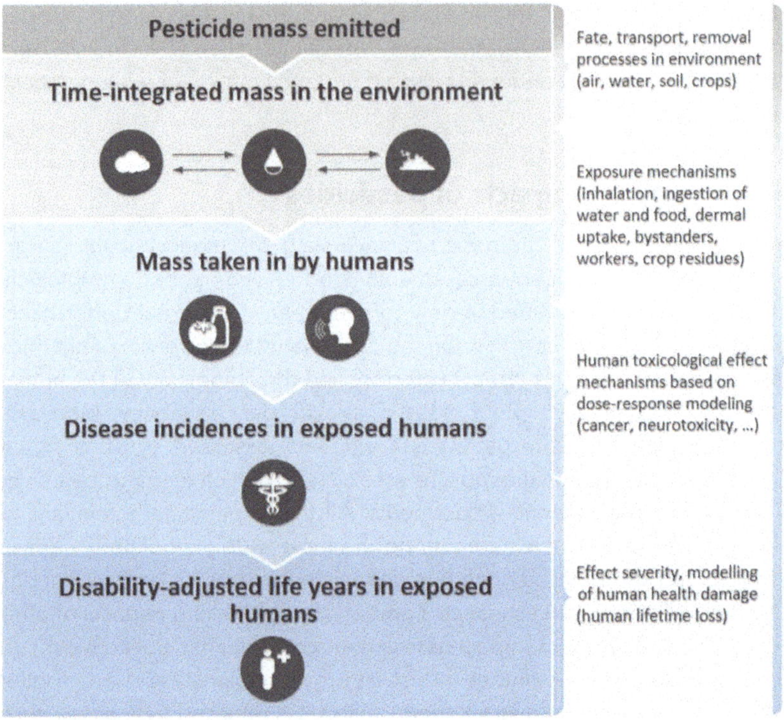

Figure 5 Generalized impact pathway for human toxicity from exposure to agricultural pesticides in environmental impact assessment (adapted from Fantke et al. 2018b).

factors at damage level are expressed as disability-adjusted life years (DALY) per kg pesticide emitted to any environmental compartment (e.g. air, water, soil), CF^h [DALY/kg$_{emitted}$]. CF^h is derived from fate factors, FF [kg$_{in\ compartment}$ per kg$_{emitted}$/d]; exposure pathway (e.g. inhalation of air, ingestion of fish) specific human exposure factors, XF^h [kg$_{intake}$/d per kg$_{in\ compartment}$]; human toxicological effect factors, EF^h [cases/d per kg$_{intake}$/d]; and a human effect severity factor, SF^h [DALY/d per cases/d] (Jolliet and Fantke, 2015):

$$CF^h = \underbrace{FF \times XF^h}_{iF} \times \underbrace{EF^h \times SF^h}_{DF^h} \tag{5}$$

As already described for ecotoxicity impact pathways (see Section 3), fate factors link the increase of pesticide mass in a given environmental compartment, that is attributable to an emission of the same pesticide in any environmental compartment, after undergoing physical and chemical transport and degradation processes. Human exposure factors link pesticide mass taken in via inhalation of air, ingestion of drinking water, ingestion of different food items and dermal exposure to pesticide mass in a given environmental compartment. Combining fate factors and human exposure factors yields the human intake fraction, $iF = FF \times XF^h$ [kg_{intake}/d per $kg_{emitted}$/d] (see Eq. 5), which is the fraction of emitted pesticide mass that is ultimately taken in by the human population via all exposure pathways (Bennett et al., 2002). Effect factors link the potential risk of developing an adverse health effect to the pesticide mass taken in by the human population. Fate, exposure and effect factors can be combined to yield human toxicity impact characterization factors expressed as cases/$kg_{emitted}$. When aggregating human toxicity impacts at damage level, severity factors are introduced which analyze differences in the severity of disabilities in terms of affecting human lifetime. Combining human toxicological effect factors and severity factors yields damage-level effect factors for human health, $DF^h = EF^h \times SF^h$ [DALY/d per kg_{intake}/d] (see Eq. 5), linking the increase of DALY in the human population to the pesticide mass taken in by the population.

Currently, the biggest challenge in environmental assessment practice for characterizing human toxicity impacts of agricultural pesticides is finding adequate pesticide property data. This might sound surprising given that pesticides are among the best-studied chemical substances requiring dedicated regulatory procedures in many countries around the world. However, some existing data is primarily reported for use in regulatory risk assessment and not readily applicable in environmental assessments without further scrutinizing and adapting the raw data, while other data is simply missing and has to be extrapolated or otherwise estimated. This is especially the case for data on biodegradation in environmental media and food crops, bioaccumulation in food products like meat and milk, and human toxicological effects. An estimation method has been proposed by Fantke et al. (2014) to derive degradation data for pesticides in crops, while degradation data in soil and water are available from other sources including the Footprint project Pesticide Properties database (Lewis et al., 2016). Effect data is usually extrapolated from animal toxicity data and between exposure routes (Huijbregts et al., 2005), and a useful data source can be the EFSA OpenFoodTox database (Dorne et al., 2017). Effect data constitutes the biggest area of uncertainty for related human toxicity impact characterization results for agricultural pesticides (Fantke et al., 2012a) and for chemical substances in general (Rosenbaum et al., 2008). This is underlined by the fact that in current human toxicity characterization methods, cancer and non-cancer effects are considered as aggregated effect metrics, and it remains unclear how health effects, such as endocrine disrupting or developmental toxicity effects (both of which are particularly relevant for many currently used agricultural pesticides), can be addressed.

4.2 Occupational exposure

Occupational exposure needs to be assessed separately for the manufacturing and use (i.e. field application) stages of agricultural pesticides, due to very different exposure environments. For occupational exposure during pesticide manufacturing, a general approach for characterizing worker exposure to chemicals has been developed based on monitoring data at industrial facilities (Kijko et al., 2015, 2016). This approach was applied to characterize occupational exposure in the US manufacturing industry including the sector 'Pesticide, Fertilizer, and Other Agricultural Chemical Manufacturing', and further to characterize worker exposure along the entire life cycle of a piece of office furniture. However, this approach is not applicable to estimate occupational exposure to agricultural pesticides during field application. While currently no operational approach is available for characterizing occupational exposure to pesticides applied in agricultural fields, a risk-based approach is used in a recent environmental assessment study for estimating worker exposure to pesticides applied in greenhouses (Boulard et al., 2012). This approach is, however, not compatible with the intake fraction framework applied in current human toxicity impact characterization and needs to be further adapted. Hence, quantifying occupational exposure to agricultural pesticide application remains a current challenge in environmental assessment. To address this challenge, the approach for characterizing bystander exposure (see Section 4.3) could be adapted, considering the different emission fractions on and beyond agricultural fields. Note, that in addition to direct exposure during pesticide field application, agricultural workers might also further expose their families at home (Vida and Moretto, 2007), which is currently not considered in environmental assessments.

4.3 Residential and non-residential bystander exposure

Bystander exposure to agricultural pesticides is currently not fully operationalized in environmental assessment, although it is acknowledged to be relevant in this context (e.g. Notarnicola et al., 2012). Several approaches are available for assessing bystander exposure (including residential bystander exposure) in risk assessment (e.g. Reiss and Griffin, 2006; Butler Ellis et al., 2010; van de Zande et al., 2010). However, these approaches are not directly applicable in an environmental assessment context based on their often conservative assumptions, such as putting bystanders in the centre of agricultural fields to estimate related exposures. By adopting spray drift curves in combination with the pesticide emission consensus model based on PestLCI 2.0 (Dijkman et al., 2012), Ryberg et al. (2018) designed the first model to estimate bystander exposure using drift-based emissions of agricultural pesticides in environmental impact assessment, which was evaluated against field measurements (Garron et al., 2009, 2012), and which can currently be used to evaluate pesticides applied to potato fields. Findings from this model show that bystanders can be exposed to a higher extent than the general population, while they represent only a small fraction of the overall population. This bystander exposure model is compatible with exposure estimates of human toxicity impact characterization, but needs to be extended to cover other crops by implementing respective drift curves, pesticide application methods, and field conditions. Furthermore, this model focuses on inhalation exposure, while dermal exposure is not currently included despite its potential relevance for bystanders. A review of human exposure pathways for chemicals in consumer products, concluded that dermal exposure models do exist that are generally compatible with the environmental impact modelling approach, although further research is required to estimate some important input parameters (Huang et al., 2017).

4.4 Exposure to residues in food crops

Initial work on characterizing human toxicity impacts from agricultural pesticides in an environmental assessment context, estimated exposure to crop residues based on measured residual pesticide concentrations, and acknowledged that this might be an important pathway (Margni et al., 2002; Humbert et al., 2007; Juraske et al., 2009; Antón et al., 2004). To include exposure to crop residues more mechanistically, the uptake, translocation and degradation in crops during the time between pesticide application and crop harvest need to be considered, and can be adapted from existing models of pesticide plant uptake developed for individual food crops for use in different assessment frameworks (e.g. Trapp, 2009). As a first attempt to include this pathway into environmental impact assessment, a single crop model for plant uptake in greenhouse tomatoes was developed (Juraske et al., 2007a), which was later adapted to compare conventional with organic farming practices (Juraske and Sanjuán, 2011). Later, the dynamic crop uptake model dynamiCROP was developed and specifically designed for evaluating multiple crops in environmental impact assessments (Fantke et al., 2011b, 2012b, 2013). In this model, the human intake fraction, iF [$kg_{intake}/kg_{applied}$], is obtained by linking the residual pesticide mass in all harvested crop components at harvest time, m_{res} [$kg_{in\ harvest}$], to the total pesticide mass applied, m_{app} [$kg_{applied}$], and by further combining this mass fraction (referred to as harvest fraction, hF) with a processing factor, PF [$kg_{intake}/kg_{in\ harvest}$], to account for losses of residual pesticides due to food processing (e.g. washing or cooking):

$$iF = \underbrace{\left(m_{res}/m_{app}\right)}_{hF} \times PF \qquad (6)$$

Note that whereas the intake fraction in Eq. 5 is based on emitted pesticide mass into a certain environmental compartment, the intake fraction for crop residues (Eq. 6) refers to applied pesticide mass to account for capturing the dynamics in crop-specific interception of pesticides. The respective intake fraction is hence combined with the applied mass fraction reaching the crop to be compatible with emission-based intake fractions (Fantke and Jolliet, 2016). The dynamiCROP model was evaluated against measured crop residues (Fantke et al., 2011a; Itoiz et al., 2012; Juraske et al., 2012; Feng et al., 2018), and applied to characterize human toxicity impacts from comparing different exposure pathways for about 900 pesticides, underlining crop residues as the predominant human exposure pathway for agricultural pesticides (Fantke and Jolliet, 2016). Based on these developments, exposure to crop residues has recently been introduced in the USEtox impact characterization model (Fantke et al., 2017b).

A major challenge for quantifying exposure to crop residues is the collection of crop degradation data, since degradation cannot be directly measured. Based on a wide set of reported dissipation data, where dissipation includes degradation and other transfer and removal processes in crops (Fantke and Juraske, 2013), an estimation model was developed to predict crop dissipation from pesticide properties, crop characteristics and environmental conditions (Fantke et al., 2014). This can be used as a proxy to derive degradation of agricultural pesticides in food crops and is available on http://half-lives.dynamicrop.org. For exposure pathways that rely on steady-state bioaccumulation processes (e.g. ingestion exposure associated with environmental emissions), related bioaccumulation factors can be found for various pesticides in Doucette et al. (2018).

4.5 The global human disease burden from pesticide exposure

The contribution of occupational and non-occupational pesticide exposure to the global human disease burden is almost certainly underestimated (Landrigan et al., 2018). A famous historical example of the negative health effects from pesticide exposure is dichloro-diphenyl-trichloroethane (known as DDT). This was widely used in agriculture as organochlorine insecticide until its worldwide ban in 2004 under the Stockholm Convention on Persistent Organic Pollutants (UNEP, 2004), and is known to be an endocrine disruptor, probably carcinogenic to humans, which affects the reproductive system. However, for many currently used pesticides, information on human dose-response, related background exposures and specific health effects is not available.

Available epidemiological and toxicological evidence for pesticides is summarized in several reviews (Sanborn et al., 2004, 2007; Bassil et al., 2007; Bolognesi, 2003), and is further detailed for selected health effects (Legler et al., 2015; Bellanger et al., 2015; Trasande et al., 2015; Bolognesi, 2003). Findings show that prenatal exposure to organophosphate pesticides in Europe alone causes the loss of 13 million IQ points and about 60000 cases of intellectual disability each year (Trasande et al., 2015). Considering that Europe makes only about 20% of the global pesticide consumption and that organophosphates are only one out of many pesticide chemical classes, a considerable human disease burden is attributable to pesticide exposure. This is supported by studies arriving at many thousands of disease cases annually due to exposure to pesticides including diseases linked to acute toxicity, carcinogenicity, reproductive and neurodevelopmental disorders, and endocrine disruption (Bourguet and Guillemaud, 2016; Fantke et al., 2012a). In an environmental sustainability assessment context, it is, however, important to separate disease burden from intentional exposure. This is because a large contributor to human disease burden from exposure to agricultural pesticides is via self-poisoning by direct pesticide ingestion, accounting for one-third of suicides globally, especially in rural Asia (Hvistendahl, 2013). These deaths are not directly attributable to any life cycle stage of pesticides and should therefore not be included in environmental sustainability assessments, where usually the disease burden from unintentional (i.e. avoidable) exposure is evaluated. Overall, characterizing human toxicity impacts associated with agricultural pesticides via all possible exposure settings is a key aspect for evaluating the environmental performance of agricultural practices, as well as for designing more sustainable pesticide formulations. Practitioners should therefore always seek to include all relevant pesticides in their assessment to avoid underestimation of related impacts on humans.

5 Addressing spatiotemporal variability

A general condition of environmental assessment studies is that – at least for processes of modelled technological background systems – emission locations and times are usually unknown or unspecific (Hauschild, 2005). However, whenever emission locations and times are known or if they are likely to contribute predominantly to the overall environmental performance profile of a studied system, accounting for spatiotemporal variability in both emission quantification and characterization modelling will be relevant.

5.1 *Aspects of spatial variability*

In the distribution of emissions and characterization of ecotoxicity and human toxicity impacts of agricultural pesticides, a regionalized assessment, that is considering geographical differences, is generally relevant, for example by using tropical species in the ecotoxicity effect assessment of agricultural pesticides in tropical regions. This concerns all environmental compartments (e.g. wind drift and long-range transport in air, or run-off from soil and subsequent mobility and transport in water networks) and organisms (e.g. ecosystems showing differences in their sensitivity to pesticides, or humans eating residues in crops after trade across national borders). For the quantification of pesticide emissions, at least during the field application, some degree of spatial differentiation is currently possible by using data on specific climate and soil regions (Dijkman et al., 2012; Fantke et al., 2017a). However, considering only the initial distribution of pesticides after application limits the role of spatial differences and only affects a few processes, such as the magnitude of wind drift (leading to air emissions) depending on differences in wind velocity and prevailing wind direction. This only leaves application scenario-specific aspects to be relevant, for example geographical differences in pesticide application (e.g. Norfaradila et al., 2014).

For characterizing ecotoxicity and human toxicity impacts of pesticides, existing impact assessment models are usually not based on site-specific data (e.g. Henderson et al., 2011; Rosenbaum et al., 2011). Especially for the effect assessment part, current impact characterization models rely on data from very few species, which have been widely tested (e.g. some temperate and subtropical or tropical species for ecotoxicity effects, or laboratory studies on rats and mice for extrapolation towards human toxicity effects). This is mainly because effect test data is mostly not available in a spatially differentiated way, which doesn't currently allow for the meaningful inclusion of spatial differentiation in related impact characterization modelling. However, it has been indicated that the sensitivity of tropical ecosystems towards several tested pesticides may be reasonably approximated by ecotoxicity data from commonly tested species (Daam and Van den Brink, 2010). Furthermore, regionalization at the level of continents and subcontinents as currently implemented in USEtox, already helps refine impact characterization results compared to using a global average if emission locations are known (Antón et al., 2014).

An overview of the variability of human toxicity impact characterization results across parameterized subcontinental regions implemented in USEtox (Kounina et al., 2014) is given in Fig. 6 for 16 common pesticides with different substance properties emitted to agricultural soil. From this example, it can be seen that variability in impact characterization results ranges over more than two orders of magnitude due to spatial differences between subcontinents. However, spatial variability is usually smaller for other emission compartments, since degradation is the main driver of variability between regions and is slowest in soil with no additional removal pathways considered. Furthermore, spatial variability is usually smaller than the variability across substances, ranging already over three orders of magnitude in Fig. 6 for only 16 pesticides emitted within the same considered region. An aspect where spatial differentiation has been assessed in impact modelling is the speciation and bioavailability of metal-based pesticides in water and soil environmental compartments (Tromson et al., 2017; Plouffe et al., 2015; Peña and Antón, 2017). In a study looking at copper-based fungicides applied to ~15 000 European vineyards, it was found that

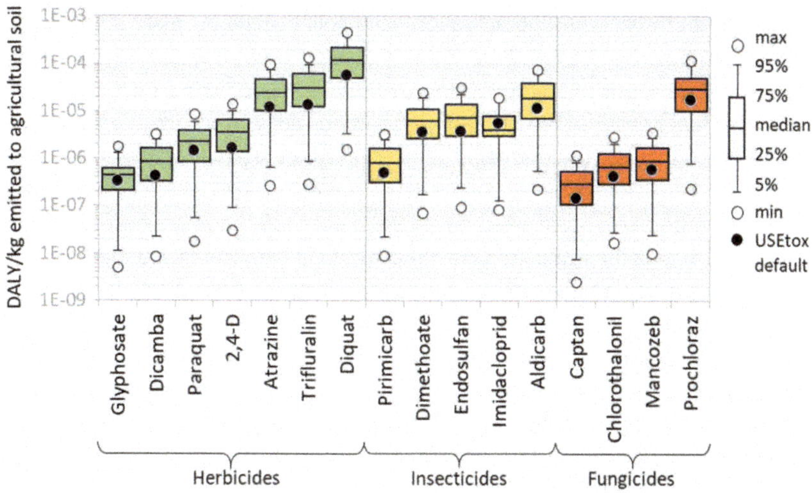

Figure 6 Variability in human toxicity impacts (expressed as disability-adjusted life years, DALY, per kg emitted to agricultural soil) across 16 sub-continental regions modelled with USEtox (http://usetox.org) for 16 common agricultural pesticides.

ecotoxicity impact characterization results vary up to two orders of magnitude as a function of spatial differences in soil parameters (Peña et al., 2017).

Overall, spatial variability can be the dominant driver of differences in emission and impact results. Given the large spatial variability of emissions and related impacts, spatialized models should hence be used whenever emission locations are known. If emission locations are unknown, it is recommended to use archetypes to represent aspects that drive variabilities in assessment results, such as field size, crop characteristics and environmental conditions.

5.2 Aspects of temporal variability

Considering the short time that needs to be considered between the application of agricultural pesticides and their almost instantaneous emission distribution, temporal dynamics are of minor relevance for quantifying related emissions. However, the temporal dimension of the considered agricultural fields, the selection of related pesticides and application methods, might vary by crop and location (Yang and Suh, 2015), These aspects are important to determine which emission distribution processes to consider in the emission inventory analysis (Rosenbaum et al., 2015). This can on the one hand just be to consider the (nearly) immediate distribution of the applied pesticide formulation within the first minutes after application, where usually only initial deposition and wind drift are relevant. On the other hand, this could also be extended to consider all subsequent processes when including several hours to days after application, for example subsurface run-off, leaching and preferential flow through the soil column, plant uptake through leaves and roots, and re-emission and volatilization from soil, water, and plant surfaces. For processes like deposition onto field soil, first-order processes are usually modelled, while wind drift requires an integration over the relevant time, which is reflected in the drift functions that are implemented in currently used emission inventory models for pesticides (Dijkman et al., 2012; Fantke et al., 2017a). As for the other life cycle stages, dynamic emission inventory modelling is possible by, for example following the approach proposed by Shimako et al. (2017) based on Tiruta-Barna et al. (2016), where the agricultural production cycle, application period and interval between pesticide applications are the most influential aspects.

As pesticides are usually applied in pulses (or rather a series of pulses per crop cycle) instead of being continuously released to agricultural fields or in greenhouses, some pathways in the impact characterization modelling cannot be modelled in a straightforward way by assuming steady-state conditions, since a steady-state might in some cases never be reached (Fantke et al., 2013; Shimako et al., 2017). The most relevant pathway in this respect is uptake of pesticides into food crops and subsequent exposure of humans to crop residues. For this pathway, the pesticide distribution dynamics in the crop-environment system are mainly a function of the time of application, crop growth and crop harvest time (Fantke et al., 2011a; Jacobsen et al., 2015). The dynamiCROP model addressing this particular pathway was developed to consider these distribution dynamics between pesticide application and crop harvest (Fantke et al., 2011b). An example of differences in the dynamics of crop residues is given in Fig. 7 for seven example pesticides that are commonly applied to field tomatoes. From this figure, it can be seen that residue dynamics vary a lot across pesticides in terms of both magnitude of the residues and time at which residues are highest for each pesticide. Nevertheless

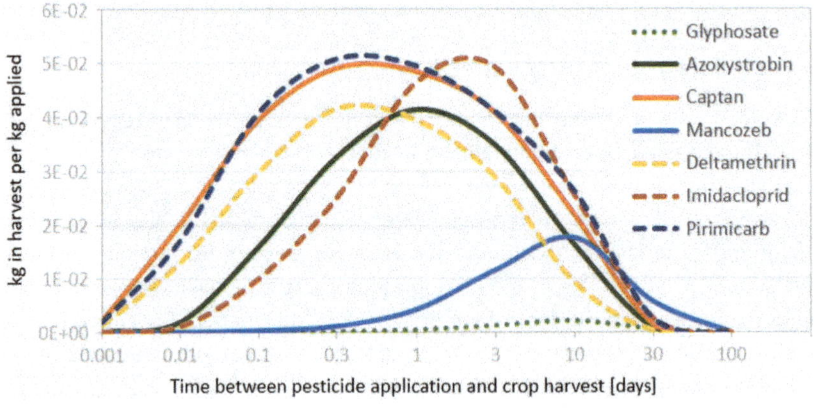

Figure 7 Differences in the tomato fruit residue dynamics of seven different pesticides (one herbicide: dotted line, three fungicides: solid lines and three insecticides: dashed lines) commonly applied to tomato crops under field conditions modelled with dynamiCROP (http://dynamicrop.org).

aligning the dynamiCROP model with the current steady-state toxicity impact characterization framework, it was parameterized as a linear model with the application-to-harvest duration as essential model input (Fantke et al., 2012b; Fantke and Jolliet, 2016). This example shows that generally, dynamics can be introduced when required for certain pathways or for the general pesticide fate model as proposed by, for example, Shimako et al. (2017). However, to increase transparency of the latter, it should be implemented as a full matrix-based solution for being consistently combined with existing matrix-based toxicity impact characterization models including USEtox (Rosenbaum et al., 2008), Impact2002 (Pennington et al., 2005), CalTOX (McKone and Enoch, 2002) or USES-LCA (van Zelm et al., 2009b), which are implemented in existing impact assessment methods.

Overall, system dynamics in the fate of pesticides will play the biggest role for metal-based pesticides that are applied in organic farming practice, since the metal component is usually entirely non-degradable. This does not mean that the metal component is infinitely mobile due to certain processes limiting its mobility over time, such as the fixation of reactive metal forms on soil particles (Crout et al., 2006). For the fate of certain metals, it has been shown that emission compartment and metal are the most important factors influencing the metal dynamics in the environment, which can reduce time-integrated freshwater ecotoxicity impacts by up to 75% for zinc-based emissions into agricultural soil considering a 100 year time horizon (Lebailly et al., 2014). It is hence recommended to consider dynamic fate at least for modelling metal-related impacts. In addition, dynamic fate models should be used for estimating impacts related to pesticide crop residues to capture the mass distribution dynamics between the time of pesticide application and crop harvest.

6 Substituting pesticides to optimize agricultural pesticide use

Different options exist to optimize the efficiency of using agricultural pesticides, including organic farming and integrated pest management (IPM) practices. These practices rely on a combination of techniques aimed at increasing overall sustainability of crop cultivation including crop rotation and mechanical or biological pest control. However, despite these options, chemical pesticides remain the predominant pest control mechanism for many farmers worldwide (Juraske et al., 2007b; Vasileiadis et al., 2017). In many countries, various pesticides are available (i.e. authorized and registered for use on specific agricultural crops) that can be used to fight one or more target pests. Pesticides all vary in their efficacy (how well they kill a target pest), application costs (of the marketed plant protection formulation containing one or more pesticide active ingredients), environmental degradation, bioaccumulation, and human toxicity and ecotoxicity potential. Therefore, there is huge potential to optimize the use of pesticides in terms of their overall costs and environmental sustainability.

Several multi-criteria decision frameworks exist and have been applied to qualitatively evaluate the sustainability of entire agricultural systems (e.g. Sadok et al., 2008; Angevin et al., 2017; Vasileiadis et al., 2017) or to evaluate risk–risk trade-offs when banning entire pesticide classes (e.g. Gray and Hammitt, 2000). In contrast, tools that focus on the substitution of individual pesticides, in support of optimizing pest control by quantitatively combining environmental sustainability and other decision support indicators relevant for farmers, are still rare. This is because the more than 100 indicators that are available for evaluating risks of pesticides (Surgan et al., 2010) are mostly applied in tools that are restricted to assessing pesticide hazards or toxicity (Labite et al., 2011) while often disregarding other aspects.

As an example, a screening-level approach was developed that combines application dose and frequency along with different hazard category indicators, such as persistence, in a single score per pesticide based on equal-weighting of all considered indicators (Juraske et al., 2007b). A similar ranking was proposed based on toxicity-related indicators but combining these with emission quantities, where it was shown that toxicity potential can be reduced up to 99% by defining adequate substitution scenarios (Fantke et al., 2011b). However, in such approaches, important indicators are missing, for example pesticide authorization, pest resistance, and application costs. All these indicators have been included in a recently released pesticide substitution tool, together with toxicity potentials and application dose and method (Steingrimsdottir et al., 2018). In this tool, pesticides effective against the same pests have been compared, indicators have been weighted according to differences in importance for farmers, and subsequently expressed in monetary values to arrive at a single quantitative score per pesticide. However, data on resistance and especially efficacy is often rarely available, which constitutes a challenge for effective pesticide substitution. Data on efficacy is usually only available to agricultural experts specialized in plant health, and is included in a technical–economic optimization tool for plant protection practices (Mghirbi et al., 2017), whereas resistance is not considered in this tool.

Overall, comparing different scoring tools shows that the ranking of pesticides strongly depends on the combination of chosen indicators (Juraske et al., 2007b). In a life cycle perspective, it is furthermore important to mention that manufacturing-related environmental impacts of agricultural pesticides can exceed impacts associated with emissions after field application, especially for application in greenhouses (Bojacá et al., 2012).

7 Case study

In this section, we evaluate in a real-life scenario selected pesticides in terms of emissions, associated human toxicity, and ecotoxicity according to available state-of-the-art emission quantification and life cycle impact characterization tools. With this case study, we aim to demonstrate how currently available data and models can be used to identify options for improving current cropping practices. An example question that we could seek to answer is 'Which of two fungicides and which of two insecticides applied to a certain crop over one year, show the least toxicity-related impacts on humans and freshwater ecosystems?'. With that, we can use the area of a single potato field as a reference unit to compare the performance of pesticides applied over one cropping season per year.

As an agricultural scenario, we chose potato as the most widely grown tuber crop worldwide. In our scenario, potatoes are grown on a humic cambisol as loamy and slightly acidic soil type according to the FAO/UNESCO Soil Map of the World. The environment was a humid continental climate with mild/cool summers according to the Köppen-Geiger climate classification, as found in central and northeastern regions within Asia, Europe and North America. Our defined square field of 9 ha area has a 1% average slope, is irrigated with 200 mm annually and has a 60 cm drainage depth with 55% being drained. Potatoes are planted after the last spring frost in mid-April and harvested at the end of August. As pesticides we select two fungicides (chlorothalonil and mancozeb) and two insecticides (methamidophos and carbofuran) commonly applied in several countries against water mould (*Phytophthora infestans*), and potato tuberworm (*Phthorimaea operculella*), respectively, as relevant potato pests. It is assumed that both fungicides and both insecticides are equally effective against their respective target pests. Hence, chlorothalonil and mancozeb are assessed as potential fungicidal substitutes for each other, while methamidophos and carbofuran are assessed as potential insecticidal substitutes for each other. For convenience, all pesticides are assumed to be applied at the end of May, that is at the end of the early crop growth stage. For the emission distribution, a land area distribution (outside the treated potato field) of 2% surface freshwater, 40% agricultural soil and 50% natural soil is assumed based on European statistics (http://ec.europa.eu/eurostat).

As the main data sources for defining the scenario, we used FAO, while information on pesticides was mainly derived from EFSA and the Pesticide Properties Database (see list of data portals in Section 9). As an emission quantification tool, we used the emission consensus model (Rosenbaum et al., 2015; Fantke et al., 2017a) that builds on PestLCI 2.0 (Dijkman et al., 2012). Finally, as impact characterization models, we used USEtox 2.1 (Rosenbaum et al., 2008) for general human toxicity and freshwater ecotoxicity impact characterization, dynamiCROP 3.07 (Fantke and Jolliet, 2016) for human toxicity impacts related to crop residues and the bystander exposure model from Ryberg et al. (2018) for impacts associated with exposure of residential bystanders. For estimating emission fractions into the environment, primary emission fractions (i.e.

considering only the initial distribution within minutes after pesticide application) were used to avoid overlaps with impact assessment models. This is because impact assessment models usually assume steady-state conditions, that is, impacts are time-integrated starting from the time of pesticide application. If longer periods (e.g. days or weeks) were considered when estimating emission fractions, this would overlap with the subsequent environmental fate modelling in the impact assessment. For estimating impacts, steady-state conditions were assumed for all pathways except for crop residues, which strongly depend on the time between pesticide application and crop harvest. An overview of data to define the case study scenario and related emission and impact characterization results is summarized in Table 3.

With treatments between 2 and 6 times per cropping season, the total annual pesticide amount applied to the potato field varies between 9 kg for carbofuran and 75.6 kg for mancozeb. Of the applied pesticide amount, only between 100 g and 700 g are emitted to freshwater bodies via direct deposition, while up to 18.2 kg and 20.4 kg are, respectively, emitted to natural soil and air. Most of the applied pesticide mass, however, is directly emitted to treated potato field soil, ranging from 47 kg for carbofuran to 389 kg for mancozeb, and to treated potato crop surface ranging from 30 kg for carbofuran to 250 kg for mancozeb.

Associated toxicity impacts of individual pesticides might be driven either by high emissions into the environment, or by high impacts per unit emission or, in the worst case, by a combination of both. This interaction is illustrated in Fig. 8, where for each emission compartment, impact scores for human toxicity (top) and freshwater ecotoxicity (bottom) are plotted on the right-side y-axes as a combination of impact characterization factors, that is, impacts per unit emission (left-side y-axes), and pesticide inventory results, that is, mass emitted over the entire cropping season per year (x-axes). Similar annual impacts across pesticides are represented along the shown diagonal equi-impact lines, where pesticides with the highest impacts are located in the top right corner of each plot. The large variation of more than seven orders of magnitude in human toxicity impacts, and more than three orders of magnitude in freshwater ecotoxicity impacts are hence attributable to the influence of variations in impact characterization and emission results. When, for example, looking at the two selected fungicides, we see that an emission of 0.5 kg chlorothalonil into freshwater with the highest ecotoxicity impact characterization factor among the selected pesticides of 5.7×10^5 PDF m^3 d/kg emitted, leads to 16 times higher freshwater ecotoxicity impacts than the slightly higher emission of 0.7 kg mancozeb, which has a 22 times lower impact characterization factor for the same emission compartment.

Overall, we can see from the results in Fig. 8 that emissions are highest in agricultural soil and crop surfaces across pesticides, while impact

Table 3 Application scenario, emission and impact characterization information of selected case study pesticides applied to potato with bold values indicating highest impact characterization result across selected pesticides per impact category and emission compartment

	Fungicides		Insecticides	
Parameter/result	Chlorothalonil	Mancozeb	Methamidophos	Carbofuran
CAS Registry Number	1897-45-6	8018-01-7	10265-92-6	1563-66-2
Target pest	Water mould (*Phytophthora infestans*)		Potato tuberworm (*Phthorimaea operculella*)	
Application dose [kg/ha][a,b]	1.5	1.4	1.0	0.5
Application count[b]	4 times	6 times	4 times	2 times
Emission fractions into compartments[c]				
Air			3.0%	
Agricultural soil[d,e]			57.1%	
Natural soil[d]			2.7%	
Surface freshwater[d]			0.1%	
Crop surface			37.1%	
Human toxicity impacts for emissions/exposure [DALY/kg][f]				
Air	3.6×10^{-6}	2.9×10^{-7}	$\mathbf{4.3 \times 10^{-5}}$	2.7×10^{-5}
Agricultural soil	4.1×10^{-7}	5.6×10^{-7}	2.0×10^{-5}	$\mathbf{2.5 \times 10^{-5}}$
Natural soil	1.7×10^{-7}	2.2×10^{-7}	$\mathbf{5.9 \times 10^{-6}}$	5.2×10^{-6}
Freshwater	1.0×10^{-5}	5.8×10^{-6}	8.2×10^{-5}	$\mathbf{1.0 \times 10^{-4}}$
Crop residues	6.3×10^{-7}	7.1×10^{-7}	$\mathbf{1.8 \times 10^{-5}}$	1.5×10^{-5}
Bystander exposure	9.3×10^{-12}	1.6×10^{-11}	$\mathbf{4.6 \times 10^{-10}}$	2.5×10^{-10}

(Continued)

Table 3 (*Continued*)

	Fungicides		Insecticides	
Parameter/result	Chlorothalonil	Mancozeb	Methamidophos	Carbofuran
Ecotoxicity impacts for emissions [PDF m^3 d/kg][g]				
Air	**2.0×10^4**	1.5×10^2	1.1×10^2	1.2×10^3
Agricultural soil	**4.1×10^3**	9.8×10^2	3.6×10^2	2.9×10^3
Natural soil[h]	**4.1×10^3**	9.8×10^2	3.6×10^2	2.9×10^3
Freshwater	**5.7×10^5**	2.6×10^4	5.0×10^3	5.6×10^4

[a] Calculated from formulation concentration of pesticide active ingredient taken from formulation product labels and recommended application rate of formulation product.
[b] Total mass applied per cropping season (i.e. per year, assuming a single cropping season per year) calculated from application dose and count according to Eq. 2 assuming a treated square field area of 9 ha for each application.
[c] Emission fractions calculated with the emission consensus model adapted and further developed from PestLCI 2.0 (Rosenbaum et al., 2015; Fantke et al., 2017a), http://qsa.man.dtu.dk/Dissemination/Pesticide-consensus, as input to arrive at emitted mass per year according to Eq. 3.
[d] Area distribution outside the treated field based on USEtox 2.1 and Eurostat with surface water (2%), agricultural soil (40%) and natural soil (50%), and the rest assumed to be built area.
[e] Emission fraction to soil includes both soil of the treated field and soil outside the treated field.
[f] Human toxicity-related impact characterization factors calculated according to Eq. 5 with USEtox 2.1 (Rosenbaum et al., 2008; Fantke et al., 2017b), http://usetox.org, with exposure to crop residues based on dynamiCROP (Fantke et al., 2011b; Fantke and Jolliet, 2016), http://dynamicrop.org, and bystander exposure based on the model of Ryberg et al. (2018).
[g] Freshwater ecotoxicity-related impact characterization factors calculated according to Eq. 4 with USEtox 2.1 (Rosenbaum et al., 2008; Fantke et al., 2017b), http://usetox.org.

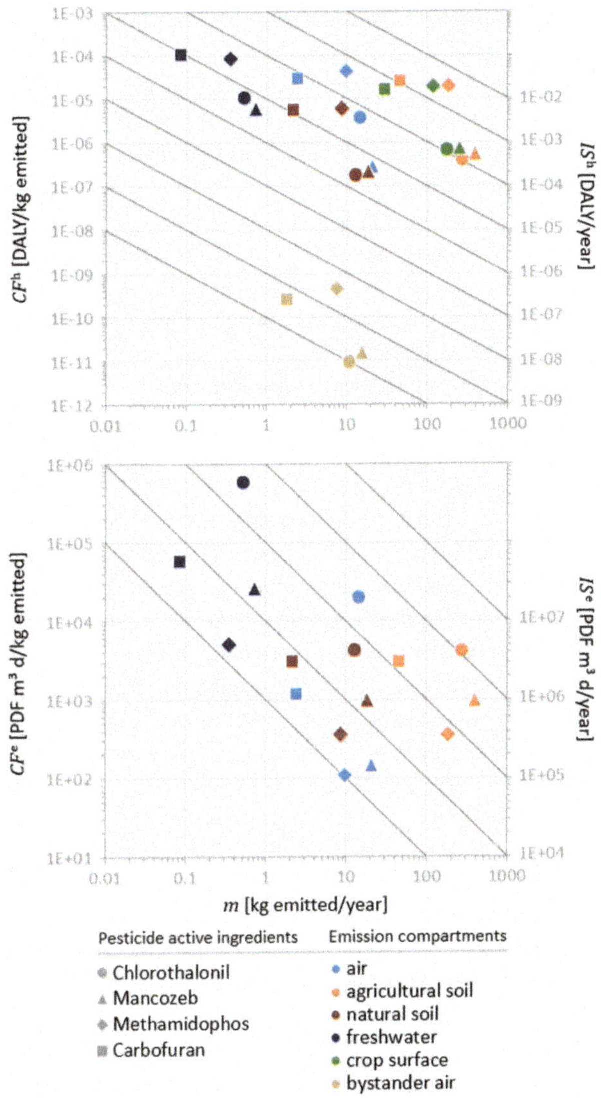

Figure 8 Impact characterization factors (left-side y-axis) for human toxicity [DALY/$kg_{emitted}$] and ecotoxicity [PDF m^3 d/$kg_{emitted}$] for emitted mass of selected pesticides applied to potato (x-axis) [$kg_{emitted}$/year] with diagonal equi-impact lines (right-side y-axis), respectively showing impact scores for human toxicity [DALY/year] and ecotoxicity [PDF m^3 d/year].

characterization factors are highest for emissions of the two insecticides into freshwater, closely followed by emissions to air, agricultural soil and crop surfaces. Related human toxicity impacts are highest for the insecticide methamidophos with 3.6×10^{-3} DALY/year for emissions to agricultural soil, and 2.2×10^{-3} DALY/year for exposure to crop residues, corresponding to an annual loss of population-level lifetime of, respectively, 1.3 and 0.8 days. At this point, it should be noted that in an environmental assessments context where land use impacts are considered together with human toxicity and/or ecotoxicity, impacts associated with direct emissions to agricultural soil might already be covered by land use impacts (e.g. De Schryver et al., 2010). In our case study, impacts via bystander exposure are several orders of magnitude lower. However, only a handful of bystanders are exposed at much higher levels compared to the general population, which might be relevant when minimizing especially high exposure levels.

For ecotoxicity, only impacts on freshwater ecosystems were considered. Impacts on freshwater ecosystems are dominated by the fungicide chlorothalonil, with 1.1×10^{6} PDF m^3 d/year for emissions to agricultural soil. However, considering the uncertainties in the emission model of at least a factor of 5 (Webb et al., 2016) and in the impact characterization models of typically around a factor of 100 (Rosenbaum et al., 2008; Fantke and Jolliet, 2016) this does not allow for a clear conclusion about any preferable pesticide in each considered target class. Consequently, either additional alternatives should be assessed, or the data and assumptions should be refined for the selected pesticides with focus on improving degradation, bioaccumulation and toxicity/ecotoxicity information. However, the present case study nevertheless indicates trends and highlights dominant pathways and therefore helps to focus further scenario and data refinement efforts to minimize related uncertainty in impact assessment results.

8 Future trends and conclusion

From the previous chapters, it is clear that considerable efforts have been made to develop frameworks for quantifying emissions and characterizing human toxicity and ecotoxicity impacts of agricultural pesticides. These have applications in environmental impact assessment, LCA and pesticide substitution assessment. As far as possible, these frameworks are science based and consistent with the general principles and conditions of comparative environmental assessment frameworks. Wherever aspects had to be addressed in emission and impact modelling that required analysis beyond current data, international consensus efforts were established to strive for consistency, harmonization and global agreement in the environmental assessment research and practitioner community (Hauschild et al., 2008; Westh et al., 2015; Rosenbaum et al., 2015; Fantke et al., 2017a). However, from environmental assessment practice, it is also clear that despite all these efforts, the existing data and models available for evaluating agricultural pesticides will have to be further improved and extended. Future efforts should therefore focus on addressing high uncertainties, cover missing compartments, organisms, pathways and toxicity-related effects, and provide additional data to avoid the (often unjustifiable) extrapolation from one pesticide with similar substance properties but different target organisms and effect mechanisms.

The Life Cycle Initiative hosted at UN Environment has established a project on 'Global Guidance for Life Cycle Impact Assessment Indicators', where task forces have been established to harmonize impact characterization modelling, and provide improved impact characterization results for – among other impact categories – human toxicity and ecotoxicity impacts associated with chemical substances including pesticides (Fantke et al., 2018a,b). Areas of focus for these task forces include introducing occupational and consumer exposure, analyzing essential metals, and improving dose-response modelling for human toxicity, as well as improving the consideration of ecosystems beyond freshwater organisms, effect modelling and substance data for ecotoxicity. Task force progress is discussed and recommendations provided in dedicated Pellston expert workshops. More specifically targeting the assessment of agricultural pesticides in LCA, the French Environmental Protection Agency (ADEME) has funded an international consortium to work over the coming years on a project for 'Operationalizing Life Cycle Assessment of Pesticides (OLCA-Pest)'. Efforts in this project build on current advances to operationalize and harmonize emission quantification and toxicity impact characterization of pesticides, including the evaluation in environmental assessment case studies. Initial results address exposure of bystanders near agricultural fields in LCA (Ryberg et al., 2018), the relevance of spatial differentiation in the emission and ecotoxicity modelling of metal-based pesticides (Peña et al., 2017), and provide data and indicators to define scenarios for the application and substitution of agricultural pesticides (Steingrimsdottir et al., 2018). Other efforts focusing on improving the environmental assessment of pesticides may exist.

Aspects currently missing or insufficiently included in environmental assessment, particularly in LCA, for agricultural pesticides along with recommendations of how to address these aspects in order to avoid underestimating related impacts in environmental assessment practice are given in Table 4. Addressing these aspects will help to identify best-in-class options working towards more sustainable agriculture, with a focus on reducing the overall load of synthetic pesticides, in the environment. Moreover, relating chemical-related impacts to absolute environmental sustainability limits is an emerging science field, which will go beyond identifying more sustainable solutions toward identifying solutions that are in fact sustainable enough (Fantke and Illner, 2019). Overall, improved assessment of impacts from pesticides encourages the development and practice of alternative farming systems, such as organic farming and IPM, which have shown that toxicity-related impacts associated with synthetic pesticides can be reduced up to 90% (e.g. Alaphilippe et al., 2013; Juraske and Sanjuán, 2011) while at the same time considering high crop yield and food safety.

Table 4 Recommendations for potential ways to address main aspects currently missing or insufficiently included in environmental assessment for agricultural pesticides to avoid underestimation of related impacts

Aspect	Affected LCA phase	How to address this aspect
Organic and inorganic pesticides used in conventional and organic farming practices that are currently not covered	LCI: current emission quantification models need to be adapted to assess metal-based inorganic pesticides (e.g. copper sulphate), non-metal inorganic pesticides (e.g. sulphur), biochemical pesticides (e.g. neem oil), microbial pesticides (e.g. *Bacillus thuringiensis*), plant-incorporated protectants in genetically modified crops and potential transformation products (metabolites) LCIA: current toxicity impact characterization models need to be adapted to assess non-metal inorganic pesticides (e.g. sulphur), biochemical pesticides (e.g. neem oil), microbial pesticides (e.g. *Bacillus thuringiensis*), plant-incorporated protectants in genetically modified crops and potential transformation products (metabolites)	Metal-based inorganic pesticides: including speciation modelling in soil and adapting environmental processes for metals based on existing models (Peña et al., 2016) Non-metal inorganic pesticides: developing fate processes based on considering relevant chemical reaction pathways (Müller and Fantke, 2017) Biochemical pesticides: building on process modelling for synthetic pesticides, but considering short-term kinetics and degradation (Chandler et al., 2011) Microbial pesticides: developing process-based models considering distribution and growth of bacteria and their interaction mechanisms with chemicals (Chandler et al., 2011) Transformation products: considering effects of transformation products in addition to the parent substance, that is directly applied pesticides (van Zelm et al., 2010), or characterizing transformation products as additional but separate substances combined with their respective transformation fraction in the environment

Aspect	Affected LCA phase	How to address this aspect
Ecosystems potentially exposed to agricultural pesticides applied in different farming practices that are currently not considered	LCI: current emission quantification models need to be adapted to assess emission fractions reaching birds, pollinators, terrestrial mammals and marine aquatic organisms LCIA: current toxicity impact characterization models need to be adapted to assess exposure and ecotoxicological effects for soil terrestrial organisms, birds, pollinators, terrestrial mammals, groundwater organisms, sediment-dwelling organisms and marine aquatic organisms	Terrestrial soil organisms: building on first approaches available for bioavailability, exposure and effects of metals (e.g. Owsianiak et al., 2015) Terrestrial birds and mammals: following and extending initial approaches for modelling the impact pathway for warm-blooded predators (Golsteijn et al., 2012) Pollinators: developing quantitative models covering the impact pathway of pollinators combining dose-related effect information with field exposure modelling of relevant insect species (Crenna et al., 2017) Groundwater organisms: incorporating a groundwater compartment based on relevant processes described in part by Geisler et al. (2004) and effect and sensitivity data for groundwater organisms where it is not feasible to extrapolate from freshwater species (Hose, 2005) Sediment-dwelling organisms: incorporating a sediment compartment and effect and sensitivity data for sediment-dwelling organisms based on initial work for metals (Pu et al., 2017) Marine aquatic organisms: adapting existing models to include exposure and effect information for marine species based on approaches developed for metals (Dong et al., 2016, 2017)

(*Continued*)

Table 4 (*Continued*)

Aspect	Affected LCA phase	How to address this aspect
Human exposure settings and routes potentially relevant for quantifying exposure to agricultural pesticides in different farming practices that are currently not operational	LCI: current emission quantification models need to be adapted to assess emissions leading to occupational exposure of agricultural field workers including exposure of worker family members, bystander exposure and crop residue exposure LCIA: current toxicity impact characterization models need to be adapted to assess occupational exposure of agricultural field workers including exposure of worker family members, bystander exposure including residential bystander exposure, and to assess dermal exposure of workers, bystanders and the general population	Occupational exposure: adapting models from occupational risk assessment in the way it was done for bystander exposure (Ryberg et al., 2018) Bystander exposure: further extending the initial approach for bystander exposure to pesticides applied to potato fields (Ryberg et al., 2018) Crop residue exposure: combining emissions to agricultural crops with respective impact characterization factors implemented in USEtox based on the approach developed by Fantke and Jolliet (2016) Dermal exposure of different population groups: developing exposure pathways and combining them with dermal effect data based on initial approach developed for exposure to cosmetic products (Ernstoff et al., 2016)
Alignment between emission and impact quantification models that are currently partly overlapping in their scope for agricultural pesticides applied in different farming practices	LCI: current emission quantification models need to be adapted to avoid any potential overlap with the environmental fate part of agricultural pesticides in impact characterization models LCIA: current toxicity impact characterization models need to be adapted to avoid any potential overlap with the emission distribution part of agricultural pesticides in emission quantification models	Overlap between emission distribution and environmental fate: aligning temporal and environmental process modelling scope between emission and toxicity impact characterization models based on initial recommendations from consensus building efforts (Rosenbaum et al., 2015; van Zelm et al., 2014)

Aspect	Affected LCA phase	How to address this aspect
Trade-offs between all life cycle stages of agricultural pesticides requiring aligned emission and impact quantification	LCI: current emission quantification models need to be adapted to comprehensively assess differences in resources use and chemical emissions associated with differences in the complexity of pesticide synthesis and manufacturing, and pesticide efficacy and specificity during pesticide application	Trade-offs between pesticide life cycle stages: increasing level of detail for environmental inputs and outputs related to pesticide synthesis based on initial work on industrial chemicals (Jimenez-Gonzalez and Overcash, 2014; Overcash, 2016), and complementing existing impact characterization models with all relevant impact pathways including occupational exposure based on work for worker exposure in different industrial sectors (Kijko et al., 2015)
	LCIA: current toxicity impact characterization models need to be adapted to comprehensively assess human toxicity impacts from occupational exposure across pesticide life cycles, and ecotoxicity and human toxicity impacts from emissions along entire pesticide life cycle	

These recommendations concern particularly LCA and its respective assessment phases, that is, life cycle emission inventory (LCI) and impact assessment (LCIA) modelling.

9 Acknowledgements

This work was financially supported by the Marie Curie project Quan-Tox (GA no. 631910) funded by the European Commission under the Seventh Framework Program, and by the OLCA-Pest project funded by the French Environment and Energy Management Agency, ADEME (GA no. 17-03-C0025).

10 Where to look for further information

Several working groups, numerous scientific and non-scientific literature and a wide range of practical guidelines, data portals and regulatory information exist that focus on one or more aspects related to pesticides, their production, their application and use in agriculture, their economic costs and benefits, and their diverse impacts on ecosystems and human health. In the present section, selected sources help to provide an overview, additional information, and in-depth reading material, for environmental assessment method developers, practitioners, agronomists and both interested experts and non-experts.

10.1 Introducing the topic

A good introduction to agricultural pesticides and their interaction with the environment including humans, is given in Matthews (2016). A more systematic introduction to different aspects around pesticides for non-experts can be found on the thematic sitemap of FAO's Plant Production and Protection Division (http://fao.org/agriculture/crops/thematic-sitemap/theme/pests) or the Organisation for Economic Co-operation and Development, OECD (http://oecd.org/env/ehs/pesticides-biocides). The most prominent international working group focusing on a wide range of pesticide-related aspects is the Pesticide Action Network (http://pan-international.org).

10.2 Academic journals and seminal articles

For those looking to gain a deeper understanding of how to assess the different aspects around application, emissions and impacts of agricultural pesticides, there is a vast body of literature available in both peer-reviewed journals as well as reference books and reports. Journals where studies related to pesticides and their interaction in the agricultural and environmental system can be found include top journals such as *Science* and *Nature*, publishing either cutting-edge topics like the current global debate around pesticides and their impacts on pollinators (e.g. Kessler et al., 2015; Woodcock et al., 2017), or advances in pesticide science of broad interest (e.g. Fenner et al., 2013). Most studies are, however, published in more specialist journals including but not limited to *Pest Management Science*, *Environmental Science and Technology*, *Japanese Journal of Pesticide Science*, *Environmental Toxicology and Chemistry*, *Chemosphere*, *Ecotoxicology and Environmental Safety*, *Science of the Total Environment*, *Crop Protection*, *Environmental Pollution*, *Journal of Environmental Management* and many more. Seminal pesticide journal articles are related to outlining impact pathways (Kovach et al., 1992), studying pesticide emissions to air (van den Berg et al., 1999), uncertainty sources in pesticide modelling (Dubus et al., 2003), addressing pesticide mixture effects (Rizzati et al., 2016), estimating damage costs of pesticide use in Europe (Fantke et al., 2012a) and societal costs of pesticide use (Bourguet and Guillemaud, 2016; Fantke et al., 2012a).

10.3 Reference books

Useful reference books providing additional in-depth knowledge around pesticides and the environment are available for application data (Turner, 2015), field application methods including application of bio-pesticides (Matthews et al., 2014), pesticide environmental fate modelling principles and processes (Richter et al., 1996), pesticide toxicology (Krieger, 2010) and uncertainty analysis in assessing pesticides (Warren-Hicks and Hart, 2010). Several open access books around pesticides, their agricultural use and their environmental effects can be found in INTECH's 'Agricultural and Biological Sciences' section (http://intechopen.com/books/subject/agricultural-and-biological-sciences).

10.4 Regular events

Dedicated conferences or workshops on pesticides and the environment are rare and include the 'International Conference on Pesticide Behaviour in Soils, Water and Air' and the 'Annual Symposium on the Sustainable Use Directive on Pesticides'. Instead, while other pesticide-related conferences rather focus on pesticide chemistry and industry workshops, aspects on pesticides in agricultural and environmental contexts are often integrated into conferences either focusing on agronomy (e.g. the 'International Conference on Agriculture & Horticulture') or environmental issues (e.g. the 'Society of Environmental Toxicology and Chemistry Annual Meeting'). An event related to pesticides and LCA is, for example the biennial 'International Conference on Life Cycle Assessment of Food'.

10.5 Data portals

Relevant data for modelling and assessing pesticides in an environmental assessment context are the EU Pesticide Database (http://ec.europa.eu/food/plant/pesticides/eu-pesticides-database), the US-EPA Office of Pesticide Programs database (http://iaspub.epa.gov/apex/pesticides/f?p=chemicalsearch:1), the pest and pesticide management portal of the Plant Production and Protection Division (AGP) of the Food and Agriculture Organization (FAO) of the United Nations (http://www.fao.org/agriculture/crops/thematic-sitemap/theme/pests), national websites on pesticides (http://oecd.org/chemicalsafety/pesticides-biocides/wwwpesticidesitesinoecdcountriesandotherorganisations.htm), the Pesticide Properties Database (http://sitem.herts.ac.uk/aeru/ppdb) and reports and databases published by EFSA (http://efsa.europa.eu).

10.6 Key research centres

Regarding information on residues and other environmental aspects of pesticides, the FAO/WHO Panel of Experts on Pesticide Management is a key research group. In the context of environmental and life cycle assessment, key research groups focusing on model and data development are the Quantitative Sustainability Assessment Division at the Technical University of Denmark (http://qsa.man.dtu.dk), and the Environmental Science group at the Radboud University Nijmegen (http://ru.nl/environmentalscience). Key environmental assessment research groups focusing on developing and application of methods and tools in practical case studies and measurement activities are, for example, the Institute for Food and Agricultural Research and Technology in Spain (http://www.irta.cat/en/), and the Agricultural Research Center at the National Agriculture and Food Research Organization in Japan (http://www.naro.affrc.go.jp/english/).

11 References

Alaphilippe, A., Simon, S., Brun, L., et al. (2013). Life cycle analysis reveals higher agro-ecological benefits of organic and low-input apple production. *Agronomy for Sustainable Development* 33: 581–92.

Angevin, F., Fortino, G., Bockstaller, C., et al. (2017). Assessing the sustainability of crop production systems: Toward a common framework? *Crop Protection* 97: 18–27.

Antón, A., Castells, F., Montero, J. I., et al. (2004). Comparison of toxicological impacts of integrated and chemical pest management in Mediterranean greenhouses. *Chemosphere* 54: 1225–35.

Antón, A., Torrellas, M., Núñez, M., et al. (2014). Improvement of agricultural life cycle assessment studies through spatial differentiation and new impact categories: Case study on greenhouse tomato production. *Environmental Science and Technology* 48: 9454–62.

Arena, M. and Sgolastra, F. (2014). A meta-analysis comparing the sensitivity of bees to pesticides. *Ecotoxicology* 23: 324–34.

Audsley, E., Alber, S., Clift, R., et al. (1997). Harmonisation of environmental life cycle assessment for agriculture. *Final Report, Concerted Action AIR3 CT94 2028.* European Commission, DG VI Agriculture, p. 139.

Barmaz, S., Croce, V., Marroncelli, S., et al. (2011). Data entry of pesticide ecotoxicology Tier 1 study endpoints in an XML schema - database. *EFSA Supporting Publications* 8: 155E.

Bassil, K. L., Vakil, C., Sanborn, M., et al. (2007). Cancer health effects of pesticides: Systematic review. *Canadian Family Physician* 53: 1704–11.

Bellanger, M., Demeneix, B., Grandjean, P., et al. (2015). Neurobehavioral deficits, diseases, and associated costs of exposure to endocrine-disrupting chemicals in the European Union. *The Journal of Clinical Endocrinology and Metabolism* 100: 1256–66.

Bennett, D. H., McKone, T. E., Evans, J. S., et al. (2002). Defining intake fraction. *Environmental Science and Technology* 36: 207A–211A.

Berthoud, A., Maupu, P., Huet, C., et al. (2011). Assessing freshwater ecotoxicity of agricultural products in life cycle assessment (LCA): A case study of wheat using French agricultural practices databases and USEtox model. *The International Journal of Life Cycle Assessment* 16: 841–7.

Bessou, C., Basset-Mens, C., Latunussa, C., et al. (2016). Partial modelling of the perennial crop cycle misleads LCA results in two contrasted case studies. *The International Journal of Life Cycle Assessment* 21: 297–310.

Birkved, M. and Hauschild, M. Z. (2006). PestLCI - A model for estimating field emissions of pesticides in agricultural LCA. *Ecological Modelling* 198: 433–51.

Bojacá, C. R., Gil, R., Casilimas, H., et al. (2012). Modelling the environmental impact of pesticides sprayed on greenhouse tomatoes: A regional case study in Colombia. *Acta Horticulturae* 957: 61–8.

Bolognesi, C. (2003). Genotoxicity of pesticides: A review of human biomonitoring studies. *Reviews in Mutation Research* 543: 251–72.

Boulard, T., Brun, R., Hayer, F., et al. (2012). Some results on environmental and human toxicity involved by greenhouse tomato production in France. *Acta Horticulturae* 952: 367–76.

Bourguet, D. and Guillemaud, T. (2016). The hidden and external costs of pesticide use. In: Lichtfouse, E. (ed). *Sustainable Agriculture Reviews: Volume 19*. Cham, Switzerland: Springer Press, pp. 35–120.

Butler Ellis, M. C., Underwood, B., Peirce, M. J., et al. (2010). Modelling the dispersion of volatilised pesticides in air after application for the assessment of resident and bystander exposure. *Biosystems Engineering* 107: 149–54.

Chandler, D., Bailey, A. S., Tatchell, G. M., et al. (2011). The development, regulation and use of biopesticides for integrated pest management. *Philosophical Transactions of the Royal Society* 366: 1987–98.

Cooper, J. S. (2015). *LCA Digital Commons Unit Process Data field Operations Work Processes and Farm Implements*. Washington DC: United States Department of Agriculture, p. 37.

Cooper, J. and Dobson, H. (2007). The benefits of pesticides to mankind and the environment. *Crop Protection* 26: 1337–48.

Crenna, E., Sala, S., Polce, C., et al. (2017). Pollinators in life cycle assessment: Towards a framework for impact assessment. *Journal of Cleaner Production* 140: 525–36.

Crout, N. M. J., Tye, A. M., Zhang, H., et al. (2006). Kinetics of metal fixation in soils: Measurement and modeling by isotopic dilution. *Environmental Toxicology and Chemistry* 25: 659–63.

Daam, M. A. and Van den Brink, P. J. (2010). Implications of differences between temperate and tropical freshwater ecosystems for the ecological risk assessment of pesticides. *Ecotoxicology* 19: 24-37.

De Schryver, A. M., van Zelm, R., Huijbregts, M. A. J., et al. (2010). Addressing land use and ecotoxicological impacts in life cycle assessments of food production technologies. In: Sonesson, U., Berlin, J. and Ziegler, F. (Eds), *Environmental Assessment and Management in the Food Industry*. Oxford, UK: Woodhead Publishing, pp. 177-206.

Dijkman, T. J., Birkved, M. and Hauschild, M. Z. (2012). PestLCI 2.0: A second generation model for estimating emissions of pesticides from arable land in LCA. *The International Journal of Life Cycle Assessment* 17: 973-86.

Dong, Y., Gandhi, N. and Hauschild, M. Z. (2014). Development of comparative toxicity potentials of 14 cationic metals in freshwater. *Chemosphere* 112: 26-33.

Dong, Y., Rosenbaum, R. K. and Hauschild, M. Z. (2016). Assessment of metal toxicity in marine ecosystems: Comparative Toxicity Potentials for nine cationic metals in coastal seawater. *Environmental Science and Technology* 50: 269-78.

Dong, Y., Rosenbaum, R. K. and Hauschild, M. Z. (2017). Metal toxicity characterization factors for marine ecosystems - considering the importance of the estuary for freshwater emissions. *The International Journal of Life Cycle Assessment*. doi:10.1007/s11367-017-1376-x.

Dorne, J. L., Richardson, J., Kass, G., et al. (2017). OpenFoodTox: EFSA's open source toxicological database on chemical hazards in food and feed. *EFSA Journal* 15: e15011.

Doucette, W., Shunthirasingham, C., Dettenmaier, E., et al. (2018). A review of measured bioaccumulation data in terrestrial plants for organic chemicals: Metrics, variability and the need for standardized measurement protocols. *Environmental Toxicology and Chemistry* 37: 21-33.

Dubus, I. G., Brown, C. D. and Beulke, S. (2003). Sources of uncertainty in pesticide fate modelling. *Science of the Total Environment* 317: 53-72.

EC. (2009). Regulation (EC) No 1107/2009 of the European Parliament and of the Council of 21 October 2009 concerning the placing of plant protection products on the market and repealing Council Directives 79/117/EEC and 91/414/EEC. European Commission, Brussels.

Ecobichon, D. J. (2001). Pesticide use in developing countries. *Toxicology* 160: 27-33.

EFSA. (2012). Compilation of a database on ecotoxicological properties of active substances and plant protection products. *EFSA Supporting Publications* 9: 364E.

Elliott, J. E., Wilson, L. K., Langelier, K. M., et al. (1997). Secondary poisoning of birds of prey by the organophosphorus insecticide, phorate. *Ecotoxicology* 6: 219-31.

Ernstoff, A. S., Fantke, P., Csiszar, S. A., et al. (2016). Multi-pathway exposure modelling of chemicals in cosmetics with application to shampoo. *Environment International* 92-93: 87-96.

Fantke, P. and Jolliet, O. (2016). Life cycle human health impacts of 875 pesticides. *The International Journal of Life Cycle Assessment* 21: 722-33.

Fantke, P. and Juraske, R. (2013). Variability of pesticide dissipation half-lives in plants. *Environmental Science and Technology* 47: 3548-62.

Fantke, P., Charles, R., de Alencastro, L. F., et al. (2011a). Plant uptake of pesticides and human health: Dynamic modeling of residues in wheat and ingestion intake. *Chemosphere* 85: 1639-47.

Fantke, P., Juraske, R., Antón, A., et al. (2011b). Dynamic multicrop model to characterize impacts of pesticides in food. *Environmental Science and Technology* 45: 8842-9.

Fantke, P., Friedrich, R. and Jolliet, O. (2012a). Health impact and damage cost assessment of pesticides in Europe. *Environment International* 49: 9-17.

Fantke, P., Wieland, P., Juraske, R., et al. (2012b). Parameterization models for pesticide exposure via crop consumption. *Environmental Science and Technology* 46: 12864-72.

Fantke, P., Wieland, P., Wannaz, C., et al. (2013). Dynamics of pesticide uptake into plants: From system functioning to parsimonious modeling. *Environmental Modelling and Software* 40: 316-24.

Fantke, P., Gillespie, B., Juraske, R., et al. (2014). Estimating half-lives for pesticide dissipation from plants. *Environmental Science and Technology* 48: 8588-602.

Fantke, P., Antón, A., Grant, T., et al. (2017a). Pesticide emission quantification for life cycle assessment: A global consensus building process. *Journal of Life Cycle Assessment, Japan* 13: 245-51.

Fantke, P., Bijster, M., Guignard, C., et al. (2017b). USEtox® 2.0 Documentation (Version 1). Kgs. Lyngby, Denmark: USEtox® Team, p. 208.

Fantke, P., Aurisano, N., Backhaus, T., et al. (2018a). Toward harmonizing ecotoxicity characterization in life cycle impact assessment. *Environmental Toxicology and Chemistry* 37: 2955-71.

Fantke, P., Aylward, L., Bare, J., et al. (2018b). Advancements in life cycle human exposure and toxicity characterization. *Environmental Health Perspectives* 126: 125001.

Fantke, P.. and Illner, N. (2019). Goods that are good enough: Introducing an absolute sustainability perspective for managing chemicals in consumer products. *Current Opinion in Green and Sustainable Chemistry* 15: 91-97.

FAO. (2014). *The International Code of Conduct on Pesticide Management*. Rome, Italy: Food and Agriculture Organization of the United Nations.

Feng, X., Wang, K., Pan, L., Xu, T., Zhang, H., and Fantke, P., (2018). Measured and modeled residue dynamics of famoxadone and oxathiapiprolin in tomato fields. *Journal of Agricultural and Food Chemistry*, doi:10.1021/acs.jafc.8b02056.

Fenner, K., Canonica, S., Wackett, L. P., et al. (2013). Evaluating pesticide degradation in the environment: Blind spots and emerging opportunities. *Science* 341: 752-8.

Garron, C. A., Davis, K. C. and Ernst, W. R. (2009). Near-field air concentrations of pesticides in potato agriculture in Prince Edward Island. *Pest Management Science* 65: 688-96.

Garron, C., Ernst, B., Julien, G., et al. (2012). Concentrations and environmental risk of chlorothalonil in air near potato fields in Prince Edward Island, Canada. *Pest Management Science* 68: 92-100.

Geisler, G., Hellweg, S., Liechti, S., et al. (2004). Variability assessment of groundwater exposure to pesticides and its consideration in life-cycle assessment. *Environmental Science and Technology* 38: 4457-64.

Geisler, G., Hellweg, S., Hofstetter, T. B., et al. (2005). Life-cycle assessment in pesticide product development: Methods and case study on two plant-growth regulators from different product generations. *Environmental Science and Technology* 39: 2406-13.

Golsteijn, L., van Zelm, R., Veltman, K., et al. (2012). Including ecotoxic impacts on warm-blooded predators in life cycle impact assessment. *Integrated Environmental Assessment and Management* 8: 372-8.

Götz, R., Sokollek, V. and Weber, R. (2012). The dioxin/POPs legacy of pesticide production in Hamburg: Part 2 - waste deposits and remediation of Georgswerder landfill. *Environmental Science and Pollution Research* 20: 1925-36.

Gray, G. M. and Hammitt, J. K. (2000). Risk/risk trade-offs in pesticide regulation: An exploratory analysis of the public health effects of a ban on organophosphate and carbamate pesticides. *Risk Analysis* 20: 665–80.

Hauschild, M. (2000). Estimating pesticide emissions for LCA of agricultural products. In: Weidema, B. P. and Meeusen, M. J. G. (Eds), *Agricultural Data for Life Cycle Assessments*. The Hague, the Netherlands: Agricultural Economics Research Institute, pp. 70–85.

Hauschild, M. Z. (2005). Assessing environmental impacts in a life-cycle perspective. *Environmental Science and Technology* 39: 81A–88A.

Hauschild, M. Z., Huijbregts, M. A. J. and Jolliet, O., et al. (2008). Building a model based on scientific consensus for life cycle impact assessment of chemicals: The search for harmony and parsimony. *Environmental Science and Technology* 42: 7032–7.

Hayashi, K., Shobatake, K., Makino, N., et al. (2012). Development of a life cycle inventory database for agricultural systems in Japan: The JALCA database. *10th Biennial International Conference on EcoBalance*, 20–23 November 2012. Yokohama, Japan.

Henderson, A. D., Hauschild, M. Z., van de Meent, D., et al. (2011). USEtox fate and ecotoxicity factors for comparative assessment of toxic emissions in life cycle analysis: Sensitivity to key chemical properties. *The International Journal of Life Cycle Assessment* 16: 701–9.

Holt, E., Von der Recke, R., Vetter, W., et al. (2008). Assessing dioxin precursors in pesticide formulations and environmental samples as a source of octachlorodibenzo-*p*-dioxin in soil and sediment. *Environmental Science and Technology* 42: 1472–8.

Hose, G. C. (2005). Assessing the need for groundwater quality guidelines for pesticides using the species sensitivity distribution approach. *Human and Ecological Risk Assessment* 11: 951–66.

Huang, L., Ernstoff, A., Fantke, P., et al. (2017). A review of models for near-field exposure pathways of chemicals in consumer products. *Science of the Total Environment* 574: 1182–208.

Huijbregts, M. A. J., Rombouts, L. J. A., Ragas, A. M. J., et al. (2005). Human-toxicological effect and damage factors of carcinogenic and noncarcinogenic chemicals for life cycle impact assessment. *Integrated Environmental Assessment and Management* 1: 181–244.

Humbert, S., Margni, M. D., Charles, R., et al. (2007). Toxicity assessment of the main pesticides used in Costa Rica. *Agriculture, Ecosystems and Environment* 118: 183–90.

Hvistendahl, M. (2013). In rural Asia, locking up poisons to prevent suicides. *Science* 341: 738–9.

Itoiz, E. S., Fantke, P., Juraske, R., et al. (2012). Deposition and residues of azoxystrobin and imidacloprid on greenhouse lettuce with implications for human consumption. *Chemosphere* 89: 1034–41.

Jacobsen, R. E., Fantke, P. and Trapp, S. (2015). Analysing half-lives for pesticide dissipation in plants. *SAR and QSAR in Environmental Research* 26: 325–42.

Jimenez-Gonzalez, C. and Overcash, M. R. (2014). The evolution of life cycle assessment in pharmaceutical and chemical applications - a perspective. *Green Chemistry* 16: 3392–400.

Jolliet, O. and Fantke, P. (2015). Human toxicity. In: Hauschild, M. and Huijbregts, M. A. J. (Eds), *Life Cycle Impact Assessment*. Dordrecht, the Netherlands: Springer Press, pp. 75–96.

Juraske, R. and Sanjuán, N. (2011). Life cycle toxicity assessment of pesticides used in integrated and organic production of oranges in the *Comunidad Valenciana*, Spain. *Chemosphere* 82: 956–62.

Juraske, R., Antón, A., Castells, F., et al. (2007a). Human intake fractions of pesticides via greenhouse tomato consumption: Comparing model estimates with measurements for captan. *Chemosphere* 67: 1102–7.

Juraske, R., Antón, A., Castells, F., et al. (2007b). PestScreen: A screening approach for scoring and ranking pesticides by their environmental and toxicological concern. *Environment International* 33: 886–93.

Juraske, R., Mutel, C. L., Stoessel, F., et al. (2009). Life cycle human toxicity assessment of pesticides: Comparing fruit and vegetable diets in Switzerland and the United States. *Chemosphere* 77: 939–45.

Juraske, R., Fantke, P., Romero Ramírez, A. C., et al. (2012). Pesticide residue dynamics in passion fruits: Comparing field trial and modeling results. *Chemosphere* 89: 850–5.

Kellogg, R. L., Nehring, R. F., Grube, A., et al. (2002). Environmental indicators of pesticide leaching and runoff from farm fields. In: Ball, V. E. and Norton, G. W. (Eds), *Agricultural Productivity: Measurement and Sources of Growth*. New York, NY: Kluwer Academic Publishers, pp. 213–56.

Kessler, S. C., Tiedeken, E. J., Simcock, K. L., et al. (2015). Bees prefer foods containing neonicotinoid pesticides. *Nature* 521: 74–6.

Kijko, G., Margni, M., Partovi-Nia, V., et al. (2015). Impact of occupational exposure to chemicals in life cycle assessment: A novel characterization model based on measured concentrations and labor hours. *Environmental Science and Technology* 49: 8741–50.

Kijko, G., Jolliet, O. and Margni, M. (2016). Occupational health impacts due to exposure to organic chemicals over an entire product life cycle. *Environmental Science and Technology* 50: 13105–14.

Kounina, A., Margni, M., Shaked, S., et al. (2014). Spatial analysis of toxic emissions in LCA: A sub-continental nested USEtox model with freshwater archetypes. *Environment International* 69: 67–89.

Kovach, J., Petzoldt, C., Degni, J., et al. (1992). A method to measure the environmental impact of pesticides. *New York's Food and Life Sciences Bulletin* 139: 1–8.

Krieger, R. I. (2010). *Hayes' Handbook of Pesticide Toxicology*, 3rd ed. London, UK: Academic Press.

Labite, H., Butler, F. and Cummins, E. (2011). A review and evaluation of plant protection product ranking tools used in agriculture. *Human and Ecological Risk Assessment* 17: 300–27.

Landrigan, P. J., Fuller, R., Acosta, N. J. R., et al. (2018). The Lancet Commission on pollution and health. *The Lancet* 391: 462–512.

Larsen, H. F. and Hauschild, M. (2007). Evaluation of ecotoxicity effect indicators for use in LCIA. *The International Journal of Life Cycle Assessment* 12: 24–33.

Lebailly, F., Levasseur, A., Samson, R., et al. (2014). Development of a dynamic LCA approach for the freshwater ecotoxicity impact of metals and application to a case study regarding zinc fertilization. *The International Journal of Life Cycle Assessment* 19: 1745–54.

Legler, J., Fletcher, T., Govarts, E., et al. (2015). Obesity, diabetes, and associated costs of exposure to endocrine-disrupting chemicals in the European Union. *The Journal of Clinical Endocrinology and Metabolism* 100: 1278–88.

Lewis, K. A., Tzilivakis, J., Warner, D. J., et al. (2016). An international database for pesticide risk assessments and management. *Human and Ecological Risk Assessment* 22: 1050–64.

Malaj, E., von der Ohe, P. C., Grote, M., et al. (2014). Organic chemicals jeopardize the health of freshwater ecosystems on the continental scale. *Proceedings of the National Academy of Sciences* 111: 9549–54.

Margni, M. D., Rossier, D., Crettaz, P., et al. (2002). Life cycle impact assessment of pesticides on human health and ecosystems. *Agriculture, Ecosystems and Environment* 93: 379–92.

Markets and Markets. (2013). Crop protection chemicals market by type, by crop types and geography: Global trends and forecast to 2018. Dallas, US: Markets and Markets, p. 403.

Matthews, G. A. (2016). *Pesticides: Health, Safety and the Environment*, 2nd ed. Chichester, UK: John Wiley and Sons.

Matthews, G. A., Bateman, R and Miller, P. (2014). *Pesticide Application Methods*, 4th ed. Chichester, UK: John Wiley and Sons.

McKone, T. E. and Enoch, K. G. (2002). CalTOX™, A Multimedia Total Exposure Model. Spreadsheet User's Guide. Version 4.0 (Beta). Berkeley, CA: Ernest Orlando Lawrence Berkeley National Laboratory, p. 45.

Mghirbi, O., Le Grusse, P., Fabre, J., et al. (2017). OptiPhy, a technical-economic optimisation model for improving the management of plant protection practices in agriculture: A decision-support tool for controlling the toxicity risks related to pesticides. *Environmental Science and Pollution Research* 24: 6951–72.

Mottes, C., Lesueur-Jannoyer, M., Le Bail, M., et al. (2014). Pesticide transfer models in crop and watershed systems: A review. *Agronomy for Sustainable Development* 34: 229–50.

Müller, N. and Fantke, P. (2017). Getting the chemicals right: Gaps and opportunities in addressing inorganics in life cycle assessment. *Society of Environmental Toxicology and Chemistry Europe 27th Annual Meeting*, 7–11 May 2017. Brussels, Belgium, p. 106.

Müller, N., de Zwart, D., Hauschild, M., et al. (2017). Exploring REACH as potential data source for characterizing ecotoxicity in life cycle assessment. *Environmental Toxicology and Chemistry* 36: 492–500.

Natalia, T. G. and Robert, M. H. (2016). Life-Cycle Assessment of Neonicotinoid Pesticides. *Journal of Fertilizers and Pesticides* 7: 165.

Nemecek, T. and Kägi, T. (2007). Life cycle inventories of Swiss and European agricultural production systems. Final report ecoinvent v2.0 No. 15. Duebendorf, Switzerland: Swiss Centre for Life Cycle Inventories, p. 360.

Nemecek, T. and Schnetzer, J. (2011). Methods of assessment of direct field emissions for LCIs of agricultural production systems. Duebendorf, Switzerland: Swiss Centre for Life Cycle Inventories, p. 34.

Nemecek, T., Bengoa, X., Lansche, J., et al. (2015). World Food LCA Database (WFLDB): Methodological Guidelines for the Life Cycle Inventory of Agricultural Products, Version 3.0. Lausanne and Zurich, Switzerland: Quantis and Agroscope, p. 84.

Nendza, M., Herbst, T., Kussatz, C., et al. (1997). Potential for secondary poisoning and biomagnification in marine organisms. *Chemosphere* 35: 1875–85.

Neto, B., Dias, A. C. and Machado, M. (2013). Life cycle assessment of the supply chain of a Portuguese wine: From viticulture to distribution. *The International Journal of Life Cycle Assessment* 18: 590–602.

Nordborg, M., Cederberg, C. and Berndes, G. (2014). Modeling potential freshwater ecotoxicity impacts due to pesticide use in biofuel feedstock production: The cases of maize, rapeseed, salix, soybean, sugar cane, and wheat. *Environmental Science and Technology* 48: 11379–88.

Nordborg, M., Davis, J., Cederberg, C., et al. (2017). Freshwater ecotoxicity impacts from pesticide use in animal and vegetable foods produced in Sweden. *Science of the Total Environment* 581-582: 448–59.

Norfaradila, J., Norela, S., Salmijah, S., et al. (2014). Life cycle assessment (LCA) for the production of palm biodiesel: A case study in Malaysia and Thailand. *Malaysian Applied Biology* 43: 53–63.

Notarnicola, B., Hayashi, K., Curran, M. A., et al. (2012). Progress in working towards a more sustainable agri-food industry. *Journal of Cleaner Production* 28: 1–8.

Overcash, M. (2016). Environmental genome of industrial products (EGIP): The missing link for human health. *Green Chemistry* 18: 3600–6.

Owsianiak, M., Rosenbaum, R. K., Huijbregts, M. A. J., et al. (2013). Addressing geographic variability in the comparative toxicity potential of copper and nickel in soils. *Environmental Science and Technology* 47: 3241–50.

Owsianiak, M., Holm, P. E., Fantke, P., et al. (2015). Assessing comparative terrestrial ecotoxicity of Cd, Co, Cu, Ni, Pb, and Zn: The influence of aging and emission source. *Environmental Pollution* 206: 400–10.

Panizzi, S., Suciu, N. A. and Trevisan, M. (2017). Combined ecotoxicological risk assessment in the frame of European authorization of pesticides. *Science of the Total Environment* 580: 136–46.

Peña, N. and Antón, A. (2017). Is copper fungicide that bad? *Acta Horticulturae* 1164: 333–8.

Peña, N. A., Antón, A. and Fantke, P. (2016). A consistent framework for modeling inorganic pesticides: Adaptation of life cycle inventory models to metal-base pesticides. *22nd SETAC Europe LCA Case Study Symposium*, 20–22 September 2016. Montpellier, France, p. 131.

Peña, N., Antón, A., Kamilaris, A., et al. (2017). Modelling ecotoxicity impacts in vineyard production: Addressing spatial differentiation for copper fungicides. *Science of the Total Environment* 616–17: 796–804.

Pennington, D. W., Margni, M. D., Ammann, C., et al. (2005). Multimedia fate and human intake modeling: Spatial versus nonspatial insights for chemical emissions in Western Europe. *Environmental Science and Technology* 39: 1119–28.

Persson, L. M., Breitholtz, M, Cousins, I. T., et al. (2013). Confronting unknown planetary boundary threats from chemical pollution. *Environmental Science and Technology* 47: 12619–22.

Plouffe, G., Bulle, C. and Deschênes, L. (2015). Assessing the variability of the bioavailable fraction of zinc at the global scale using geochemical modeling and soil archetypes. *The International Journal of Life Cycle Assessment* 20: 527–40.

Plouffe, G., Bulle, C. and Deschênes, L. (2016). Characterization factors for zinc terrestrial ecotoxicity including speciation. *The International Journal of Life Cycle Assessment* 21: 523–35.

Pu, Y., Laratte, B. and Ionescu, R. E. (2017). Freshwater sediment characterization factors of copper oxide nanoparticles. *IOP Conference Series: Earth and Environmental Science* 51: 012020.

Räsänen, K., Mattila, T., Porvari, P., et al. (2015). Estimating the development of ecotoxicological pressure on water systems from pesticides in Finland 2000–11. *Journal of Cleaner Production* 89: 65–77.

Reiss, R. and Griffin, J. (2006). A probabilistic model for acute bystander exposure and risk assessment for soil fumigants. *Atmospheric Environment* 40: 3548–60.

Renaud-Gentié, C., Dijkman, T. J., Bjørn, A., et al. (2015). Pesticide emission modelling and freshwater ecotoxicity assessment for Grapevine LCA: Adaptation of PestLCI 2.0 to viticulture. *The International Journal of Life Cycle Assessment* 20: 1528–43.

Richter, O., Diekkrüger, B. and Nörtersheuser, P. (1996). *Environmental Fate Modelling of Pesticides: From the Laboratory to the Field Scale*. Weinheim, New York, Basel, Cambridge and Tokyo: VCH Verlagsgesellschaft.

Rizzati, V., Briand, O., Guillou, H., et al. (2016). Effects of pesticide mixtures in human and animal models: An update of the recent literature. *Chemico-Biological Interactions* 254: 231–46.

Rosenbaum, R. K. (2015). Ecotoxicity. In: Hauschild, M. Z. and Huijbregts, M. A. J. (Eds), *Life Cycle Impact Assessment*. Dordrecht, the Netherlands: Springer Press, pp. 139–62.

Rosenbaum, R. K., Bachmann, T. M., Gold, L. S., et al. (2008). USEtox - The UNEP-SETAC toxicity model: Recommended characterisation factors for human toxicity and freshwater ecotoxicity in life cycle impact assessment. *The International Journal of Life Cycle Assessment* 13: 532–46.

Rosenbaum, R. K., Huijbregts, M. A. J., Henderson, A. D., et al. (2011). USEtox human exposure and toxicity factors for comparative assessment of toxic emissions in life cycle analysis: Sensitivity to key chemical properties. *The International Journal of Life Cycle Assessment* 16: 710–27.

Rosenbaum, R. K., Anton, A., Bengoa, X., et al. (2015). The Glasgow consensus on the delineation between pesticide emission inventory and impact assessment for LCA. *The International Journal of Life Cycle Assessment* 20: 765–76.

Ryberg, M. W., Rosenbaum, R. K., Mosqueron, L., et al. (2018). Addressing bystander exposure to agricultural pesticides in life cycle impact assessment. *Chemosphere* 197: 541–9.

Sadok, W., Angevin, F., Bergez, J.-É., et al. (2008). *Ex ante* assessment of the sustainability of alternative cropping systems: Implications for using multi-criteria decision-aid methods. A review. *Agronomy for Sustainable Development* 28: 163–74.

Sanborn, M., Cole, D., Kerr, K., et al. (2004). *Systematic Review of Pesticide Human Health Effects*. Toronto, Ontario: The Ontario College of Family Physicians, 188.

Sanborn, M., Kerr, K. J., Sanin, L. H., et al. (2007). Non-cancer health effects of pesticides. *Canadian Family Physician* 53: 1712–20.

SCBD. (2010). *Global Biodiversity Outlook 3*. Montréal, Canada: Secretariat of the Convention on Biological Diversity.

Schmidt Rivera, X. C., Bacenetti, J. and Fusi, A., et al. (2017). The influence of fertiliser and pesticide emissions model on life cycle assessment of agricultural products: The case of Danish and Italian barley. *Science of the Total Environment* 592: 745–57.

Shimako, A. H., Tiruta-Barna, L. and Ahmadi, A. (2017). Operational integration of time dependent toxicity impact category in dynamic LCA. *Science of the Total Environment* 599–600: 806–19.

Steingrimsdottir, M. M., Petersen, A. and Fantke, P. (2018). A screening framework for pesticide substitution in agriculture. *Journal of Cleaner Production* 192: 306–15.

Surgan, M., Condon, M. and Cox, C. (2010). Pesticide risk indicators: Unidentified inert ingredients compromise their integrity and utility. *Environmental Management* 45: 834–41.

Tiruta-Barna, L., Pigné, Y., Navarrete Gutiérrez, T., et al. (2016). Framework and computational tool for the consideration of time dependency in Life Cycle Inventory: Proof of concept. *Journal of Cleaner Production* 116: 198–206.

Trapp, S. (2009). Bioaccumulation of polar and ionizable compounds in plants. In: Devillers, J (Ed.), *Ecotoxicology Modeling*. Dordrecht, the Netherlands: Springer Press, pp. 299–353.

Trasande, L., Zoeller, R. T., Hass, U., et al. (2015). Estimating burden and disease costs of exposure to endocrine-disrupting chemicals in the European Union. *The Journal of Clinical Endocrinology and Metabolism* 100: 1245–55.

Tromson, C., Bulle, C. and Deschênes, L. (2017). Including the spatial variability of metal speciation in the effect factor in life cycle impact assessment: Limits of the equilibrium partitioning method. *Science of the Total Environment* 581-582: 117–25.

Tsiropoulos, I., Faaij, A. P.C, Lundquist, L., et al. (2015). Life cycle impact assessment of bio-based plastics from sugarcane ethanol. *Journal of Cleaner Production* 90: 114–27.

Turner, J. A. (2015). *The Pesticide Manual*, 17th ed. Hampshire, UK: BCPC Publications, British Crop Protection Council.

UNEP. (2004). *Stockholm Convention on Persistent Organic Pollutants (POPs)*. Geneva, Switzerland: United Nations Environment Programme.

USDE. (2012). *U.S. Life Cycle Inventory Database*. Washington DC: United States Department of Energy.

US-EPA. (1995). *Compilation of Air Pollutant Emission Factors, Volume 1: Stationary Point and Area Sources*, 5th ed. Washington DC: Food and Agricultural Industries and United States - Environmental Protection Agency.

van den Berg, F., Kubiak, R., Benjey, W. G., et al. (1999). Emission of pesticides into the air. *Water, Air and Soil Pollution* 115: 195–218.

van de Zande, J. C., Wenneker, M and Michielsen, J. M.G. P. (2010). Risk estimation of bystander and residential exposure from orchard spraying based on measured spray drift data. International Advances in Pesticide Application. *Aspects of Applied Biology* 99: 149–56.

van Zelm, R., Huijbregts, M. A. J., Posthuma, L., et al. (2009a). Pesticide ecotoxicological effect factors and their uncertainties for freshwater ecosystems. *The International Journal of Life Cycle Assessment* 14: 43–51.

van Zelm, R., Huijbregts, M. A. J. and van de Meent, D. (2009b). USES-LCA 2.0 - a global nested multi-media fate, exposure, and effects model *The International Journal of Life Cycle Assessment* 14: 282–4.

van Zelm, R., Huijbregts, M. A. J. and van de Meent, D. (2010). Transformation products in the life cycle impact assessment of chemicals. *Environmental Science and Technology* 44: 1004–9.

van Zelm, R., Larrey-Lassalle, P. and Roux, P. (2014). Bridging the gap between life cycle inventory and impact assessment for toxicological assessments of pesticides used in crop production. *Chemosphere* 100: 175–81.

Vasileiadis, V. P., Dachbrodt-Saaydeh, S., Kudsk, P., et al. (2017). Sustainability of European winter wheat- and maize-based cropping systems: Economic, environmental and social ex-post assessment of conventional and IPM-based systems. *Crop Protection* 97: 60–9.

Verweij, W., Blonk, B. and Kraak, M. (2015). *How to Protect Groundwater Organisms to Toxic Substances?* Luxembourg, UK: EU Working Group Groundwater.

Vida, P. and Moretto, A. (2007). Pesticide exposure pathways among children of agricultural workers. *Journal of Public Health* 15: 289–99.

Warren-Hicks, W. J. and Hart, A. (2010). *Application of Uncertainty Analysis to Ecological Risks of Pesticides*. Boca Raton, FL: CRC Press.

Webb, J., Hutchings, N., Amon, B., et al. (2016). *EMEP/EEA Air Pollutant Emission Inventory Guidebook 2016: Agriculture Other Including Use of Pesticides*. Luxembourg, UK: European Environment Agency.

Weber, R., Aliyeva, G. and Vijgen, J. (2013). The need for an integrated approach to the global challenge of POPs management. *Environmental Science and Pollution Research* 20: 1901–6.

Westh, T. B., Hauschild, M. Z., Birkved, M., et al. (2015). The USEtox story: A survey of model developer visions and user requirements. *The International Journal of Life Cycle Assessment* 20: 299–310.

Woodcock, B. A., Bullock, J. M., Shore, R. F., et al. (2017). Country-specific effects of neonicotinoid pesticides on honey bees and wild bees. *Science* 356: 1393–5.

Yang, Y. (2015). A note on pesticide-related toxicity impacts of crops in the USA. *The International Journal of Life Cycle Assessment* 20: 1604–6.

Yang, Y. and Suh, S. (2015). Changes in environmental impacts of major crops in the US. *Environmental Research Letters* 10: 094016.

Zhang, W. (2016). *China Pesticide Trade*. Beijing, China: Institute for the Control of Agrochemicals, Ministry of Agriculture, P. R. China.

Zhang, C., Hu, R., Shi, G., et al. (2015a). Overuse or underuse? An observation of pesticide use in China. *Science of the Total Environment* 538: 1–6.

Zhang, Y., Liu, X., Xiao, R., et al. (2015b). Life cycle assessment of cotton T-shirts in China. *The International Journal of Life Cycle Assessment* 20: 994–1004.

Chapter 8

Assessing socio-economic impacts of agriculture

Bo P. Weidema, Aalborg University, Denmark

1 Introduction

This chapter is dedicated to impacts that are related to social and economic pressures, such as underpayment of labour, illegitimate resource acquisition and control, and inadequate work conditions, in contrast to the impacts that are related to physical pressures (use of land, fossil fuels, freshwater etc., or emissions of e.g. nutrients and pesticides), issues that have been largely covered in the preceding chapters of this book.

However, it is worth noting that it is only possible to maintain the distinction between (bio-) physical and socio-economic when referring to the level of pressures, i.e. the direct interactions or flows between human activities and the environment. Once the impact pathway is modelled further from these pressures, it turns out that all biophysical pressures eventually also have impacts on the social and economic environment (Weidema, 2018).

http://dx.doi.org/10.19103/AS.2018.0044.10

To perform a holistic socio-economic assessment, the integration of different scientific traditions is required, each with their own strengths and weaknesses. From the economic discipline we have cost-benefit analysis with a strong quantitative tradition, developed quite isolated from the rest of the social sciences where a more qualitative social impact assessment has been developed. Both of these disciplines have developed in isolation from the quantitative ecological science that has largely left it to economists to perform the valuation of the ecological impacts. A parallel development has occurred in the health sciences, where health economics is a relatively new discipline building on the quantification of risk factors for different health outcomes. Another example of the compartmentalisation of science can be found in the early development of energy analysis of product supply chains into what is now known as life cycle assessment, a development that took place in isolation from economic disciplines of input-output analysis and cost-benefit analysis. This compartmentalisation is now slowly being overcome, as described in this chapter.

This chapter describes the development and state of the art of cost-benefit assessment and social impact assessment in agriculture (Sections 2–6) and their integration into models and tools with an economy-wide supply chain perspective (Section 7). A broad picture of socio-economic impacts of agricultural development is treated in Sections 8–11. Section 12 considers the role of certification and fair-trade schemes.

This chapter does not intend to explicitly cover all socio-economic impacts of agriculture. For example, no descriptions are provided on how to measure impacts on animal welfare or on the role of agriculture in preservation of cultural heritage.

2 Cost-benefit analysis and assessment in agriculture

Economic assessment in the form of cost-benefit or cost-effectiveness assessments remains the dominating form of socio-economic assessment in agriculture. This can partly be attributed to the central role of economic income in determining wellbeing of producers and consumers, partly the relative ease of obtaining quantitative data for their private (internal) costs and benefits, and partly its ability to express external costs and benefits in the same monetary units as the internal costs and benefits, thus promising a holistic assessment. As expressed by Richards and Nicholls (2015: p. 10): 'by attaching an assessment of relative importance, where financial proxies provide monetary representations of the worth of social outcomes, we are better able to appreciate their value. This is because we have a consistent unit of measurement to compare outcomes with one another, and the costs of their production. As a result, we are better able to make decisions for the maximisation of social value.'

The origins of cost-benefit analysis is often ascribed to the 1936 Flood Control Act, which contain the wording, 'the Federal Government should improve or participate in the improvement of navigable waters or their tributaries, including watersheds thereof, for flood-control purposes if the benefits to whomsoever they may accrue are in excess of the estimated costs.' In response to this, the U.S. Army Corps of Engineers developed some ad hoc methods that were discovered by academic economists in the 1950s and improved into a rigorous system for the economic analysis of public investments (Watkins, s.a.). Early cost-benefit analysis was mainly concerned with the direct and indirect economic costs and benefits to different actors, and less with environmental externalities. It was not until the 1960s and early 1970s, with the background of rising environmental awareness, that the valuation of nature and health became an important issue in cost-benefit analysis (Krutilla, 1967; Rice, 1967). From then on it is more correct to use the term cost-benefit assessment, rather than analysis, to emphasise on that valuations and trade-offs are unavoidable when resources need to be allocated between different impacts and/or actors, or when some impacts are beneficial and others are detrimental.

Cost-benefit assessment has been criticised for the limitations of its practical implementations. The inclusion of externalities has often been incomplete, and the methods applied for the valuation of biophysical and social externalities have been uncertain, without this resulting in sufficient caveats on the conclusions. In comparison to life cycle assessments and input-output analyses, the impacts in the supply chain have often been insufficiently included. In the early cost-benefit tradition, it was common to exclude impacts outside of the national boundaries. The use of discounting (utility-weighting applied over temporally distinct populations) has often been unjustified, and in stark contrast to the general lack of utility-weighting of the immediate impacts on present stakeholders, in spite of this topic being included in standard textbooks on cost-benefit assessment (e.g. Boardman et al., 2006; Florio, 2014).

However, the criticism of cost-benefit assessments is mainly a criticism of inadequate implementation than a criticism of the fundamental principles and ambitions to provide a holistic, quantitative social assessment expressed in a single numeraire. Much of the early criticism has slowly been adopted and integrated in the form of significant improvements in the consistency and completeness of the economic assessment techniques (see Kulig et al., 2010 for a condensed historical overview). What remains is mainly a concern that not all impacts can (or should) be quantified and expressed in monetary units. This concern is especially expressed and addressed in the social impact assessment community that can be seen as seeking to complement the cost-benefit assessment technique by including issues that are still not sufficiently or adequately covered in the practical implementations of this technique.

3 Social impact assessment

Social impact assessment has formally developed as a part of environmental impact assessment (EIA). In spite of the definition of EIA as 'the process of identifying, predicting, evaluating and mitigating the biophysical, social, and other relevant effects of development proposals prior to major decisions being taken and commitments made' (IAIA and IEA, 1999), the practical implementation has been strongly biased towards the biophysical effects on ecosystems and to a lesser extent human health. Thus, social impact assessment originally developed as the 'lesser sibling' of EIA (Hildebrandt and Sandham, 2014), but the social impacts are now increasingly being integrated into EIA, to the extent that the technique has been re-branded by some (IFC, 2012; Dendena and Corsi, 2015) with the somewhat pleonastic term 'Environmental and Social Impact Assessment'. Esteves et al. (2012) provide a good summary of the history and state of the art of social impact assessment, while Burdge and Vanclay (1996) give a more detailed account of its early history and the challenges it faced.

With its innovative combination of process- and content-focus, the ISO 26000 guidance standard on Social Responsibility (ISO, 2010) provides an effective management framework for social impact assessment (Hahn, 2012) along with the IFC social performance standards (IFC, 2012).

The development of social impact assessment has suffered from the schisms within the social sciences, both between economics and sociology and within each of these disciplines. For example, the concept of impact pathways or cause–effect chains, linking pressures (economic, biophysical or social outputs of the economic activities under study) via midpoint impact indicators (quantifying the output of biophysical or social processes) to final damage (quantifying the detrimental or beneficial impact as perceived and valued by humans) is well described in the cost–benefit literature (ExternE, 1995, 2005), while the concept was only recently introduced in social impact assessment, and even when attempting to separate pressure indicators, midpoint indicators, and damage indicators, these are often confused (van Schooten et al., 2003).

The most severe schism has been between a qualitative, interpretive, or constructivist tradition on the one hand, with a focus on process, involvement, narratives, power relations, and local conditions where no general rules appear applicable, and a quantitative and analytical tradition on the other hand, with a focus on outcomes, generalizable theory, and comparability across local contexts. Burdge and Vanclay (1996: p. 66) point out that 'social science traditions, especially sociology, tend to be critical and discursive, rather than predictive and explanatory' and that this contrasts with the objective of social impact assessment, which requires 'conceptual frameworks or valid measures for testing the interrelationships among variables'. Scoones (2009: p. 181) notes that 'livelihoods approaches, coming as they did from a complex disciplinary parentage that emphasised the local, have not been very good at dealing with big shifts in the state of global markets and politics. In the frameworks, these were dumped in a box labelled 'contexts'. But what happens when contexts are the most important factor, over-riding the micro-negotiations around access to assets and the finely-tuned strategies of differentiated actors?' Burdge and Vanclay (1996: p. 72) provide one plausible reply to this question, noting that results of social impact assessments that 'pay more attention to local concerns over the concerns of the broader distant public' may be outweighed in the ultimate decision process.

The qualitative or constructivist position has been especially vocal in defending that no impacts assessment occurs in a power vacuum, only providing 'objective' knowledge. In the words of Howitt (1989: p. 154): 'the impact assessment process itself is part of a process of social change with significant social and political consequences.' Therefore, sustainable development practitioners need to understand and become involved in the power game around decision-making, providing empowerment to stakeholders that have the more sustainable solutions but currently have too low power to have them implemented. One powerful tool in this game is to call for due diligence by the more powerful players that have the less sustainable solutions. Because these players are powerful, it may be necessary to find ways to temporarily compensate their losses, to ensure that the best possible long-term compromises can be implemented.

The qualitative/quantitative schism has been exaggerated to the extent that the two approaches have been described as distinct and opposing paradigms in what comes close to being caricatures of the opposing positions (see e.g., Aledo-Tur and Domínguez-Gómez, 2017), pointing to the schism being rather a sociological convention than a practically or epistemologically distinguishable separation (Hanson, 2008). Other authors have stressed the complementarity of quantitative and qualitative techniques and the options for integrating them, taking advantage of their respective strengths and their ability to compensate for each other's weaknesses (Granovetter, 1973; Coleman, 1988; Carvalho and White, 1997; Parthasarathy, 2002; Becker, 2003). For example, quantitative indicators have the advantage of being generalizable, reproducible, additive, and amenable to verifiable cause-effect modelling with explicit uncertainty, while qualitative techniques can identify missing perspectives and indicators and provide richer local context and a deeper understanding of the motivations, culture, values, power relations, and change potentials of the affected societal groups (Vanclay, 2002), which may work to accelerate or sidetrack a specific development pathway. In the context of the EU Impact Assessment system, a number of FP6 and FP7 research projects have developed attempts at integrated sustainability assessment methods; reviewed in Kasperczyk and Knickel (2006).

A seminal contribution to social impact assessment is Amartya Sen's concept of capabilities (Sen, 1984, 1985), focusing on the effective access and freedom to choose between sets of valuable functionings ("what people are actually able to be and do", in the words of Nussbaum, 2000: p. 5), rather than their pecuniary wealth or income.

Writing in the interpretive or constructivist tradition, Sen did not provide clear practical guidelines on how to assess or identify capabilities, arguing that being prescriptive would 'deny the possibility of fruitful public participation on what should be included and why' (Sen, 2004: p. 77). Hoon et al. (1997: p. 15) challenges this position: 'Examples of basic capabilities are leading a life free of avoidable morbidity, being informed and educated, well nourished and having access to satisfactory levels of resource and asset bases. Being able to function on the basis of these essential capabilities is an objective, observable phenomenon; it is not a matter of subjective perception, nor is it culture specific.'

4 Sustainable livelihoods and the concept of capitals

In their report to the World Commission on Environment and Development (the 'Brundtland commission'), the Advisory Panel on Food Security, Agriculture, Forestry and Environment (APFSAFE, 1987) merged Sen's capability approach with the concept of 'Sustainable livelihoods security'. This idea was further developed by Chambers and Conway (1991), defining a 'livelihood' as a combination of 'capabilities, assets (stores, resources, claims and access) and activities required for a means of living' (Chambers and Conway, 1991: p. 6). Bernstein (1992: p. 24) then formulated some questions that should always be asked in a livelihoods analysis: 'Whose livelihoods? What are the social relations (and social, cultural, political and institutional factors) that affect: who owns what? who does what? who gets what? what do they do with it?'. The further development of the 'sustainable livelihoods' approach to social impact assessment is nicely summarised by Solesbury (2003) and in a very personal insider account by Ian Scoones in the Journal of Peasant Studies (Scoones, 2009).

An important aspect of this development was its attempt at bridging the gap between economics and sociology by using the language of economics. Note that Bernstein's above-cited questions are questions about the social relations that govern the distribution of property, work, income, consumption, and accumulation. As pointed out by Bernstein (1992: p. 25) these are terms associated with economic life. The livelihoods perspective expands the narrow economic analysis of these terms by also stressing the importance of social relations: 'Considering social relations always involves asking 'whose livelihoods?', and reminds us that economic inequalities (of property, work, income, consumption, accumulation) typically incorporate and express social, cultural, political and institutional inequalities as well.' In the words of Scoones (2009: pp. 176–7): 'In the notionally trans-disciplinary subject area of development, making sense to economists is a must. With economists only recently having discovered institutions ... social relations and culture ... a moment had opened up to generate some productive conversation, even if largely on disciplinary economics' terms.'

The economic concept of 'capital' has played an important role in this context. From originally signifying only a 'stock' of manufactured goods used as a production factor, the concept of capital has slowly grown to be used as a synonym for any factor of production, which again is used synonymously with the concepts of 'resources' and 'wealth', and the value of capital thus being measurable as the net present value of the future stream of income expected from employing that capital. The three production factors of classical economics, stocks, land, and labour, which today would be called manufactured, natural and human capital, were first supplemented by 'social capital' by Coleman (1988), recognizing the value of social network relations[1] in facilitating action and reducing externalities, an idea taken up by Paul Ekins in his *Four-Capital Model of Wealth Creation* (Ekins, 1992). Viederman (1994) and Bebbington (1999) expanded this to five capitals by adding `cultural capital', although without providing a clear definition to distinguish this from human capital, and without obtaining much following. A more successful 'Five Capitals' model including 'financial capital' as the fifth part of a pentagon is often attributed to Jonathon Porritt (or Forum for the Future, the organization that he co-founded with Sara Parkins and Paul Ekins), although the pentagon of five capitals first appeared in Carney (1998) as part of the Sustainable Rural Livelihoods framework. Latest, the International Integrated Reporting Council (IIRC, 2011) has suggested a 'Six Capitals' model, by separating intangible manufactured capital out as a separate 'intellectual capital', a development that may have been prompted by the five capitals model apparently limiting the account of manufactured capital to 'physical capital'.

The value of a capital comes from its potential to supply a future stream of income and is measured in terms of the net present value of these expected future streams, typically accounted on an annual basis. This stresses the mutual relationship between capitals and the annual flow accounts of income and consumption, in which the net accumulation of capital provides the link between the capital balance of one year and the next. Sustainability of a capital is then defined as a zero or positive trend in the capital balance over time.

1 When used as a complement to the other three capitals, 'social capital' should more precisely be called 'social network capital', since 'social' is here used in the narrow sociological sense of 'pertaining to the interactions between people', rather than in the sense of 'aggregated at the level of society', which is the common usage in economics since Marx (1885) and von Böhm-Bawerk (1889).

The contribution of human activities to sustainable capital development can thus be seen as the extent to which they contribute to building, maintaining, consuming, degrading, and/or depleting the different types of capital. While such changes to the capital balance are sometimes included in cost-benefit assessments, they are often neglected, for example the influence of human activities on education and skills (the fact that capabilities are upgraded - or downgraded - while performing an activity), distributional issues, as well as influences on our time usage, e.g. the amount of time spent commuting or on different household activities. Although belonging to the economic sphere, these issues are in practice often better covered by social impact assessments (van Schooten et al., 2003).

Social network capital is different from the other forms of capital in that it cannot be bought or sold, and like a public good it does not diminish when consumed. Yet it can still be built, maintained, consumed, degraded, and depleted. And its value can be determined in parallel to that of the other capitals, in its ability to increase productivity and reduce transaction costs. This comes about in three ways, as distinguished by Coleman (1988): in the form of the insurance provided by the mutual obligations and expectations created in the deep, frequent, long-lasting, trust-based relationships of narrow community networks (named 'bonding' social capital by Gittell and Vidal, 1998); In the form of information sharing, mainly in between-group relations where the diversity of relations is of importance (named 'bridging' social capital by Putnam, 1995); in the form of the safety and security provided by norms and sanctions, normally embedded in an institutionalised power structure - named 'linking' social capital in World Bank (2000). Coleman (1988) noted that social network capital not only facilitates certain actions, it also constrains others. Depending on how the capital is used, it can have net beneficial or net detrimental effects on overall social wellbeing. However, this is not specific to social network capital: all capitals can be used or misused. It is the actions taken that determine the contribution to social wellbeing, and thus indirectly the social value of the capital, which may thus be a net value from potential beneficial and detrimental actions.

Financial capital is different from the other forms of capital in that it does not have any value in itself but simply represents the value of that part of the other capitals that is currently used by another actor than the one that owns it, for which reason its value is listed as a debit in the user's financial balance sheet and a credit in the owner's financial balance sheet. Financial capital may also be seen as a temporal transfer of consumption options for the owner and user, with the credit representing postponed consumption for the owner balanced by the preponed consumption of the user. Since the debit and the credit counterbalances, the amount of financial capital at the global societal level is zero. However, the user of financial capital usually pays an economic rent to the owner, in addition to transaction costs. Even though the amount of financial capital at the global societal level is zero, its distribution over different actors at different points in time is essential for understanding developments in economic activity and inequality (Godley and Lavoie, 2007; Caiani et al., 2019).

5 Applications of the capitals concept, critique, and a way beyond

Although Sen (1984, 1985) wrote in opposition to the contemporary utilitarian approach to cost-benefit assessment, which at that time was very much focused on physical resources and commodities, and without any equity considerations, his concept of capabilities can nevertheless be seen as a forerunner for what is now made more explicit with the detailed indicators for human and social capital. And the above-cited definition of 'livelihood' by Chambers and Conway (1991) closely mirrors the economic production function with inputs (capabilities and assets) converted by activities into outputs ('a living'). For the specific agricultural context, the United Nations Food and Agricultural Organisation has published very useful practitioners' and field guides to socio-economic assessment based on the livelihoods approach (FAO, 2011).

A very comprehensive set of sustainable development indicators has been proposed by the Conference of European Statisticians Recommendations on Measuring Sustainable Development (UNECE, 2014). It provides a very clear description of the relationship between these indicators and the national accounting framework. The report particularly points out that for each aspect to be measured, both a geographical (imports/exports) and a temporal (capital transfer to future generations) perspective need to be covered. Statisticians from more than 50 countries have endorsed the recommendations. The conceptual foundation and the potential indicators suggested in the UNECE publication serve as a good starting point for further harmonisation of the measurement systems and development of a set of indicators that could be used for comparison across countries, and probably with some adaptation also across individual organisations and products.

However, in their overview of the different aspects of social network capital, Woolcock and Narayan (2000: p. 235) provide a word of warning of a too aggregated, statistical approach: "In providing broad statistical evidence for the importance of social capital, the subtlety, richness, and enormous variation gleaned from case studies of individual countries and communities is lost, as are the voices of those most directly affected by weak public institutions: the poor." Also Scoones (2009: p. 178), in spite of his above-cited recognition of the need to make sense to economists, laments the central role that the pentagon of five capitals obtained within the livelihoods research, shifting the focal point from institutions to assets: 'In some respects the focus on the 'asset pentagon' and the use of the 'capitals' metaphor was an unfortunate diversion' compared to the previous emphasis on 'in-depth qualitative understandings of power, politics and institutions'.

While the qualitative/quantitative schism in social science may indeed largely be a 'social construct', as argued by Hanson (2008), there is a fundamental difference in focusing on the indirect contribution to wellbeing from human and social capital as factors of production, and the broader understanding implicit in Sen's capability approach, where human and social capabilities are seen *both* as a capital factor of production and as what could be called a 'factor of consumption', i.e. as a direct contributor to wellbeing. This is made explicit in the editorial by Sen (1997), where he illustrates the distinction with the example of education: 'If education makes a person more efficient in commodity production, then this is clearly an enhancement of human capital. This can add to the value of production in the economy and also to the income of the person who has been educated. But even with the same level of income, a person may benefit from education, in reading, communicating, arguing, in being able to choose in a more informed way, in being taken more seriously by others, and so on. The benefits of education, thus, exceeds its role as human capital in commodity production. The broader human-capability perspective would record – and value – these additional roles (Sen, 1997: p. 1959).'

These warnings on the overemphasis on a narrow economics perspective are still valid. The focus on the capital concept can easily lead to undesired commodification and a reductionist trap of looking only at what is easily measured.

What is needed is a more complementary understanding of means and ends, rather than the irrelevant and futile contradiction of quantitative/qualitative. An example of such a more constructive approach is the 'capitals and capabilities' framework developed by Bebbington (1999) for his rural livelihoods and poverty studies in the Andes, again drawing on Sen's classic work. He saw assets as 'vehicles for instrumental action (making a living), hermeneutic action (making living meaningful) and emancipatory action (challenging the structures under which one makes a living)' (Bebbington, 1999: p. 2022). Concordantly, from the context of life cycle sustainability assessment, Sala et al. (2013: p. 1658) comment on the four and five capitals models and find that these capitals 'are not purely of instrumental value but represent an appropriate framework within which particular endpoints of intrinsic value can be identified'.

To operationalise this view, the 'areas of protection' concept of the UNEP/SETAC Working Group on Impact Assessment (Jolliet et al., 2003) could be expanded to reflect both the instrumental 'capital' or production aspect and the intrinsic 'enjoyment' or final consumption aspect of each asset type; see Table 1.

Table 1 Areas of protection with both intrinsic enjoyment value and instrumental capital value

Objects considered:		Manufactured		Human		
Ends/means values:	Natural	Physical	Intellectual	capabilities	Social networks	Financial
Ends: Intrinsic value (for enjoyment / final consumption)	Natural heritage (incl. biodiversity)	Physical consumption goods; physical cultural heritage	Intellectual consumption goods; intellectual heritage	Leisure time; autonomy; health; skills	Participation and influence; safety and security; cultural heritage	-
Means: Instrumental value (for production)	Natural resources (incl. ecosystems)	Manufactured physical capital (biotic and abiotic)	Intellectual capital	Human capital	Social network capital	Financial capital

6 Valuation of impacts

Valuation, understood as the expression of relative preferences for marginal changes in the different areas of protection, is generally applicable within both intrinsic and instrumental areas of protection. Valuation serves the purpose of facilitating comparisons of across different impacts, by applying weights (values) that reflect their relative importance. Without valuation it becomes impossible to recommend the best decision when the alternative options score best on different impacts measured in different units.

Welfare economics posits some fundamental premises that must be fulfilled for valuations to be valid. These include that values should be statistically representative for the affected population, that the wellbeing of each individual should be weighted equally, and that any systematic irrational bias caused by the contextual and informational setting should be corrected for. Much critique of valuation - especially monetary valuation - is due to the non-fulfilment of these premises in (unfortunately the majority) of current practical valuation studies and policy recommendations. Another source of critique comes from the confusion of the marginal values applied in trade-offs between alternative resource allocations and the absolute moral values (like the value of democracy or the value of human life as such) that cannot be subject to quantified measurements and trade-offs.

Valuation is in its essence anthropocentric, since its purpose is to support human decision-making. Any concern for other species (or for that matter for any other group than the one that has the power to take the decision) must necessarily come as a concession from those who perform the valuation. However, the fact that it appears very difficult - or rather impossible - to design a truly non-anthropocentric valuation scheme does not make it unimportant to raise the issue and seriously contemplate its relevance when deciding on the design of a valuation method. It should also be noted that an anthropocentric valuation does not necessarily imply a low valuation of nature; nature does have high value for humans, both use value (today often referred to as the value of ecosystem services) and non-use values (existence value and bequest value). In fact, from the current best estimates of the monetary valuation of nature (Murakami et al., 2018), it can be deduced as socially optimal to halt any further human-induced species loss.

If specific groups are affected more than other groups, it is important that they are represented in the population that performs the valuation, and additionally it may be required that such specifically affected groups give their informed consent to the valuation procedure or its specific results. This issue is largely parallel to the issue of minority protection in democracy.

Especially in the context of sustainable development, which has an inherent concern for equality between and within generations (World Commission on Environment and Development, 1987), it is important to ensure that the valuation methodology gives equal weight to the values of each individual. For this purpose, utility-weights can be applied, which weights values in relation to the utility that each individual obtains from an additional unit of income. Utility-weights (W) can be calculated as the proportion between the global average income and the income of the affected sub-group raised to the power of the elasticity of marginal utility of income (δ):

$$W = \left(\frac{\text{Average Income}}{\text{Subgroup Income}} \right) \wedge \delta \tag{1}$$

where both incomes are corrected for purchasing power. The value of δ can be empirically determined from surveys of wellbeing and income and is typically between 1.1 and 1.4 (see e.g., Layard et al., 2008), which implies a larger weight to the same impact when it affects a person with low income compared to a person with higher income. Alternatively, δ can be normatively set to 1, in which case the same impact will be weighted equally across all levels of income. Utility-weighting over time is obtained by multiplying the elasticity of marginal utility of income (δ) by the expected annual increase in real income, which gives an annual discount rate that is consistent with the intra-generational utility-weights.

It is a widespread critique of valuation methods that they assume that participants exhibit rational, utility-maximising behaviour when making valuations, while empirical evidence shows that people do not exhibit this rational behaviour, neither in normal market transactions nor in experimental settings, but are influenced by the framing of the decision situation. A large body of literature on behavioural economics suggests improvements to the survey techniques to control and adjust for the systematic biases caused by the contextual and informational setting of the valuation (see e.g. Angner and Loewenstein, 2017). One important example is that of the endowment effect, which causes larger weight to a change that is framed as a loss rather than as a gain, which must therefore be taken into account when interpretation values are solicited under these two frames. Another example is the issue of scaling, which may cause large singular instances of impacts (catastrophes) to be systematically over-weighted relative to the same impact over a larger space or time, while empirical studies systematically controlling for this bias show that neither experts nor lay people are particularly catastrophe averse. By making such biases explicit, it is possible to adjust for them.

The individualistic nature of values and valuation has been criticised for not adequately taking into account that some values only occur as 'group values' and that individual values may be affected by the discussions and opinions given in a group context. For many complex environmental goods, respondents may not have well-defined preferences prior to the process of value solicitation, but their latent preferences are developed during the solicitation process itself. This criticism has often been voiced in connection to the general critique of the assumptions made in neo-classical economics. While classical valuation techniques can take into account the (group) context of the solicited individual values, the values of the group as a whole are seen as the sum of the individual values.

This limitation in classical valuation procedures has led to the development of additional procedures in valuation under the name of 'deliberative valuation' (Kenter, 2015). In deliberative valuation, the solicitation of values takes place in a process of reasoned social discourse and learning, which increases the participants' awareness of the issue and of the perceptions of other participants. Participants can discuss considerations such as equity, fairness, rights, and responsibilities, alongside discussions of costs, benefits and trade-offs, and uncertainties and risks, in order to come to a more meaningful constitution of their contextual values (Kenter, 2015), thus taking more explicitly into account that preferences are socially constructed.

Deliberative valuation can be used both before a valuation survey, to test the survey design (e.g. wording and comprehension of questions, validate the information content, help identify design biases) and after a valuation survey, to validate the survey result. When combined with recording of the evolution of participants' values, attitudes, beliefs, and norms during the course of the deliberation, the procedure offers the potential for increasing both the interpretative and the explanatory depths of the results (understanding the meaning of the responses and the reasons behind the responses, and how the complexities, uncertainties, and risks influence the values solicited).

However, deliberative valuation is limited by the resources and time required to resolve complex issues, especially where competing value systems or beliefs are present. If used as a stand-alone technique, rather than as an additional or integrated part of a more traditional valuation study, the representativeness of the results may be challenged due to the typically small numbers of people who can be involved and the difficulty of adequately eliminating influence from power-relationships.

Deliberative valuation may also be used as a technique to overcome the incommensurability problem, i.e. that some individuals reject a valuation completely because they do not accept that the items to be valued are comparable. A claim for incommensurability is valid when the items to be valued belong to different logical levels (e.g. 'What is largest: Mont Blanc or the roar of a lion?'), but is not a viable position when comparing items for which trade-offs between alternative resource allocations are in reality being made, in which case the problem of choice is unavoidable (Beckerman and Pasek,

1997). Participants that outright reject the valuation questions can be invited to participate in a deliberative valuation session instead, where the concerns about incommensurability are taken serious and additional context of the valuation is provided. The results from the deliberative valuation will then add to the representativeness of the overall valuation.

Valuation is also a part of multi-criteria decision-making methods, where the relative preferences can be solicited either prior to or as a progressive part of the decision-making process. However, the values applied are often those of a specific decision-maker or a smaller panel, thus not fulfilling the above-stated premises of welfare economics (statistically representativeness, equal weighting of wellbeing of each affected individual, and correction for systematic irrational biases). This also applies to other valuation methods that apply the values solicited from smaller panels of experts or politicians, which is therefore sensitive to the same critique as raised for deliberative valuation as a stand-alone technique.

Distance-to-target weighting is sometimes applied as an alternative to valuation of different impacts. The target can be politically determined or revealed from political decisions. Increasingly, there is an interest in determining 'science-based' targets, based on the concept of dose–response curve thresholds, beyond which an additional impact will have much more severe impacts than below the threshold (e.g. popularised as 'planetary boundaries'). The approach applies the relative distance between the current state and a target state, implying that the further below a target an impact is, the lower the weight, and the further above the target, the higher the weight. The approach conflicts with rational decision making, because the same damage is not valued equally across impact pathways, across time, and across affected population groups. The approach has become increasingly popular, probably because of the important and often necessary role that targets play in policymaking and management, both for expressing and communicating intent and for monitoring progress. This has resulted in targets being misunderstood as ends in themselves rather than means to achieve a politically determined reduction in damage. The more logical choice for comparing impacts is a valuation of the damage itself, rather than a specific targeted level of damage (Pizzol et al., 2017; Weidema, 2016). Targets could better be understood and applied as absolute constraints within which marginal valuation is then applied to identify the socially optimal solution.

Also marginal abatement costs are sometimes applied as an alternative to valuation. However, since the socially optimal abatement of an impact occurs at the point where the marginal abatement cost equals the value of the marginal damage, a separate valuation of the damage is anyway required. Often, abatement costs are instead determined at a politically determined target, which then violates the premises of welfare economics.

Valuations of predicted social impacts are often expressed in monetary (currency) units, as also recommended by The European Commission's Impact Assessment Guidelines (European Commission, 2009). However, this is mainly a question of convenience. There is no need for using monetary units during value solicitation and there is no need to express the final single-score in monetary units. Since valuations are simply cardinal expressions of relative preferences, they can be expressed in any unit of convenience, e.g., ecopoints, happiness points, Quality Adjusted Life Years, or currency units. When communicating values, the most appropriate unit should be chosen, depending on the audience. The use of monetary units for communicating values should be limited to those situations where it is desired by the audience.

Quality-Adjusted person-Life-Year (QALY) and Disability-Adjusted Life-Year (DALY) are concepts originating in health economics, where they combine a measure of severity (the Adjustment) with an extent (person-Life) and a duration (Year). The severity is obtained from surveys that use simple pairwise comparison questions, where respondents are asked to indicate which person they would regard as healthier out of two hypothetical individuals characterized by different functional limitations. These simple questions permit input from a broad cross-section of respondents spanning diverse cultural, environmental, and demographic circumstances (Salomon, 2010). Simple analytic procedures allow the ordinal responses by individuals to be translated into cardinal severity weights of diseases at the population-level on a scale from 0 (no loss of health) to 1 (equivalent to death). QALYs are essentially derived in a similar manner, but with 0 as death and 1 as a state of perfect wellbeing, thus intended to cover all aspects of the 'Ends/Intrinsic value' top row in Table 1, and thus more directly amenable to monetary valuation.

7 The product life cycle perspective

The modelling of social impacts is complicated by the initial unequal distribution of all forms of capital and the interconnectedness of human activities in global supply chains. To understand how human activities are related and to support decisions, economic models and tools have been developed, notably IO, general equilibrium, LCA, and agent-based models. These models, with adaptations to the language of different decision makers, have different levels of detail for assessing policies, projects, specific products, or activities.

In spite of their different levels of detail, the different models all share the same basic structure, namely that of a directed graph, where the nodes are activities, e.g. a dairy herd, and the edges are input and output flows, e.g. fodder in and milk and meat out. Such a directed graph, which can also be represented as a matrix, is amenable to simple mathematical operations, such as the Leontief inverse, that provide an answer to the question: 'To supply each specific output, what is the amount needed of each activity in the system?' Summing over all activities in the system then allows answering many different questions, such as 'For this specific output, what is the total material and labour input needed, the total amount of waste and CO_2 produced, and the total economic cost?' If the system includes environmental mechanisms and valuations, also 'For this specific output, what is the total impact on human well-being?'

'Input-output' (IO) is a somewhat imprecise term for a group of models based on national supply-use tables, largely maintaining a steady-state accounting perspective with fixed and homogeneous prices, constant returns to scale, no input or capacity constraints, and no substitution among inputs (Miller and Blair, 2009). The main advantage of these models is that they are firmly grounded in empirical data with a high level of detail, while remaining simple enough to maintain their transparency. Because of the low importance placed on constraints, IO models are most relevant for analyses with a long-time horizon where constraints are generally of less importance.

Similar to IO models, General Equilibrium (GE) models are based on national supply-use tables. However, they add an extra layer of optimising algorithms to reflect the balancing of supply and demand with price changes. However, the empirical basis for the additional algorithms is often limited, and even when based on econometric estimation on long time series, a substantial element of subjectivity or context-dependency is introduced by the selection of which relationships to model and which not. Therefore, the utmost prudence is required when interpreting the results. Furthermore, the large number of additional assumptions and parameters included can make it very difficult to maintain transparency of the models. Perali and Scandizzo (2018) provide an optimistic description of the current developments in GE modelling.

In contrast to IO and GE models that seek to include the entire economy top-down, Product Life Cycle Assessment (LCA) has its origin in energy and engineering process models, where each process is analysed for its inputs, outputs, and efficiency, and inputs are traced through the supply chain to arrive at the cumulative energy and material requirements for the product. This technique was expanded in the early 1970s (Hunt et al., 1992) to include environmental emissions and became known as a Resource and Environmental Profile Analysis (REPA). Further developments in the 1980s and beginning of the 1990s added methods for impact assessment of the emissions. From a modelling perspective, process models are similar to IO models (Heijungs and Suh, 2002), the only significant difference being their relative incompleteness due to cut-offs (Lenzen, 2000). In 1996, LCA was standardised through the ISO 14040-series, adding a completeness requirement that effectively can only be fulfilled by combining process and IO-modelling. However, the requirements of the ISO 14040-standards are not yet widely applied for LCA studies. Hybrid modelling, which combines the technical detail of LCA and the completeness of IO models, is an area currently in rapid development. One area that still requires more attention is to integrate uncertainty propagation in these models (Suh and Nakamura, 2007).

Agent-based (AB) models add an extra dimension of actors (agents) as operators of the different activities in the economy and as explicit owners of assets. Actors can have properties and decision-rules that can be affected by the levels of their assets. AB models thereby come closer to a realistic modelling of the actual interactions between inhomogeneous decision-makers with limited information and bounded rationality. Especially the AB models with stock-flow consistency (Caiani et al., 2016) are explicit about 'Who owns what and what do they do with it?' – thus integrating the role of financial capital, which is particularly important in explaining inequality and fluctuations in economic activity (Godley and Lavoie, 2007; Caiani et al., 2019).

Compared to contemporary cost–benefit assessments that are still notoriously incomplete when it comes to covering the full supply chain, an integration with IO, GE, LCA, and/or AB models would be an important step forward. Ideally, the best aspects of these different models should be combined. However, even when having ensured completeness in terms of modelling the full economy, there needs also to be a procedure for ensuring that all potential impacts are considered (Burdge and Vanclay, 1996). This can be obtained by expanding the balancing principles from the IO models and the stock-flow consistent financial models with the comprehensive assets concept from Table 1. Table 2 shows the structure of a transaction and stock change matrix that is exhaustive in both rows (includes all asset types) and columns (includes all types of activities). A third dimension could be added in the form of balanceable properties, which could be, e.g. economic value, dry mass, water mass, person-time, or specific elements of interest, such as carbon. Each such 'property layer' can then be balanced, considering that each row and each column must sum to zero, since every output of an object in one column must be an input in the same row but for another activity. Each of the rows and columns in Table 2 should of course be disaggregated to the appropriate level for a specific analysis. Columns can furthermore be assigned to specific actors, allowing the same actor to perform different activities. Such an expanded Social Accounting Matrix (SAM) could then form the basis for consistent analyses of the impacts of specific decisions across all asset types, at different levels of detail.

Assessments cannot draw conclusions on a more specific level than that the disaggregation level of the data allows. Therefore, it is important to have an adequate geographical differentiation of data and to differentiate population groups and households by income, gender, age, migratory/refugee status, ethnicity, and particular disabilities (poverty in all asset dimensions).

Table 2 Transaction and stock change matrix including all five (six) asset types

	Causing activities:				
Affected objects	Ecosphere activities	Productive activities	Leisure activities	Export/import	Stock changes
Natural	+ Natural ecosystem services	− Natural ecosystem services	− Ecosystem services, incl. natural heritage	± Biogeochemical flows	± Natural resources
Manufactured - physical	+ Deposition on cultivated areas and manufactured structures	± Manufactured physical goods	− Manufactured physical goods, incl. physical cultural heritage	± Manufactured goods	∓ Manufactured goods
Manufactured - intellectual	-	± Intellectual capital	− Intellectual capital, incl. heritage	± Intellectual capital	∓ Intellectual capital
Human capabilities	± Vectors of human health	− Work-hours ± Human capabilities + Occupational health vectors ∓ Lost hours	− Leisure hours ± Human capabilities, incl.: − autonomy, health and skills ± Lost hours	± Migration	∓ Human capabilities
Social networks	-	± Social networks, incl.: − license to operate	± Social networks, incl.: − cultural heritage	± Cross-border networks	± Social networks
Financial capital	-	− Labour pay − Interest ± Taxes ∓ Savings ∓ Voluntary transfers ∓ Untaxed rent ∓ Underpaid labour	+ Labour income + Interest ∓ Savings ∓ Voluntary transfers ± Untaxed rent ± Underpaid labour	± Foreign investments ± Voluntary transfers	+ Credit - Debt

By adding the necessary completeness of instrumental and intrinsic 'areas of protection' and by applying the modern advances within economics to provide better justified and less uncertain methods for valuation, some of the limitations of traditional cost-benefit assessment can be overcome.

8 Socio-economic impacts of agricultural development

The importance of agriculture in sustainable development stems from its large share (more than 80%) of the original economies as compared to the share in developed economies (less than 1%, e.g. in Germany). This development, driven by increasing agricultural labour-productivity through technological innovation and dissemination, also implies change in technology and education in the rest of society. The social impacts of such a major economy-wide technological change depend crucially on how the benefits of the increase in agricultural productivity is distributed, both between regions, between groups in society, and within individual households. This distribution is affected by the prior distribution of assets (natural, manufactured, human, and social) and how well the markets for labour, capital, insurance, information, and products are regulated: Who have surplus capital to invest and at what time in the development? Who owns the capital, especially the land, and is the ownership secure? Are markets freely accessible for all? Are economic rents appropriately taxed and invested?

A very detailed assessment of the prerequisites for sustainable agricultural development is provided by the 'Competitive Commercial Agriculture for Africa' study, published in a synthesis report (Morris et al., 2009) and a large number of background reports from which I would especially point to Chapter 8 of De Muro et al. (2007).

In the following sections, some of these issues will be treated in more detail. Most of these issues can only be effectively addressed by well-functioning public institutions. In Section 12, it is discussed as to what extent certification and fair-trade schemes might address these issues in the absence of public governance.

9 Inequality

From a simple equilibrium model, Eswaran and Kotwal (1986) show how - in absence of economies of scale and under imperfect supervision of agricultural labourers - an initial unequal distribution of capital will lead to larger farm sizes and a decrease in the labour-to-land ratio with a consequent decrease in social welfare for the rural population. The socially optimal situation, with efficient labour-intensive small-holders and the highest agricultural wages, occurs when access to capital is more distributed, providing a theoretical rationale for the benefits of credit institutions that accept other forms of collateral than owned land-holdings. Henderson and Isaac (2017) introduce a modern monopsony situation into the model from Eswaran and Kotwal (1986) and show how this shifts the optimal towards larger farm sizes, increases overall welfare (before utility-weighting) due to the larger prices paid to the small-holders, while having detrimental effects on the wages for the landless agricultural workers. Large-scale labour-extensive farms are only cost-effective in situations of significant economies of scale, insufficient rule of law that allows monopsony, and when agricultural labour productivity is below that of urban jobs at the same skill level (Morris et al., 2009: p. 9).

Especially when labour, capital, and insurance markets are unregulated, with unequal access to and insufficient taxation of capital gains, an initial inequality in the distribution of assets can be perpetuated and amplified when technological innovations provide new options for profitable investments to first-adopters. If the change in agricultural labour-productivity happens too fast or too unexpectedly the consequent necessary adjustments in the agricultural labour force (investment in new technology, education, and/or migration to other employment) can lag behind and create a gap between those that have sufficient assets to adjust and those that remain in increasingly lower-paid agricultural employment (Timmer, 2009). The intra-household distribution between men, women, and children may also be affected (Stockbridge, 2007). Initial inequality can be perpetuated over generations through unequal investment in education (Banerjee and Newman, 1993; Galor and Zeira, 1993).

Including utility-weighting of outcomes is therefore essential for socio-economic assessments in agriculture. Socio-economic assessments should also consider any impacts on labour, capital, and insurance markets, since well-functioning markets can alleviate some of the causes of inequality.

10 Land tenure, ownership, rents, and taxation

The increased agricultural productivity and the increasing urban populations lead to higher land values, and thus to higher land rents. For sustainable agricultural development, it is important that these land rents are appropriately taxed, and the revenue invested in further capital development, especially investing in physical and social infrastructures and in human capital by improving society-wide access to health and education. Untaxed land rents imply a socially inefficient transfer from the productive economy to passive rentiers and should be specified as such in socio-economic assessments.

Secure and transferable land tenure rights are important as an incentive to invest in increasing land productivity, for land to be used by those who can use it most productively, and as a basis for taxation of the land value. For many agriculturalists, the land they own is also the most immediate collateral that they can offer to obtain financial credit. Secure property rights are therefore often seen as a prerequisite for pro-poor growth and for facilitating migration to other employment (Morris et al., 2009).

However, secure and transferable land tenure rights do not necessarily imply private property rights. The disruption of communal ownership may in fact work against a sustainable development due to higher costs and corruptibility of land reforms and the difficulty of completely capturing the implicit social security in communal land tenure (Platteau, 1996). Platteau (1996) also finds scarce evidence for any effect of land ownership on access to credit, suggesting that collateral substitutes are often sufficient, such as future crops, third party or group guarantees, and the threat of loss of future borrowing opportunities, most of which involve effective trust-building and group sanction mechanisms. The use of communally owned land can also be taxed at least as effectively as privately owned land.

11 Food prices and price instability

For sustainable agricultural development it is necessary that food markets are well functioning, so the increase in agricultural productivity leads to lower food prices and thereby also benefits the steadily increasing amount of poor urban households (former agriculturalists) that spend a significant part of their income on staple foods while establishing themselves in their new employment. Stability of food prices is also important to avoid periods of food insecurity for poor households and to avoid that poor, risk-averse farmers underinvest in productivity-enhancing technology. Instability of food prices is partly caused by natural cycles (harvests are concentrated in time and yields are dependent on weather and pests) and partly by over-interpretation of short-term price fluctuations caused by hoarding or when many market actors make production decisions simultaneously (Galtier and Vindel, 2013). In their thorough study on which price stabilisation instruments work best in different contexts, Galtier and Vindel (2013) suggest that price stabilisation for staple foods may be cheaper and more effective than targeted aid, when implemented as predictable and realistic interventions when prices deviate from a price band around the medium-term market trend.

12 Certification and fair-trade schemes

It is fundamental to life cycle thinking and assessments that production and its impacts are driven by consumer demand and that these impacts can therefore be changed via a change in consumer choices. The concentration of global consumer goods trade through supermarkets has led to increased demands on product quality as well as fulfilment of labour and environmental standards, often combined with technical support and better payment to farmers, although the latter may be eroded due to monopsony. Alongside – or on top of – this development, more specific 'fair trade' labels add their own standards that may include (even) better payment of farmers, market access for small producers, price stability, micro-credit, community empowerment projects, and advocacy.

Stockbridge (2007) points out that the increasing demands from global supermarket value chains to meet quality, labour, and environmental standards while keeping prices low may indeed lead to higher wages and better conditions for skilled, high-productive workers, but that this comes at the expense of less-skilled labour and women, who often depend upon the flexibility of casual labour.

The 'fair trade' labels have been criticised for being an expensive form of aid, due to the overhead required for certification (de Janvry et al., 2015), product traceability, and marketing, as well as the many options for value-drain and corruption in the chain (Johannessen and Wilhite, 2010). Some of these problems may be overcome with the introduction of distributed ledger technologies. The few studies that have been performed on the actual impacts of 'fair trade' labels (for recent reviews, see Darko et al. (2017) and Dragusanu et al. (2014)) have not compared them to the impacts of similar-sized direct project donations, and this is also complicated by the large diversity of 'fair trade' labels and specific conditions.

Standards of supermarkets or 'fair trade' labels may play a beneficial role in agricultural development if formulated with a specific concern for credit options for low-skilled labour and women, but it will take quite a lot of ingenuity to develop a consumer-driven approach to, e.g. capture and reinvest land rents. Likewise, while the price stabilisation provided by some 'fair trade' labels is a safeguard for producers, it does not address the problem of food price volatility for the urban population in developing countries.

13 Where to look for further information

13.1 Seminal works

Bebbington, A. 1999. Capitals and capabilities: a framework for analyzing peasant viability, rural livelihoods and poverty. *World Development* 27(12), 2021–44.

Carney, D. (Ed.). 1998. *Sustainable Rural Livelihoods: What Contribution Can We Make?* Department for International Development, London.

Chambers, R. and Conway R. 1991. Sustainable rural livelihoods: practical concepts for the 21st century. IDS discussion paper 296. Available at: https://www.ids.ac.uk/files/Dp296.pdf.

Coleman, J. S. 1988. Social capital in the creation of human capital. *The American Journal of Sociology* 94, S95–120.
Sen, A. K. 1984. *Resources, Values and Development*. Basil Blackwell, Oxford.
Sen, A. K. 1985. *Commodities and Capabilities*. North-Holland, Amsterdam and Elsevier, New York.

13.2 Key organisations

- Food and Agriculture Organization of the United Nations (www.fao.org)
- International Association for Impact Assessment (www.iaia.org)
- World Bank (www.worldbank.org)

13.3 Key journals and conferences

- *Journal of Agrarian Change*
- *Environmental Impact Assessment Review*
- *Impact Assessment and Project Appraisal*

13.4 Research centres

- Institute of Development Studies (www.ids.ac.uk)
- International Food Policy Research Institute (www.ifpri.org)

14 References

Aledo-Tur, A. and Domínguez-Gómez, J. A. 2017. Social Impact Assessment (SIA) from a multidimensional paradigmatic perspective: challenges and opportunities. *Journal of Environmental Management* 195(1), 56–61. doi:10.1016/j.jenvman.2016.10.060.
Angner, E. and Loewenstein, G. F. 2017. Behavioral economics. In: Mäki, U. (Ed.), *Handbook of the Philosophy of Science: Philosophy of Economic*. Elsevier, Amsterdam, pp. 641–90.
APFSAFE. 1987. *Global Policies for Sustainable Agriculture. A Report of the Advisory Panel on Food Security, Agriculture, Forestry and Environment to the World Commission on Environment and Development*. Zed Books, London.
Banerjee, A. V. and Newman, A. F. 1993. Occupational choice and the process of development. *The Journal of Political Economy* 101(2), 274–98. doi:10.1086/261876.
Bebbington, A. 1999. Capitals and capabilities: a framework for analyzing peasant viability, rural livelihoods and poverty. *World Development* 27(12), 2021–44. doi:10.1016/S0305-750X(99)00104-7.
Becker, H. A. 2003. Theory formation and application in social impact assessment. In: Becker, H. A. and Vanclay, F. (Eds), *The International Handbook of Social Impact Assessment: Conceptual and Methodological Advances*. Edward Elgar Publishing, Cheltenham and Northampton, pp. 129–42.
Beckerman, W. and Pasek, J. 1997. Plural values and environmental valuation. *Environmental Values* 6(1), 65–86. doi:10.3197/096327197776679202.

Bernstein, H. 1992. Poverty and the poor. In: Bernstein, H., Crow, B. and Johnson, H. (Eds), *Rural Livelihoods: Crises and Responses*. Oxford University Press, Oxford, pp. 13–26.

Boardman, A. E., Greenberg, D. H., Vining, A. R. and Weimer, D. L. 2006. *Cost-Benefit Analysis: Concepts and Practice* (3rd edn.). Pearson, Upper Saddle River.

Burdge, R. J. and Vanclay, F. 1996. Social impact assessment: a contribution to the state of the art series. *Impact Assessment* 14(1), 59–86. doi:10.1080/07349165.1996.9725886.

Caiani, A., Godin, A., Caverzasi, E., Gallegati, M., Kinsella, S. and Stiglitz, J. E. 2016. Agent based-stock flow consistent macroeconomics: towards a benchmark model. *Journal of Economic Dynamics and Control* 69, 375–408. doi:10.1016/j.jedc.2016.06.001.

Caiani, A., Russo, A. and Gallegati, M. 2019. Does inequality hamper innovation and growth? An AB-SFC analysis. *Journal of Evolutionary Economics* 29(1), 177–228. doi:10.1007/s00191-018-0554-8.

Carney, D. (Ed.). 1998. *Sustainable Rural Livelihoods: What Contribution Can We Make?* Department for International Development, London.

Carvalho, S. and White, H. 1997. Combining the quantitative and qualitative approaches to poverty measurement and analysis: the practice and the potential. World Bank technical paper no. 366. World Bank, Washington DC.

Chambers, R. and Conway, R. 1991. Sustainable rural livelihoods: practical concepts for the 21st century. IDS discussion paper 296. Available at: https://www.ids.ac.uk/files/Dp296.pdf.

Coleman, J. S. 1988. Social capital in the creation of human capital. *The American Journal of Sociology* 94 (Suppl.), S95–S120. doi:10.1086/228943.

Darko, E., Lynch, A. and Smith, W. 2017. *The Impact of Fairtrade: A Review of Research Evidence 2009–2015*. Overseas Development Institute, London.

de Janvry, A., McIntosh, C. and Sadoulet, E. 2015. Fair Trade and free entry: can a disequilibrium market serve as a development tool? *The Review of Economics and Statistics* 97(3), 567–73. doi:10.1162/REST_a_00512.

De Muro, P., Bocci, R., Gorgoni, S., Lombardo, L., Martone, E., Silici, L. and Russo, L. 2007. *Mozambique, Nigeria and Zambia Case Studies. Social and Environmental Impact Assessment*. Department of Economics, Università degli Studi Roma Tre, Roma.

Dendena, B. and Corsi, S. 2015. The Environmental and Social Impact Assessment: a further step towards an integrated assessment process. *Journal of Cleaner Production* 108(A), 965–77. doi:10.1016/j.jclepro.2015.07.110.

Dragusanu, R., Giovannucci, D. and Nunn, N. 2014. The economics of Fair Trade. *Journal of Economic Perspectives* 28(3), 217–36. doi:10.1257/jep.28.3.217.

Ekins, P. 1992. A four-capital model of wealth creation. In: Ekins, P. and Max-Neef, M. (Eds), *Real-Life Economics: Understanding Wealth Creation*. Routledge, London and New York, pp. 147–55.

Esteves, A. M., Franks, D. and Vanclay, F. 2012. Social impact assessment: the state of the art. *Impact Assessment and Project Appraisal* 30(1), 34–42. doi:10.1080/14615517.2012.660356.

Eswaran, M. and Kotwal, A. 1986. Access to capital and agrarian production organization. *The Economic Journal* 96(382), 482–98. doi:10.2307/2233128.

European Commission. 2009. *Impact Assessment Guidelines*. SEC(2009) 92. Available at: http://ec.europa.eu/governance/impact/commission_guidelines/docs/iag_2009_en.pdf.

ExternE. 1995. *Volume 2: Methodology. ExternE: Externalities of Energy*. Office for Official Publication of the European Communities, Luxembourg.

ExternE. 2005. *Methodology, 2005 Update. ExternE: Externalities of Energy*. Office for Official Publication of the European Communities, Luxembourg.

FAO. 2011. *Social Analysis for Agriculture and Rural Investment Projects. Vol. II and III: Practitioners Guide and Field Guide*. The Food and Agricultural Organisation of the United Nations, Rome.

Florio, M. 2014. *Applied Welfare Economics: Cost-Benefit Analysis of Projects and Policies*. Routledge, New York.

Galor, O. and Zeira, J. 1993. Income distribution and macroeconomics. *The Review of Economic Studies* 60(1), 35–52. doi:10.2307/2297811.

Galtier, F. and Vindel, B. 2013. *Managing Food Price Instability in Developing Countries. A Critical Analysis of Strategies and Instruments*. Agence Française de Développement, Paris. (À Savoir 17).

Gittell, R. and Vidal, A. 1998. *Community Organizing: Building Social Capital as a Development Strategy*. Sage Publications, Newbury Park.

Godley, W. and Lavoie, M. 2007. *Monetary Economics: an Integrated Approach to Credit, Money, Income, Production and Wealth*. Palgrave Macmillan, Basingstoke and New York.

Granovetter, M. S. 1973. The strength of weak ties. *American Journal of Sociology* 78(6), 1360–80. doi:10.1086/225469.

Hahn, R. 2012. Standardizing social responsibility? New perspectives on guidance documents and management system standards for sustainable development. *IEEE Transactions on Engineering Management* 59(4), 717–27. doi:10.1109/TEM.2012.2183639.

Hanson, B. 2008. Wither qualitative/quantitative?: Grounds for methodological convergence. *Quality and Quantity* 42(1), 97–111. doi:10.1007/s11135-006-9041-7.

Heijungs, R. and Suh, S. 2002. *Computational Structure of Life Cycle Assessment*. Kluwer Publishers, Dordrecht.

Henderson, H. and Isaac, A. G. 2017. Modern value chains and the organization of agrarian production. *American Journal of Agricultural Economics* 99(2), 379–400. doi:10.1093/ajae/aaw092.

Hildebrandt, L. and Sandham, L. A. 2014. Social Impact Assessment: the lesser sibling in the South African EIA process? *Environmental Impact Assessment Review* 48, 20–6. doi:10.1016/j.eiar.2014.04.003.

Hoon, P., Singh, N. and Wanmali, S. 1997. Sustainable livelihoods: concepts, principles and approaches to indicator development. Paper presented at the workshop Sustainable Livelihood Indicators, Social Development and Poverty Eradication Division, UNDP, New York.

Howitt, R. 1989. Social impact assessment and resource development: issues from the Australian experience. *Australian Geographer* 20(2), 153–66. doi:10.1080/00049188908702987.

Hunt, R. G., Sellers, J. D. and Franklin, W. E. 1992. Resource and Environmental Profile Analysis: a life cycle environmental assessment for products and procedures. *Environmental Impact Assessment Review* 12(3), 245–69. doi:10.1016/0195-9255(92)90020-X.

IAIA and IEA. 1999. *Principles of Environmental Impact Assessment Best Practice*. International Association for Impact Assessment, Fargo and Institute of Environmental Assessment, Lincoln.

IFC. 2012. *Performance Standards on Social and Environmental Sustainability*. International Finance Corporation, Washington DC.

IIRC. 2011. *Towards Integrated Reporting. Communicating Value in the 21st Century*. International Integrated Reporting Council, London.

ISO. 2010. *Guidance on Social Responsibility*. International Organization for Standardization, Geneva.

Johannessen, S. and Wilhite, H. 2010. Who really benefits from fairtrade? An analysis of value distribution in fairtrade coffee. *Globalizations* 7(4), 525–44. doi:10.1080/14747731.2010.505018.

Jolliet, O., Brent, A., Goedkoop, M., Itsubo, N., Mueller-Wenk, R., Peña, C., Schenk, R., Stewart, M. and Weidema, B. P. 2003. *Final Report of the LCIA Definition Study. Life Cycle Impact Assessment Programme of the UNEP/SETAC Life Cycle Initiative*. United Nations Environmental Programme, Paris. Available at: https://lca-net.com/p/1100.

Kasperczyk, N. and Knickel, K. 2006. *Final Case Study Report. Deliverable 15 of the SustainabilityA-Test project. Advanced Techniques for Evaluation of Sustainability Assessment Tools*. Institute for Environmental Studies (IVM), Amsterdam.

Kenter, J. O. 2015. Deliberative monetary valuation: An overview. Working Papers 2015-04. Laurence Mee Centre for People and the Sea, Oban.

Krutilla, J. V. 1967. Conservation reconsidered. *American Economic Review* 57, 777–86.

Kulig, A., Kolfoort, H. and Hoekstra, R. 2010. The case for the hybrid capital approach for the measurement of the welfare and sustainability. *Ecological Indicators* 10(2), 118–28. doi:10.1016/j.ecolind.2009.07.014.

Layard, R., Mayraz, G. and Nickell, S. 2008. The marginal utility of income. *Journal of Public Economics* 92(8–9), 1846–57. doi:10.1016/j.jpubeco.2008.01.007.

Lenzen, M. 2000. Errors in conventional and input-output–based life–cycle inventories. *Journal of Industrial Ecology* 4(4), 127–48. doi:10.1162/10881980052541981.

Marx, K. (Ed. Engels F.). 1885. *Das Kapital. Kritik der Politischen Oekonomie. Zweiter Band, Buch II: Der Cirkulationsprocess des Kapitals*. O. Meissner, Hamburg.

Miller, R. E. and Blair, P. D. 2009. *Input-Output Analysis: Foundations and Extensions* (2nd edn.). Cambridge University Press.

Morris, M., Binswanger, H., Byerlee, D., Savanti, P. and Staatz, J. 2009. *Awakening Africa's Sleeping Giant. Prospects for Commercial Agriculture in the Guinea Savannah Zone and Beyond*. World Bank, Washington DC.

Murakami, K., Itsubo, N., Kuriyama, K., Yoshida, K. and Tokimatsu, K. 2018. Development of weighting factors for G20 countries. Part 2: estimation of willingness to pay and annual global damage cost. *The International Journal of Life Cycle Assessment* 23(12), 2349–64. doi:10.1007/s11367-017-1372-1.

Nussbaum, M. 2000. *Women and Human Development: the Capabilities Approach*. Cambridge University Press, New York.

Parthasarathy, D. 2002. Measuring or defining sustainability? Impact assessment of an agricultural technology. *Impact Assessment and Project Appraisal* 20(4), 293–8. doi:10.3152/147154602781766537.

Perali, F. and Scandizzo, P. L. 2018. General equilibrium modelling: the integration of policy and project analysis. In: Perali, F. and Scandizzo, P. L. (Eds), *The New Generation of Computable General Equilibrium Models*. Springer, Heidelberg. Chapter 1.

Pizzol, M., Laurent, A., Sala, S., Weidema, B. P., Verones, F. and Koffler, C. 2017. Normalisation and weighting in life cycle assessment: quo vadis? *The International Journal of Life Cycle Assessment* 22(6), 853–66. doi:10.1007/s11367-016-1199-1.

Platteau, J.-P. 1996. The evolutionary theory of land rights as applied to sub-Saharan Africa: a critical assessment. *Development and Change* 27(1), 29–86. doi:10.1111/j.1467-7660.1996.tb00578.x.

Putnam, R. D. 1995. Tuning in, tuning out: the strange disappearance of social capital in America. *PS: Political Science and Politics* 28(4), 664–83. doi:10.1017/S1049096500058856.

Rice, D. P. 1967. Estimating the costs of illness. *American Journal of Public Health and the Nation's Health* 57(3), 424–40. doi:10.2105/ajph.57.3.424.

Richards, A. and Nicholls, J. (Eds). 2015. *A Discussion Document on the Valuation of Social Outcomes*. Social Value International, Liverpool.

Sala, S., Farioli, F. and Zamagni, A. 2013. Progress in sustainability science: lessons learnt from current methodologies for sustainability assessment: part 1. *The International Journal of Life Cycle Assessment* 18(9), 1653–72. doi:10.1007/s11367-012-0508-6.

Salomon, J. A. 2010. New disability weights for the global burden of disease. *Bulletin of the World Health Organization* 88(12), 879. doi:10.2471/BLT.10.084301.

Scoones, I. 2009. Livelihoods perspectives and rural development. *The Journal of Peasant Studies* 36(1), 171–96. doi:10.1080/03066150902820503.

Sen, A. K. 1984. *Resources, Values and Development*. Basil Blackwell, Oxford.

Sen, A. K. 1985. *Commodities and Capabilities*. North-Holland, Amsterdam and Elsevier, New York.

Sen, A. K. 1997. Editorial: human capital and human capability. *World Development* 25(12), 1959–61. doi:10.1016/S0305-750X(97)10014-6.

Sen, A. K. 2004. Capabilities, lists, and public reason: continuing the conversation. *Feminist Economics* 10(3), 77–80. doi:10.1080/1354570042000315163.

Solesbury, W. 2003. Sustainable livelihoods: a case study of the evolution of DFID policy. Working paper 217. Overseas Development Institute, London.

Stockbridge, M. 2007. *All-Africa Review of Experiences with Commercial Agriculture: Social Impacts. Background Report to the Competitive Commercial Agriculture in Sub-Saharan Africa (CCAA) Study*. World Bank, Washington DC.

Suh, S. and Nakamura, S. 2007. Five years in the area of input-output and hybrid LCA. *The International Journal of Life Cycle Assessment* 12(6), 351–2. doi:10.1065/lca2007.08.358.

Timmer, C. P. 2009. *A World without Agriculture: the Structural Transformation in Historical Perspective*. AEI Press, Washington DC.

UNECE. 2014. Conference of European Statisticians Recommendations on Measuring Sustainable Development. United Nations, New York and Geneva.

Vanclay, F. 2002. Conceptualising social impacts. *Environmental Impact Assessment Review* 22(3), 183–211. doi:10.1016/S0195-9255(01)00105-6.

van Schooten, M., Vanclay, F. and Slootweg, R. 2003. Conceptualizing social change processes and social impacts. In: Becker, H. A. and Vanclay, F. (Eds), *The International Handbook of Social Impact Assessment: Conceptual and Methodological Advances*. Edward Elgar Publishing, Cheltenham and Northampton, 74–91.

Viederman, S. 1994. Five capitals and three pillars of sustainability. *The Newsletter of PEGS (Committee on the Political Economy of the Good Society)* 4(1), 5,7,11–2.

von Böhm-Bawerk, E. 1889. *Positive Theorie des Kapitales*. Verlag der Wagner'schen Universitäts-Buchhandlung, Innsbruck.

Watkins, T. (s.a.). *Introduction to Cost Benefit Analysis*. Available at: http://www2.sjsu.edu/faculty/watkins/cba.htm (accessed on 29 November 2018).

Weidema, B. P. 2016. Valuation procedures for Product Environmental Footprint (PEF). In: Christiansen, K. (Ed.), *Nordic Environmental Footprint Seminar on LCA/PEF Issues*. Chapter 5. Available at: http://lca-net.com/p/2551.

Weidema, B. P. 2018. Towards a taxonomy for social impact pathway indicators. *Proceedings of the 6th International Conference on Social Life Cycle Assessment*, 10–12 September 2018. Springer, Heidelberg.

Woolcock, M. and Narayan, D. 2000. Social capital: implications for development theory, research, and policy. *The World Bank Research Observer* 15(2), 225–49. doi:10.1093/wbro/15.2.225.

World Bank. 2000. *World Development Report 2000/01*. Oxford University Press, New York.

World Commission on Environment and Development. 1987. *Our Common Future*. Oxford University Press.

Part 3

Improvement options

Chapter 9

Improvement options for agricultural crop production for food, feed and bioenergy

Benoît Gabrielle, AgroParisTech and Université Paris-Saclay, France; and Pietro Goglio, Wageningen University, The Netherlands and Cranfield University, UK

1 Introduction: the case for improving the performance of crop production systems

As underlined in previous chapters of this book, agriculture is increasingly pointed out as both the source of a range of pressing environmental issues and a part of the solution to mitigate those. Climate change may serve as a prime example of this dual challenge. On the one hand, mitigation measures based on land management (in particular cropland and grassland) feature in all scenarios complying with the 2°C warming limit agreed on in the 2015 Paris Conference of the Parties (Smith et al., 2016). On the other hand, the agricultural sector accounts for about 24% of greenhouse gas (GHG) emissions worldwide, half of which are due to land-use changes driven by the expansion of cropland and grassland (Pellerin et al., 2017). Biodiversity loss is another prominent issue with agriculture, generally viewed as a trade-off with food supply (Foley et al., 2011), not to mention non-food applications which are growing in particular to substitute GHG-intensive fossil resources (Smith et al., 2016).

Increasing the performance of agricultural systems, in particular for crops, has been highlighted as a promising route to reconcile the need for agriculture to provide more biomass while reducing its environmental footprint

http://dx.doi.org/10.19103/AS.2018.0044.11

(Paustian et al., 2016; Tilman et al., 2002; Foley et al., 2011). Environmental performance is generally expressed in terms of input use efficiency regarding nutrients, pesticides, water and energy, with also a particular emphasis on land resources. It may be measured through various frameworks, methods and indicators, reflecting the wide range of environmental and ecological issues that agriculture interacts with. Improvements may be gained at different scales, from individual cultivated fields to landscape or regional level, and encompass one growing season to decades of crop sequences or carry-over effects of land-use changes (Goglio et al., 2018b). They involve changes in single management practices, such as tillage or fertilizer management, to a more complex systems approach taking into account the inter-linkages between crops in a given crop sequence, and the intrinsic logics of cropping systems (Goglio et al., 2018b).

This chapter reviews these possible options to improve the performance of cropping systems, starting with some methodological background on performance metrics before discussing practices one at a time and later in the context of a system's approach. It wraps up with a summary of these options and suggests avenues for future research whether into a better estimation of performance or towards the implementation of low-impact crop production systems. It concludes on the necessary linkage with the demand side to achieve sustainable land-use and management.

2 Methodological background

2.1 Measuring the environmental performance

For agricultural production systems, compliance with sustainable development may be viewed from various perspectives, given the multiple functions these systems may be assigned: contributing jobs and economic revenues on the farm, maintaining landscape amenities by tending to farmland, or producing biomass for food or non-food purposes. Fully encompassing these various functions and their overall effects on the environment calls for a systems integration approach, as put forth by Liu et al. (2015), and frameworks such as the evaluation of ecosystem services or planetary boundaries, at global level.

However, at the local scale of the decision-making process (i.e. on the farms), assessing the impact of a particular management practice or cropping system is most commonly done using life-cycle assessment (LCA), a generic, holistic framework used to inform policy-making, product development or consumer choices (Hellweg and Mila i Canals, 2014). Originally, its application to agriculture focused on liquid biofuels in the early 1990s (Cherubini, 2010), but quickly proceeded to include other value-chains and in particular all types of foodstuff (Poore and Nemecek, 2018). The details of this ISO-standardized methodology are covered elsewhere and only summarized here. In the rest of this chapter, readers are assumed to be familiar with LCA and in particular the impact categories it addresses.

The objective of LCA is to estimate the impacts resulting from the use of a particular product through its entire life cycle 'from cradle to grave' (i.e. from the extraction of raw materials to the recycling or disposal of the product considered).

The results of LCA are expressed relative to a measure reflecting the usefulness of the product system, called 'functional unit'. The functional unit is defined as the quantified performance of a product system for use as a reference unit (ISO, 2006). These choices are directly related to the goals of the assessment, which may be diverse: benchmarking a given product against an alternative or reference product (e.g. bio-based vs fossil-derived, or first- vs. second-generation biofuels), identifying 'hot spots' in the life-cycle of a product to improve its environmental performance, or reporting the environmental footprint of a product or a portfolio of products to consumers or regulation authorities. LCA evaluates a wide range of impact categories, encompassing climate change, human health, toxicity for ecosystems (ecotoxicity), the eutrophication of fresh water or sea water bodies, or air pollution. This multi-criteria characteristic is one of the strong points of LCA (Klöpffer and Curran, 2014; Lo et al., 2005).

The metric commonly associated with the LCA of crop production systems amounts to a ratio between the magnitude of environmental impacts (I) for all the processes within the system boundary and the system's outputs, whether in the form of feed, food or other bio-based products (O). Minimizing this I:O ratio amounts to achieving the best possible performance. This is generally portrayed as a quest for efficiency, sometimes referred to as 'eco-efficiency' (Kulak et al., 2013).

As an example, Fig. 1 reports an early study on the energy efficiency of wheat grown with various rates of fertilizer inputs in the long-term classical experiments of Rothamsted in the United Kingdom (Brentrup et al., 2004). The performance indicator is the life-cycle fossil energy demand in MJ per tonne of grains produced. Since mineral fertilizers are energy-intensive inputs, this ratio is quite sensitive to the rates of N applied, and it appears that a mid-range value (96–144 kg N/ha) would be optimal from this point of view. Note that wheat yields peak at a higher N input value (192 kg N/ha), implying that this 'energy' optimum may not correlate with the economic one. This example uses grain yields as a measure of the system's output, in relation to its most obvious function.

Overall, there are considerable variations in I:O ratios across the world and systems: for instance, a recent meta-analysis showed that the GHG emissions of 1 kg of rice grains ranged from 0.4 to 2.3 kg CO_2-eq. (Poore and Nemecek,

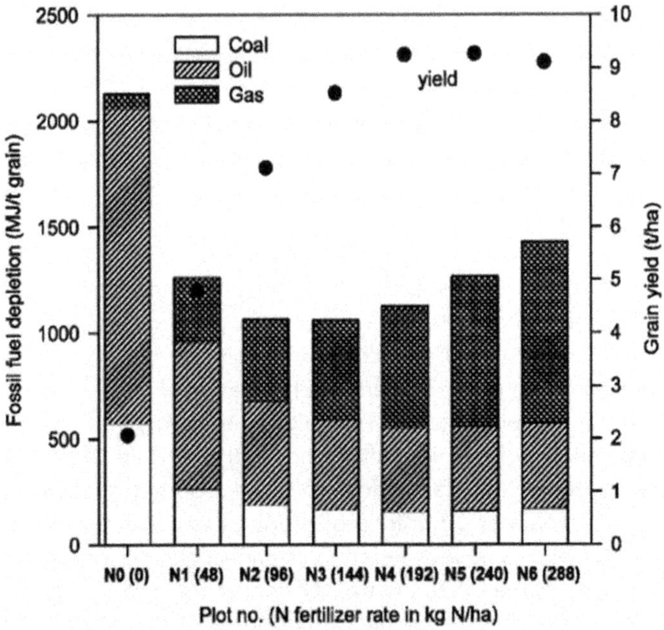

Figure 1 Combined consumption of fossil fuels (MJ/ton grain, bars) and yields (t/ha, dots) at increasing N-fertilizer rates, for wheat grown in Rothamsted, UK. Source: Adapted from Brentrup et al. (2004).

2018). Variation ranges were even larger for other impact categories, such as water eutrophication, acidification or water withdrawal. As is frequently the case with LCA, some of this variability may stem from differences in methodological assumptions (Poore and Nemecek, 2018), but according to the authors most of it was due to local context and production systems. Unravelling the effects of biophysical factors (climate, soil properties) and crop management practices to evidence strategies that consistently improve crops' performance is the core subject of the following subsections.

Key elements when assessing the environmental performance of agricultural systems include a proper accounting of greenhouse gases and of the C, N and P cycles, which play a pivotal role in the outcomes of the LCA (van Stappen et al., 2018; Goglio et al., 2015, 2017, 2018b; Nemecek et al., 2014). Indeed, the C cycle affects the soil CO_2 and CH_4 emissions responsible for the impact on climate change, several reactive N species related to the N cycles (NH_3, N_2O, NO) affect several impacts including climate change, eutrophication and acidification (De Klein et al., 2006; Brady and Weil, 2002). Finally, the P cycles are largely affecting eutrophication (Brady and Weil, 2002).

Before we proceed, note that a series of methods other than LCA are available to assess environmental sustainability, which are complementary to LCA or simply rely on alternative system boundaries or impact metrics. For instance, the ecological footprint expresses all impacts and resource consumptions by a given system relative to a single metric, the use of global hectares of biosphere which both supply energy and raw materials for this system and act as sinks for the pollutants it emits (de Oliveira et al., 2005). Its measure of environmental performance is therefore much simpler to grasp than that of LCA, amounting to a single figure (the footprint in hectares). The ecosystem services framework appears particularly relevant to investigate the impact of agriculture on the status and functioning of ecosystems worldwide. It complements LCA in that it can evaluate the effect of land management on the capacity of ecosystems to sustain biomass production but also to regulate key biogeochemical cycles or support crop growth by preserving soil resources in the long term. However, its use in the context of environmental performance assessment for agricultural systems is still challenging (de Souza et al., 2018).

2.2 Designing cropping systems and management scenarios

A crucial step when comparing agricultural systems and seeking to improve them is the way these systems are designed. Studies usually start with a 'baseline' or reference system which is supposed to represent current agricultural practices in a given supply area. Various sources may be mobilized and possibly combined to put together management and crop yield data: farmers' surveys, agricultural statistics, trials in experimental farms or expert knowledge from local scientists or advisors. Implementing variants with respect to this baseline system, to evidence improvements and quantify their potential to abate adverse environmental impacts (and thereby increase the performance) is a daunting exercise given the vastness of possible alternative practices and their combinations. In fact, designing systems ex ante has become the core subject of agronomy as a science, interweaving concepts and knowledge from a variety of disciplines (Kropff et al., 2001). Readers are referred to the latter reference for more details, since a thorough discussion would go beyond the scope of this chapter. Ultimately, design and assessment are difficult to separate since candidate systems are compared and sorted on the basis of their performance, whether economic, environmental or technical (Loyce et al., 2002). For the sake of simplicity, and in line with the current practices in agricultural LCA, we will only contrast incremental, single factor management alternatives with approaches at cropping system level.

In single management factor approaches management practices are varied one at a time (e.g. crop cultivar, fertilizer N input rate or irrigation), whereas with the systems approach cropping systems are assigned different sets of objectives, which impact a wide range of management practices. As an example of the systems approach, Colnenne-David et al. (2017) sought to halve the GHG emissions of a cropping system representative of the greater Paris (France) area, compared with a productive, high-performance counterfactual. Practices selected for the low-GHG system were intended to favour C sequestration in soils, such as the systematic use of cover crops, and legumes were introduced to save inorganic N inputs.

2.3 Playing around with scales to expand your options

In terms of spatial scale, performance is primarily assessed for an individual field characterized by a given sequence of management practices and outcomes (yields and environmental emissions). Some degree of representativeness is usually sought, be it implicit (e.g. an actual field trial which would follow regional trends in terms of management – Colnenne-David et al., 2017), or more explicitly based on statistics, surveys or a more formal typology of production systems. For instance, Renaud-Gentié et al. (2014) proposed a multidimensional data analysis method to derive generic types of management options for the farming of wine grapes in the Loire Valley in France. The objective was to apply LCA to these five types subsequently.

Management options and opportunities for change are also constrained by the fact that crops are part of larger systems and availability of non-utilized biomass material from local sources: farms, value chains which collect feedstock and interact with other systems (e.g. waste water treatment which needs land to spread biosolids on). Substitution of mineral inputs with organic forms is interesting to improve the performance of crops, but is difficult to generalize in the absence of local sources of such potential inputs, which cannot be transported over long distances for logistical reasons (Kulak et al., 2013) as a large portion of farms is plant based only. In regions highly specialized in arable crops with little livestock farming, other options will have to be sought. At value-chain level, Magrini et al. (2016) argues that the strong decline in the area cropped to pea in France was partly the result of the agri-food industry setting standards unfavourable to low-input crops, and ill-adapted logistics systems.

On the positive side, synergies may be sought with local industrial operations, for instance to valorize the heat co-generated by an anaerobic digestion facility or turn exogenous organic waste from household industries into organic amendments or fertilizers. Thus, expanding the boundaries of the agricultural system considered may generate a wider range of options and improvement routes, along with a more realistic assessment of their actual potential.

3 Improvement options

3.1 Incremental approaches or the quest for efficiency

Individual practices for crop production tested include fertilizer management (in particular N input rates, or fertilizer forms), irrigation, crop disease and weed management, soil tillage and crop residue management. Comparison of crops is also common for non-food biomass production (e.g. lignocellulosic crops or trees), and will be addressed although it should fall under the 'systems' approach since it entails changes for a number of practices. Crop sequences are covered in the next section. Interestingly, very few studies have been conducted on genotypes, except for emergent lignocellulosic crops such as *Miscanthus* or biomass sorghum (Serra et al., 2017).[1] In a similar context dealing with ethanol production from sugarcane, Otto et al. (2016) argued that genotypes delivering high yields with low N demand were paramount to improve the sustainability of biofuels in Brazil. However, breeders tend to focus on resistance to insects or drought and biomass quality characteristics, rather than nitrogen use efficiency - but also on quantitative yields. Environmental performance is directly related to this parameter since it is measured as the ratio of environmental impacts to crop yields, which should be minimized. So, even if very few studies mention genotypes, they clearly factor in genetic improvements by using yield data and trends that include these evolutions (see e.g. Otto et al. (2016) for sugarcane over the past 40 years). It is also further discussed in Kulak et al. (2013) in the context of low-input systems.

A few systematic reviews of literature are available regarding the above practices and their performance assessment following life-cycle principles, as proposed above. Kulak et al. (2013) is probably the most extensive one, focusing on low-input systems in the context of Western Europe but drawing on more general literature. Readers are referred to this article for a more thorough discussion of improvement options and their potential for adoption. The following will build on such reviews and complement them with individual studies and attempt to put them in perspective when possible, although their outcome is highly context-dependent (Bessou et al., 2011). There are clearly strong interactions between individual practices (e.g. between soil tillage and weed management), but they will be discussed separately here for the sake of simplicity, and addressed in the next subsection. This also reflects the approach taken by studies pertaining to the 'incremental' route. For instance, Goglio et al. (2018a) showed soil C and soil N_2O emissions management results are the main priorities for potential improvements of practices. Paustian et al. (2016)

1 A search done in July 2018 on the Web of Science (scientific data base) with the keywords 'genotype' and 'environmental assessment' or 'performance' yielded fewer than 10 references relevant to the topic of this chapter.

reported that improving *tillage management* and soil management would allow the sequestration of nearly 0.2 Gt of C/year at global scale. To mitigate climate change, optimal tillage and residue management both reduces GHG emissions and enhances soil carbon storage, and appears necessary to meet the Paris Agreement terms (Williamson, 2016). Enhanced soil management practices include no-till, reduce tillage and strip tillage (Brady and Weil, 2002). Conservation tillage (CT) which is now widely adopted worldwide, primarily for soil conservation reasons, reduces the use of farm machinery and favours the build-up of soil organic matter in soils (Holland, 2004). Compared to conventional tillage involving deep ploughing, life-cycle energy savings per kg of harvested crop biomass ranges from 5% to 15% depending on the tillage practices considered (Holland, 2004; Labreuche et al., 2011), also entailing GHG savings from reduced fuel use by tractors. The effect of CT on soil organic matter stocks and its potential to sequester C has been extensively researched and debated in the literature, but appears sizeable among the panel of land-based climate mitigation options (Smith et al., 2016). Its magnitude varies according to soil type, climate conditions, the type of tillage implements used and the frequency of tillage (e.g. in minimum tillage vs. no-till systems), altogether with crop yields. According to an early study, no-till generated enough C sequestration in the soil to fully offset the GHG emissions incurred by the management practices of a corn–soybean–wheat rotation in the Midwest United States, compared to the conventional tillage counterfactual (Robertson et al., 2000). Thus, the no-till rotation ended up as being almost carbon-neutral, thanks to its having created a soil sink of 1100 kg CO_2-eq./year/ha. A recent review of the current scientific evidence concluded to a 0–1100 kg CO_2-eq./year/ha range for no-till, with an average of 550 CO_2-eq./year/ha, a more conservative average sink strength of 360 CO_2-eq./year/ha for occasional ploughing, and a zero-sink for minimum tillage (Chenu and Butault, 2013).

LCA results are also reflected in these findings, see for instance, Fig. 2 comparing the impacts of various crops grown in the Walloon region of Belgium (van Stappen et al., 2018). It emphasizes the key contribution of fertilizers to the overall emissions of GHG, with a 20–70% share across crops, equally split between the production and transport of inputs and the losses of various forms of N and P after fertilizer application (often referred to as 'direct' or 'field emissions' – Perrin et al., 2014). Other research included soil-borne GHG emissions in the assessment showing that the impact of soil emissions represent the largest contribution towards global warming potential for cropping systems established in Alberta (Goglio et al., 2014; see Table 1). Direct N_2O emissions resulting from the application of fertilizer N were up to 2-fold larger than GHG emissions from the use of agricultural machinery, while the soil's sink or source strength for atmospheric CO_2 was on a par with these emissions.

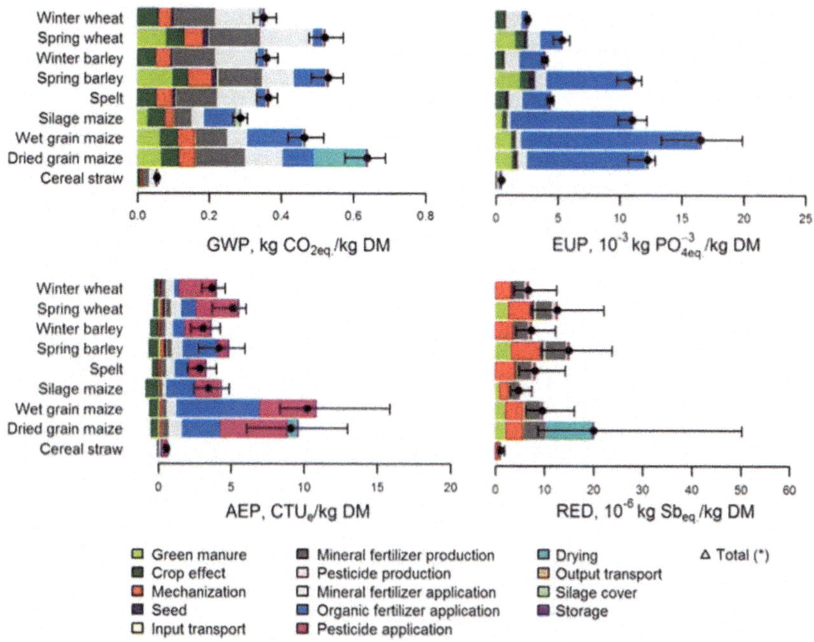

Figure 2 Life-cycle impacts for cereals grown in the Walloon region of Belgium, per kg of dry matter (DM) produced, and breakdown across cultivation steps. Whiskers represent the confidence interval of the population ($P = 0.05$). Key to impact categories: GWP: global warming potential; EUP: eutrophication potential; AEP: aquatic ecotoxicity potential; RED: renewable energy development. Source: Adapted from van Stappen et al. (2018).

Other impact indicators than global warming respond to specific management practices, beyond fertilizer inputs. This is an important conclusion as many studies tend to focus on only a limited set of indicators, in particular these GHG emissions (termed GWP on Fig. 2). Fossil energy demand is also widely used in the assessment of bioenergy systems (Muench and Guenther, 2013). While this focus may be justified from a policy-making perspective, whereby biofuels find their justification through their potential to mitigate climate change or decrease reliance on energy imports, it only conveys a restricted picture of their overall impacts and ignores trade-offs which typically occur in terms of local effects on eutrophication or biodiversity (Koh and Ghazoul, 2010).

Coming back to the case of crop production, Fig. 2 shows eutrophication potential to be mostly sensitive to fertilizer management, and especially organic manure application, which tends to result in higher direct emissions of reactive forms of N (e.g. NH_3, NO or N_2O) than mineral forms. Input production only plays a minor role. This pattern is similar with the aquatic ecotoxicity potential (AEP), although pesticides feature more prominently than fertilizers, as could be

Table 1 Percentage contribution to the global warming potential (100-year horizon) on a hectare basis for arable cropping systems at two locations in Alberta, Canada

			M[b] (%)	CO_2[c] (%)	Direct N_2O emissions (%)	CH_4[c] (%)	Indirect N_2O from NH_3 (%)	Indirect N_2O from NO_x (%)	Indirect N_2O from NO_3^- (%)
Breton (B)	50	Residue removal (0r)	48.5	−20.0	68.0	−2.0	2.4	0.1	2.9
		Residue retained (1r)	47.6	−42.7	90.1	−2.4	2.8	0.2	4.5
	75	Residue removal (0r)	42.2	−23.0	75.0	−1.6	4.4	0.3	2.8
		Residue retained (1r)	35.5	−32.1	89.5	−1.6	4.6	0.3	3.8
Ellerslie (E)	50	Residue removal (0r)	29.6	25.8	44.5	−2.5	2.1	0.1	0.4
		Residue retained (1r)	24.5	19.1	56.0	−2.4	2.1	0.1	0.7
	75	Residue removal (0r)	20.7	24.8	52.1	−1.5	2.9	0.1	0.9
		Residue retained (1r)	17.0	20.4	60.1	−1.5	2.8	0.1	1.0

[a] RM, residue management.
[b] M, Machinery, including fuel and lubricating oil use, transport and production, machinery production, maintenance repairs, transport of machinery and of raw materials for machinery manufacture.
[c] Negative values indicate a GWP reduction, due to CO_2 uptake and/or CH_4 oxidation. Soil CO_2 emissions are due to soil carbon dynamics.
Source: Adapted from Goglio et al. (2014).

expected. Negative emissions appear for this indicator (under the 'crop effect' item of the balance), that arise from the net export of heavy metals from soils in the harvested biomass, and create a benefit by extracting toxic substances from soils. Note that the heavy metals contained in harvested biomass may later be released to water bodies or other environmental compartment, an issue that may not be adequately covered in this study. Lastly, the indicator related to non-renewable resource depletion (RED) closely mirrors GHG emissions, with a slightly higher emphasis on mechanization, an energy-intensive component of the production system.

When looking at the hot spots of the crops assessed by van Stappen et al. (2018), *fertilizer management* appears as an obvious culprit to improve upon: it features prominently in most of the indicators, and is involved both in the input subsystems and the field emissions. However, alternatives to current management of fertilizer are related to the availability of organic fertilizer manure which is dependent on the presence of livestock farms. The environmental impact of organic fertilizer is also dependent on management. Increasing nutrient use efficiency (NUE: the ratio of harvested biomass to input rates) is an obvious target, as already emphasized by Otto et al. (2016), especially since it appears to be rather low in some parts of the world, but could 'tunnel' rapidly to higher, asymptotic values pending the appropriate combination of technologies and management practices (Zhang et al., 2015). In India or China, a doubling of NUE (from 30% to 60%) could for instance be considered – this would entail a performance increase by at least 50%, since fertilizers typically account for around 50% of the impacts of crops. In regions that have already achieved an optimal NUE (e.g. in parts of Europe or the United States), decreasing the reliance on mineral forms of fertilizers is a promising avenue, involving enhanced recycling of organic waste streams (including urban and industrial waste), and the introduction of legumes that can fix atmospheric nitrogen from the air. A study aiming at reducing GHG emissions from agriculture in France (Pellerin et al., 2017) estimated that improved N management could save 2 Mt CO_2-eq./year over the whole of France, offsetting 2% of the overall emissions of this sector. From an LCA perspective, some trade-offs may appear with the use of organic forms of fertilizers, since they result in larger eutrophication or acidification impacts (Fig. 2). However, field emissions upon spreading may be controlled by appropriate practices such as soil incorporation or injection for the application of liquid manure to reduce ammonia volatilization. Further appropriate timing will also contribute in the abatement of field emissions, especially for N_2O (Saggar, 2010; Snyder et al., 2009).

Practices involved in *crop disease and weed management* contribute little to global warming but feature prominently in indicators related to human health or ecotoxicity. While these impacts are known to be fraught with high uncertainties because of the hypotheses they involve in the emission to the damage chain, a 'consensus model' was developed for aquatic ecosystems (Hauschild et al., 2008). It was used by Berthoud et al. (2011) to estimate the impacts from pesticide use for wheat crops from real practices data collected by a network of cooperatives on about 50 000 ha in northeastern France. Pesticide application programmes differed according to soil types, and hence their overall impacts, but only to a slight extent (Fig. 3). More interestingly, the authors identified for each soil type the active ingredient that contributed most to these impacts, and substituted with a pesticide presenting a better environmental profile. Simply changing one ingredient in the programme allowed a 2-fold reduction in the ecotoxicity impacts of wheat, an abatement which actually corresponds to the objectives of the French government in terms of pesticide use between 2008 and 2018 (Hossard et al., 2017) – achieved with a much more modest adjustment in farming practices, when pesticide use failed to decrease overall during the targeted period.

This example followed the incremental approach and focused here solely on input substitution and the choice of active, chemical ingredients, whereas disease and weed management options usually follow more systemic guidelines. Such comparisons, for instance between organic and conventional agricultural systems, will be addressed in the next section.

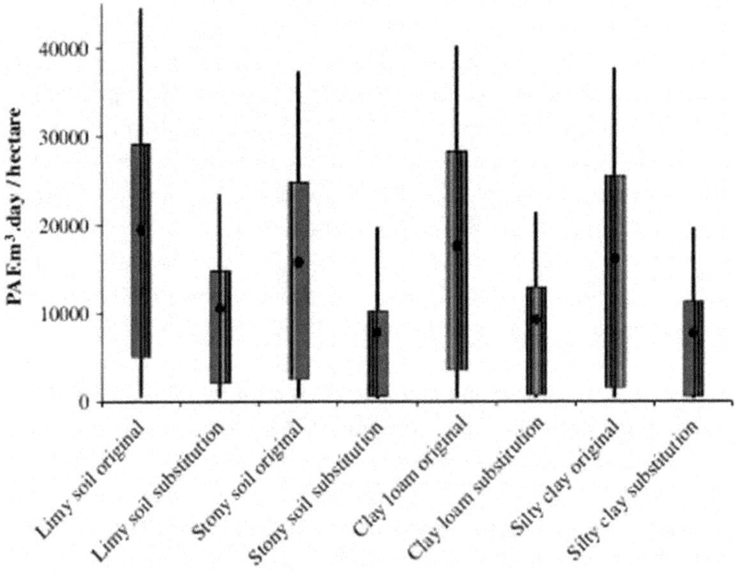

Figure 3 Active ingredients substitution effect on freshwater ecotoxicity of pesticides used on winter wheat in France (P10, Q1, average, Q3, P90). The impacts are expressed on a hectare basis, as a fraction of potentially affected species (PAF) integrated over time (day) and volume of water (m³). Source: Adapted from Berthoud et al. (2011).

In close relation to tillage, the *residue management* may also provide benefits or incur drawbacks. Frequent residue returns to soils increases soil organic matter and can create a sink of a magnitude close to no-till – with which it is often combined, if slightly higher, compared to systematic removal. Saffih-Hdadi and Mary (2008) reported a 285–1400 kg CO_2-eq./year range for the difference in soil organic C build-up between removal and return of straw in crop rotations dominated by cereals. These pertained to a set of long-term experiments over Europe and Thailand. Residue return also impacts N dynamics and other impact categories than global warming, but only to a minor extent (Gabrielle and Gagnaire, 2008).

Overall, a meta-analysis on LCA for soybean-derived bio-diesel in the United States concluded that no-till systems achieved 3-fold lower GHG emissions than their conventional counterparts. This highlights the importance of this management parameter via in particular its effect on soil C variations and sequestration (Sieverding et al., 2015).

Irrigation comes out as a similarly important practice in areas where water is a limiting factor for crop growth. Alleviating this constraint results in large productivity gains, but comes at the expense of extra energy use and costly infrastructure. It is interesting to see whether the associated impacts are compensated for by yield differences. A study on sorghum in Portugal pointed to a 3.5-fold difference between rainfed and irrigated crops (Figueiredo et al., 2017). When compared on a kg basis, both systems had similar life-cycle GHG emissions and eutrophication potential, while irrigation entailed a 2- to 3-fold increase in acidification, due to the use of fertilizer P and N in this system. In the South of France, irrigation accounts for about a third of the life-cycle energy demand of maize (Willmann et al., 2014).

Lastly, Fig. 2 highlighted large *differences between crop types* in terms of environmental profiles. The GHG emissions of grain maize are twice larger than those of silage maize, on the basis of harvested biomass. One may argue that there are differences in quality and nutritional or economic value between these two products, and that they are not directly comparable. Energy content is a commonly used metric to correct for these differences, but even so variations across crops generally persist across a whole rotation due to contrast in input requirements, productivity or susceptibility to pests and diseases. Within a sequence of crops that follow each other on the same arable field, the attribution of impacts to each specific crop is nevertheless difficult since temporal boundaries are hard to draw for the presence of complex interactions between crops (Goglio et al., 2017). For instance, it is difficult to distinguish the effect of a particular crop on the dynamics of GHG emissions for a cropping system, as the previous crop and its management affects the following crop in the sequence (Goglio et al., 2017, 2018b). In the wake of the push for agro-ecology, the agronomic question regarding crop selection is rather that of fostering the combination of a wider diversity of species over time and space (using intercrops for instance), to create synergies between organisms and enhance the resilience of agro-ecosystems (Altieri et al., 2017). This opens up the range of options available in the design of cropping systems and definitely calls for a wider approach as presented in the next subsection.

3.2 From single practices to a full-blown systems approach

Acknowledging the caveat expressed in the previous section that it is difficult to disentangle the effect of a single crop within a longer sequence (Goglio et al., 2017), and in line with concepts framed decades ago by agronomists regarding the design and drivers of these sequences (Kulak et al., 2013), the LCA community has recently tackled the organization level of cropping systems. Indeed a dual approach which considers cropping systems and crops at the same time has been proposed (Goglio et al., 2017). For the latter case, Goglio et al. (2017) proposed a series of recommendations: (1) the environmental burden of cropping system issues easily related to the crop should be fully attributed to the corresponding crops. For instance, for a crop management practice (pesticide treatment), the emissions of the sprayer should be attributed to the corresponding sprayed crop. Indeed the effect of the pesticide treatment is limited to the crop which is sprayed (Goglio et al., 2017). (2) For the environmental burden of cropping system issues which need to be attributed to a crop, combined approaches using specific attribution criteria should be adopted where possible. For instance, this approach could be used when assessing nitrate leaching with regard to eutrophication potential for a cropping system containing wheat (*Triticum aestivum* L.), clover (*Trifolium* sp.) and oilseed rape (*Brassica napus* L.). In this case, nitrate leaching can be associated to the fertilizer events as part of each crop management. (3) If the previous options are not possible, allocation approaches with generic attribution criteria should be applied (Goglio et al., 2017). This could be in the case of long-term effects such as soil carbon dynamics or the evaluation of companion cropping. Cropping systems can have variable composition, a different level of complexity, and are generally composed by a series of crops (Goglio et al., 2017; Malézieux, 2012). Performance for these systems is no longer assessed on the basis of the products of a single crop on the whole sequence of crops, which may span 6 years or more in case of highly diversified systems.

A first step, close to the single management factor approach, consists of comparing sequences of arable crops that differ by *one major management variant*: for instance CT versus traditional inversion tillage, the inclusion of legumes (vs. an absence of such crops) in the sequence, or the management of bare soil cover in between two main crops of interest with catch crops or cover crops. It is interesting to revisit tillage options, which were already discussed in the previous section, with the systems approach. Using data from a 10-year-long cropping systems experiment involving an intensive tillage versus no-till comparison and a rotation of cereals with field bean, Prechsl et al. (2017) concluded that the absence of tillage only led to a slight, non-significant reduction of GHG emissions of up to 5%. This abatement potential is lower than reported in the previous assessments, and factors in yield differences between the systems which mitigate the 10% reduction achieved by the no-till variant on a hectare basis. However, it does not include soil organic variations, in contrast with the Robertson et al. (2000) study cited earlier – for lack of a reliable measurement method to monitor variations over less than 10 years.

Another interesting outcome of such studies regards the selection of species when designing crop sequences, as touched upon in the previous section. The effects of including legumes, which is promoted by the agro-ecology literature (Altieri et al., 2017) were tested by Nemecek et al. (2008) in four locations across Europe. Compared to systems that use high-input rates, grain legumes achieved a significant reduction of energy demand, global warming potential, ozone formation and acidification, due to lower N-fertilizer requirements, conditions more favourable to reduced tillage techniques, and a greater diversification of the crop rotations which facilitate weed and disease management. However, there appeared a trade-off with nitrate leaching, which was higher than the baseline system. This may be mitigated by the use of cover crops.

Goglio et al. (2018b) examined the effect of including a 4-year perennial legume (alfalfa) in a rotation of cereals, oilseed rape (canola) and faba beans in Canada. In terms of global warming potential per GJ of harvested biomass they found highly significant differences between both cropping systems, and also between annual crops and alfalfa, and between legumes and non-legume crops. The magnitude of these differences varied according to the various estimations tested (field observations vs. a number of modelling approaches), but alfalfa consistently presented lower GHG emissions than the other crops, down to negative net values (i.e. turning into a CO_2 sink).

Systems that differ overall in terms of input use and intensity represent another, more integrated brand of systems comparison: *low-input versus high-input systems* (usually the baseline) have been extensively tested in agronomical trials (see Kulak et al., 2013 for a review), in particular for systems that include non-food/feed crops. By construction they involve a clear trade-off between the productivity per unit area (in relation to input levels) and the impacts incurred by the use of inputs; a key question ensuing is where the compensations occur and whether highly productive systems are more efficient overall than low-productivity ones - and whether there may be an optimum somewhere down the 'extensification' road (as discussed around Fig. 1 for N input rates and energy consumption). An important issue there is that reducing productivity per hectare compared to the current situation may entail indirect effects on land-use, which are hard to assess and factor into the environmental assessments (De Rosa et al., 2016). In principle this warrants a consequential approach in LCA to extend system boundaries and capture these effects, but it is still an open question due to the lack of scientific consensus on the methodology (see Section 5). Aside from this question, Kulak et al. (2013) concluded that 'reducing farm-external inputs may lead to either an improvement or reduction in eco-efficiency, depending on the crop, initial level of inputs, and environmental impacts considered'. In addition to the individual management practices reviewed in the previous section, they emphasized intercropping (i.e. mixing crop species on a given parcel of land) and biomass recycling (directly on the field or via the processing of exogenous waste streams) as major avenues for progress.

The comparison of systems based on *labels or production standards* (e.g. organic vs. conventional agriculture, or integrated pest management) may be viewed as a more sophisticated or integrated level to differentiate production options. A study on nuts, wine grapes and berries in the United States found mixed results since organic production had higher GHG emissions than conventional systems in 7 out of the 12 crop cases assessed (Venkat, 2012). In Switzerland, in the trial mentioned earlier by Prechsl et al. (2017), the performance of two organic systems included in the analysis were also contrasted: one of them had 30% lower GHG emissions than the conventional systems, on the basis of their biomass production level, but the other had twice larger emissions because of its poor yields. Organic systems had much larger eutrophication and acidification impacts because of the higher emissions of reactive N associated with manure inputs, but they entailed 4-times lower aquatic ecotoxicity impacts.

While production standards focus on the means of crop cultivation and not on their outcomes, the latter may be explicitly stated to design systems that seek to meet *specific objectives* or comply with *particular constraints* in terms of resource use (Colnenne-David et al., 2017). Objectives may have to do with the reduction of the systems' carbon footprint or water consumption. This is more challenging since one needs a capacity to predict these variables ex ante and to measure them when trialling the designed systems in real life.

Results from such a trial conducted in the Paris (France) area are given in Table 2. A system which was supposed to halve GHG emissions with regard to a reference system representative of the regional agricultural systems only achieved a 29.5% reduction in practice. However the confidence intervals between the Orid and 50% GHG systems were overlapping.

Table 2 Impact assessment with GJ of grain output as functional unit for the two cropping systems of the cropping systems under constraints trial in Grignon (France): a reference system termed 'Productive and High Environmental Performance (PHEP)', and the 50% GHG system aiming at halving GHG emissions with respect to PHEP

		50% GHG		PHEP	
Impact category	Unit (G/J)	Values	CI	Values	CI
Cumulative energy demand	MJ-eq.[a]	263	0	287	0
Global warming (100-year horizon)	kg CO_2-eq.	16.7	9.7	23.7	11.2
Acidification	g SO_{2-}eq.	153.3	0.3	162.6	0.3
Eutrophication	kg PO_4^{-3}-eq.	0.66	0.01	5.11	0.02
Human toxicity air	m^3	2.27E+05	9.59E+01	2.50E+05	7.29E+01
Human toxicity water	m^3	1.13E+03	0	1.13E+03	0
Human toxicity soil	m^3	2.42E+00	0	2.53E+00	0
Ecotoxicity water chronic	m^3	4.93E+04	0	4.45E+04	0
Ecotoxicity water acute	m^3	1.90E+03	0	1.90E+03	0
Ecotoxicity soil chronic	m^3	1.12E+04	0	1.24E+04	0

[a] 1 MJ-eq. corresponds the amount of 1 MJ consumed from different primary source.
CI: 95% confidence intervals estimated in relation to the variability of reactive N and CO_2 emission between rotations, which is mostly due to the interactions between climate and crop management.

To conclude, a few words on *agro-ecology*, generally deemed the route to sustainable agriculture and the beacon to ultimate performance. Beyond the benefits brought about by its underlying principles, very few assessments of agro-ecological systems have been carried out so far, at least regarding their performance as defined in this chapter (Altieri et al., 2017).[2] A quantification following the ecosystem services framework may be more suitable than the output-to-impact ratio used so far, since the principles of agro-ecology may be broadly translated as a maximization of these services. However, there are possible connections between this framework and that of impact assessment, as approached thus far by LCA, and used in policy-making (De Souza et al., 2018). In particular, both methods use the same inputs in their assessment, such as crop management practices or type of land-use impacted by the system considered. For instance, the study by Petit et al. (2015) showing that herbicide use could be reduced by resorting to alternative practices to control weeds, which contribute complementary ecosystem services of interest. The environmental sustainability of these strategies was not assessed by LCA but by indicators such as N input rates or fuel consumption, which are important items in the inventory of life-cycle emissions and provide surrogates for the performance indicators associated. Based on these metrics, integrated weed management out-performs both conventional and organic managements, thanks to its ability to maintain the higher yields achieved by the pesticide-intensive system.

4 Summary

The search for efficiency in agricultural production has been put forward for several decades now as an inescapable option to reconcile increasing biomass supply with the mitigation of global environmental changes. From its original focus on external inputs such as fertilizers, water and pesticides (Tilman et al., 2002), it has expanded its scope to include both a larger set of impacts and environmental issues (e.g. freshwater ecotoxicity), geographical coverage and management options, culminating with a full-blown systems approach in line with agronomic principles.

The rapid growth in the body of scientific literature available on this topic has produced a rich but complex image of performance for crop production: options which appear promising and relatively easy to implement, such as reduced tillage, were recently questioned in terms of climate mitigation benefits (see Section 3.1). Others may be very efficient in one given location but perform poorly in others (e.g. for organic agriculture – see Section 3.2), or result in trade-offs between environmental issues. Although some patterns are

2 A search done on 24 July 2018 on the Web of Science data base on 'agro-ecology' and 'LCA' or 'environment* assessment' yielded only three papers relevant to the topic at hand.

more consistent in terms of performance (e.g. the introduction of legumes in crop rotations), the general conclusion may be that performance optimization should be done on a case-by-case basis, carefully combining the options that have shown some potential in this particular context.

Further mapping of the evidence regarding these options is therefore warranted to provide better guidance to the relevant stakeholders in that respect. Integrated modelling of cropping systems on the one hand, to make this knowledge available to stakeholders, and participatory approaches such as co-design workshops on the other (Lesur-Dumoulin et al., 2018) appear a promising route in the quest for performance and overall sustainability.

5 Future trends in research

There is a growing body of work on improving environmental assessment methodologies to make them more relevant and reliable for agricultural systems. Consequential LCA for instance seeks to track the environmental implications of the deployment of product systems, while the more widespread attributional approach aims to ascribe impacts to the same particular products but ignores the potential effects of its development on other products, markets or economic sectors (Ben Aoun et al., 2013). As far as bio-based products are concerned, consequential LCA has typically been used to factor in indirect land-use change (LUC) effects related to the development of biofuels, and to address food/feed/fuel competition (Muench and Gunther, 2013). Land-use change effects are difficult to track and carry a large degree of uncertainty, but also affects significantly the GHG balances (El Akkari et al., 2018). Progress on these aspects of LCA may be expected in a near future as these effects are increasingly investigated. Meta-analyses such as that by El Akkari et al. (2018) on LUC effects on the GHG abatement potential of bioenergy are helpful to identify trends and quantify them in terms of performance. The modelling of LUC dynamics, in the context of LCA has made good progress recently, despite relying on a diverse set of approaches (De Rosa et al., 2016).

There is a lack of meta-analysis or overall review on some practices, cropping systems or impacts. Further, there is a need of standardization in the application of the LCA methodology to allow comparisons across management practices and geographical regions. For instance, conservation agriculture (which usually focuses on tillage practices and crop diversification) has been extensively assessed from the point of view of GHG emissions and crop yields (Pittelkow et al., 2015), but little is known as yet on water use, soil chemical quality or air pollution. 'Mapping' the current evidence is one way forward, whether from the point of view of geographical coverage, technologies tested or environmental issue addressed (McKinnon et al., 2015).

This is especially crucial for systems that follow the 're-design' approach of agro-ecology, that is that bear little resemblance to currently dominating systems (e.g. input-intensive mono-cultures in the Midwest United States). Because they are experimental, far from being standardized and highly variable, the resulting systems are hard to document and hence assess. The same is true for practices that appear promising on paper, but have not been fully researched and appraised so far (e.g. biochar application – Kulak et al., 2013).

Lastly, adoption by farmers of innovative practices or systems is likely to be hampered by various barriers, economic but also related to a lack of know-how or advisory services, risk aversion or sociological factors. Co-design with local stakeholders appears a promising route, with scientists acting as facilitator and providing insights into the performance and impacts of candidate cropping systems (Lesur-Dumoulin et al., 2018).

6 Where to look for further information

Review articles for non-specialists:

- from the reference list: Foley et al. (2011), Tilman et al. (2002), Poore and Nemecek (2018) and Kulak et al. (2013).

Societies or professional organizations:

- the Society of Environmental Toxicology and Chemistry (SETAC), 'dedicated to the study, analysis and solution of environmental problems, the management and regulation of natural resources' (www.setac.org)

Journals and conferences:

- *Journal of Cleaner Production, Agricultural Systems, Field Crops Research*
- LCA Food: an international Conference on Life Cycle Assessment in the Agri-Food Sector, which takes place every 2 years

International research projects:

- the AgriBalyse project/data base on the LCA of agricultural products (focused on France): https://www.ademe.fr/agribalyse-r
- Task 38 of the International Energy Agency (IEA) on the GHG balances of bioenergy (ask38.ieabioenergy.com/about)
- the International GBEP initiative on the sustainability of bio-based products (www.globalbioenergy.org)
- the Life-Cycle initiative of the UN Environment Programme (https://www.lifecycleinitiative.org/)
- US Global Change Research program (https://www.globalchange.gov/)
- Global Research Alliance on Agricultural Greenhouse Gases (https://globalresearchalliance.org/about/)

Research centres:

- LCA platform of the Joint Research Centre of the European Commission: eplca.jrc.ec.europa.eu
- the AgroScope research institute on food and environment (Switzerland), (https://www.agroscope.admin.ch/agroscope/en/home.html)

7 References

Altieri, M. A., Nicholls, C. I. and Montalba, R. (2017). Technological approaches to sustainable agriculture at a crossroads: An agroecological perspective. *Sustainability*, 9, 3722–43.

Ben Aoun, W., Gabrielle, B. and Gagnepain, B. (2013). The importance of land use change in the environmental balance of biofuels. *OCL - Oilseeds and fats Crops and Lipids*, 20, D505.

Berthoud, A., Maupu, P., Huet, C. and Poupart, A. (2011). Assessing freshwater ecotoxicity of agricultural products in life cycle assessment (LCA): A case study of wheat using French agricultural practices databases and USEtox model. *International Journal of Life Cycle Assessment*, 16, 841–7.

Bessou, C., Ferchaud, F., Gabrielle, B. and Mary, B. (2011). Biofuels, greenhouse gases & climate change. A review. *Agronomy for Sustainable Development*, 31, 1–79.

Brady, N. and Weil, R. (2002). *The Nature and Properties of Soils*, 13th ed. Prentice Hall, Upper Saddle River, NJ.

Brentrup, F., Küsters, J., Lammel, J., Barraclough, P. and Kuhlmann, H. (2004). Environmental impact assessment of agricultural production systems using the life cycle assessment (LCA) methodology II. The application to N fertilizer use in winter wheat production systems. *European Journal of Agronomy*, 20, 265–79.

Chenu, C. and Butault, J. (2013). Développer les techniques culturales sans labour pour stocker du carbone dans les sols. Technical report, INRA, Paris. URL: http://inra-dam-front-resources-cdn.brainsonic.com/ressources/afile/246624-bd3bc-resource-etude-ges-2013-rapport-partie-2-action-3-non-labour.html (accessed 24 July 2018).

Cherubini, F. (2010). GHG balances of bioenergy systems – Overview of key steps in the production chain and methodological concerns. *Renewable Energy*, 35, 1565–73.

Colnenne-David, C., Grandeau, G., Jeuffroy, M. H. and Doré, T. (2017). Ambitious environmental and economic goals for the future of agriculture are unequally achieved by innovative cropping systems. *Field Crops Research*, 210, 114–28.

De Klein, C., Novoa, R. S. A., Ogle, S., Smith, K. A., Rochette, P., Wirth, T. C., McConkey, B. G., Mosier, A., Rypdal, K., Walsh, M. and Williams, S. A. (2006). Chapter 11: N_2O emissions from managed soils, and CO_2 emissions from lime and urea application. In: Gytarsky, M., Hiraishi, T., Irving, W., Krug, T. and Penman, J. (Eds), *IPCC Guidelines for National Greenhouse Gas Inventories*. IPCC International Panel on Climate Change, Geneva, pp. 11.1–11.54.

De Oliveira, M. E. A. D., Vaughan, B. E. and Rykiel Jr., E. J. (2005). Ethanol as fuel: Energy, carbon dioxide balances, and ecological footprint. *BioScience*, 55, 593–602.

De Rosa, M., Knudsen, M. T. and Hermansen, J. E. (2016). A comparison of land use change models: Challenges and future developments. *Journal of Cleaner Production*, 113, 183–93.

De Souza, D. M., Lopes, G. R., Hansson, J. and Hansen, K. (2018). Ecosystem services in life cycle assessment: A synthesis of knowledge and recommendations for biofuels. *Ecosystem Services*, 30, 200–10.

El Akkari, M., Réchauchère, O., Bispo, A., Gabrielle, B. and Makowski, D. (2018). A meta-analysis of the greenhouse gas abatement of bioenergy factoring in land use changes. *Scientific Reports*, 8, 8563–7.

Figueiredo, F., Castanheira, E. G. and Freire, F. (2017). Life-cycle assessment of irrigated and rainfed sunflower addressing uncertainty and land use change scenarios. *Journal of Cleaner Production*, 140, 436–44.

Foley, J. A., Ramankutty, N., Brauman, K. A., Cassidy, E. S., Gerber, J. S., Johnston, M., Mueller, N. D., O'Connell, C., Ray, D. K., West, P. C., Balzer, C., Bennett, E. M., Stephen Carpenter, R., Hill, J., Monfreda, C., Polasky, S., Rockström, J., Sheehan, J., Siebert, S., Tilman, D. and Zakset, D. P. M. (2011). Solutions for a cultivated planet. *Nature*, 478, 337–42.

Gabrielle, B. and Gagnaire, N. (2008). Life-cycle assessment of straw use in bio-ethanol production: A case-study based on deterministic modelling. *Biomass and Bioenergy*, 32, 431–41.

Goglio, P., Grant, B. B., Smith, W. N., Desjardins, R. L., Worth, D. E., Zentner, R. and Malhi, S. S. (2014). Impact of management strategies on the global warming potential at the cropping system level. *Science of the Total Environment*, 490, 921–33.

Goglio, P., Smith, W. N., Grant, B. B., Desjardins, R. L., McConkey, B. G., Campbell, C. A. and Nemecek, T. (2015). Accounting for soil carbon changes in agricultural life cycle assessment (LCA): A review. *Journal of Cleaner Production*, 104, 23–39. https://doi.org/10.1016/j.jclepro.2015.05.040.

Goglio, P., Brankatschk, G., Knudsen, M. T., Williams, A. G. and Nemecek, T. (2017). Addressing crop interactions within cropping systems in LCA. *International Journal of Life Cycle Assessment*, 23, 1735–43. https://doi.org/10.1007/s11367-017-1393-9.

Goglio, P., Smith, W., Grant, B., Desjardins, R., Gao, X., Hanis, K., Tenuta, M., Campbell, C., McConkey, B., Nemecek, T., Burgess, P. and Williams, A. (2018a). A comparison of methods to quantify greenhouse gas emissions of cropping systems in LCA. *Journal of Cleaner Production*, 172, 4010–17.

Goglio, P., Smith, W. N., Worth, D. E., Grant, B. B., Desjardins, R. L., Chen, W., Tenuta, M., McConkey, B. G., Williams, A. and Burgess, P. (2018b). Development of Crop.LCA, an adaptable screening life cycle assessment tool for agricultural systems: A Canadian scenario assessment. *Journal of Cleaner Production*, 172, 3770–80. https://doi.org/10.1016/j.jclepro.2017.06.175.

Hauschild, M. Z., Huijbregts, M., Jolliet, O., Macleod, M., Margni, M., van de Meent, D., Rosenbaum, R. K. and McKone, T. E. (2008) Building a model based on scientific consensus for life cycle impact assessment of chemicals: The search for harmony and parsimony. *Environmental Science & Technology*, 42, 7032–7.

Hellweg, S. and Milà i Canals, L. (2014). Emerging approaches, challenges and opportunities in life cycle assessment. *Science*, 344, 1109–13.

Holland, J. M. (2004). The environmental consequences of adopting conservation tillage in Europe: Reviewing the evidence. *Agriculture, Ecosystems & Environment*, 103, 1–25.

Hossard, L., Guichard, L., Pelosi, C. and Makowski, D. (2017). Lack of evidence for a decrease in synthetic pesticide use on the main arable crops in France. *Science of the Total Environment*, 575, 152–61.

ISO. 2006. *SS-EN ISO 14040 Environmental Management- Life Cycle Assessment, Principles and Framework*. International Organization for Standardization, Geneva.
Klöpffer, W. and Curran, M. (2014). *Background and Future Prospects in Life Cycle Assessment. Series: LCA Compendium - The Complete World of Life Cycle Assessment*. Springer, Dordrecht, The Netherlands.
Koh, L. P. and Ghazoul, J. (2010). Spatially explicit scenario analysis for reconciling agricultural expansion, forest protection, and carbon conservation in Indonesia. *Proceedings of the National Academy of Sciences*, 107, 11140–4.
Kropff, M., Bouma, J. and Jones, J. (2001). Systems approaches for the design of sustainable agro-ecosystems. *Agricultural Systems*, 70, 369–93.
Kulak, M., Nemecek, T., Frossard, E. and Gaillard, G. (2013). How eco-efficient are low-input cropping systems in Western Europe, and what can be done to improve their eco-efficiency? *Sustainability*, 5, 3722–43.
Labreuche, J., Lellahi, A., Malaval, C. and Germon, J. (2011). Impact of no-tillage agricultural methods on the energy balance and the greenhouse gas balance of cropping systems. *Cahiers Agricultures*, 20, 204–15 (in French).
Lesur-Dumoulin, C., Laurent, A., Reau, R., Guichard, L., Ballot, R., Jeuffroy, M. and Loyce, C. (2018). Co-design and ex ante assessment of cropping system prototypes including energy crops in Eastern France. *Biomass and Bioenergy*, 116, 205–15.
Liu, J., Mooney, H., Hull, V., Davis, S. J., Gaskell, J., Hertel, T., Lubchenco, J., Seto, K. C., Gleick, P., Kremen, C. and Li, S. (2015). Systems integration for global sustainability. *Science*, 347, 963.
Lo, S.-C., Ma, H. and Lo, S.-L. (2005). Quantifying and reducing uncertainty in life cycle assessment using the Bayesian Monte Carlo method. *Science of the Total Environment*, 340, 23–33. doi:10.1016/j.scitotenv.2004.08.020.
Loyce, C., Rellier, J. P. and Meynard, J. M. (2002). Management planning for winter wheat with multiple objectives (2): Ethanol-wheat production. *Agronomy for Sustainable Development*, 72, 33–57.
Magrini, M.-B., Anton, M., Cholez, C., Corre-Hellou, G., Duc, G., Jeuffroy, M.-H., Meynard, J.-M., Pelzer, E., Voisin, A.-S. and Walrand, S. (2016). Why are grain-legumes rarely present in cropping systems despite their environmental and nutritional benefits? Analyzing lock-in in the French agrifood system. *Ecological Economics*, 126, 152–62.
Malézieux, E. (2012). Designing cropping systems from nature. *Agronomy for Sustainable Development*, 32, 15–29. doi:10.1007/s13593-011-0027-z.
McKinnon, M. C., Cheng, S. H., Garside, R., Masuda, Y. J., Miller, D. C. (2015). Map the evidence. *Nature*, 528, 185–7.
Muench, S. and Guenther, E. (2013). A systematic review of bioenergy life cycle assessments. *Applied Energy*, 112, 257–73.
Nemecek, T., von Richthofen, J.-S., Dubois, G., Casta, P., Charles, R. and Pahl, H. (2008). Environmental impacts of introducing grain legumes into European crop rotations. *European Journal of Agronomy*, 28, 380–93.
Nemecek, T., Bengoa, X., Lansche, J., Mouron, P., Rossi, V. and Humbert, S. (2014). *Methodological Guidelines for the Life Cycle Inventory of Agricultural Products*. Version 2.0. World Food LCA Database (WFLDB), Quantis and Agroscope, Lausanne and Zurich, Switzerland.
Otto, R., Castro, S. A. Q., Mariano, E., Castro, S. G. Q., Franco, H. C. J. and Trivelin, P. C. O. (2016). Nitrogen use efficiency for sugarcane-biofuel production: What is next? *BioEnergy Research*, 9, 1272–89.

Paustian, K., Lehmann, J., Ogle, S., Reay, D., Robertson, G. P., Smith, P. (2016). Climate-smart soils. *Nature*, 532, 49–57. doi:10.1038/nature17174.

Pellerin, S., Bamière, L., Angers, D., Béline, F., Benoit, M., Butault, J.-P., Chenu, C., Colnenne-David, C., De Cara, S., Delame, N., Doreau, M., Dupraz, P., Faverdin, P., Garcia-Launay, F., Hassouna, M., Hénault, C., Jeuffroy, M.-H., Klumpp, K., Métay, A., Moran, D., Recous, S., Samson, E., Savini, I. and Pardon, L. (2017). Identifying cost-competitive greenhouse gas mitigation potential of French agriculture. *Environmental Science & Policy*, 77, 130–9.

Perrin, A., Basset-Mens, C. and Gabrielle, B. (2014). Life cycle assessment of vegetable products: A review focusing on cropping systems diversity and the estimation of field emissions. *International Journal of Life Cycle Assessment*, 19, 1247–63.

Petit, S., Munier-Jolain, N., Bretagnolle, V., Bockstaller, C., Gaba, S., Cordeau, S., Lechenet, M., Meziere, D. and Colbach, N. (2015). Ecological intensification through pesticide reduction: Weed control, weed biodiversity and sustainability in arable farming. *Environmental Management*, 56, 1078–90.

Pittelkow, C. M., Liang, X., Linquist, B. A., van Groenigen, K. J., Lee, J., Lundy, M. E., van Gestel, N., Six, J., Venterea, R. T. and van Kessel, C. (2015). Productivity limits and potentials of the principles of conservation agriculture. *Nature*, 517, 365–8.

Poore, J. and Nemecek, T. (2018). Reducing food-environmental impacts through producers and consumers. *Science*, 360, 987–92.

Prechsl, U. E., Wittwer, R., van der Heijden, M. G., Lüscher, G., Jeanneret, P. and Nemecek, T. (2017). Assessing the environmental impacts of cropping systems and cover crops: Life cycle assessment of FAST, a long-term arable farming field experiment. *Agricultural Systems*, 157, 39–50.

Renaud-Gentié, C., Burgos, S. and Benoit, M. (2014). Choosing the most representative technical management routes within diverse management practices: Application to vineyards in the Loire Valley for environmental and quality assessment. *European Journal of Agronomy*, 56, 19–36.

Robertson, G. P., Paul, E. A. and Harwood, R. R. (2000). Greenhouse gases in intensive agriculture: Contributions of individual gases to the radiative forcing of the atmosphere. *Science*, 289, 1922–5.

Saffih-Hdadi, K. and Mary, B. (2008). Modeling consequences of straw residues export on soil organic carbon. *Soil Biology and Biochemistry*, 40, 594–607.

Saggar, S. (2010). Estimation of nitrous oxide emission from ecosystems and its mitigation technologies. *Agriculture, Ecosystems & Environment*, 136, 189–91. https://doi.org/10.1016/j.agee.2010.01.007.

Serra, P., Giuntoli, J., Agostini, A., Colauzzi, M. and Amaducci, S. (2017). Coupling sorghum biomass and wheat straw to minimise the environmental impact of bioenergy production. *Journal of Cleaner Production*, 154, 242–54.

Sieverding, H. L., Bailey, L. M., Hengen, T. J., Clay, D. E. and Stone, J. J. (2015). Meta-analysis of soybean-based biodiesel. *Journal of Environmental Quality*, 44, 1038–48.

Smith, P., Davis, S. J., Creutzig F., Fuss S., Minx J., Gabrielle B., Kato, E., Jackson, R. B., Cowie, A., Kriegler, E., van Vuuren, D. P., Rogelj, J., Ciais, P., Milne, J., Canadell, J. G., McCollum, D., Peters, G., Andrew, R., Krey, V., Shrestha, G., Friedlingstein, P., Gasser, T., Grübler, A., Heidug, W. K., Jonas, M., Jones, C. D., Kraxner, F., Littleton, E., Lowe, J., Moreira, J. R., Nakicenovic, N., Obersteiner, M., Patwardhan, A., Rogner, M., Rubin, E., Sharifi, A., Torvanger, A., Yamagata, Y., Edmonds, J. and Yongsung, C.

(2016). Biophysical and economic limits to negative CO_2 emissions. *Nature Climate Change*, 6, 42–50.

Snyder, C. S., Bruulsema, T. W., Jensen, T. L. and Fixen, P. E. (2009). Review of greenhouse gas emissions from crop production systems and fertilizer management effects. *Agriculture, Ecosystems & Environment*, 133, 247–66. https://doi.org/10.1016/j.agee.2009.04.021.

Tilman, D., Cassman, K. G., Matson, P. A., Naylor, R. and Polasky, S. (2002). Agricultural sustainability and intensive production practices. *Nature*, 418, 671–7.

van Stappen, F., Mathot, M., Loriers, A., Delcour, A., Stilmant, D., Planchon, V., Bodson, B., Léonard, A. and Goffart, J.-P. (2018). Sensitive parameters in local agricultural life cycle assessments: The illustrative case of cereal production in Wallonia, Belgium. *International Journal of Life Cycle Assessment*, 23, 225–50. doi:10.1007/s11367-017-1325-8.

Venkat, K. (2012). Comparison of twelve organic and conventional farming systems: A life cycle greenhouse gas emissions perspective. *Journal of Sustainable Agriculture and the Environment*, 36, 620–49.

Williamson, P. (2016). Emissions reduction: Scrutinize CO_2 removal methods. *Nature*, 530, 153–5. https://doi.org/10.1038/530153a.

Willmann, S., Dauguet, S., Tailleur, A., Schneider, A., Koch, P. and Lellahi, A. (2014). LCIA results of seven French arable crops produced within the public program AGRIBALYSE ® – Contribution to better agricultural practices. In *Proceedings of the 9th International Conference on Life Cycle Assessment in the Agri-Food Sector*, 8–10 October 2014, San Francisco, USA. ACLCA, Vashon, WA, pp. 1541–50.

Zhang, X., Davidson, E. A., Mauzerall, D. L., Searchinger, T. D., Dumas, P. and Shen, Y. (2015). Managing nitrogen for sustainable development. *Nature*, 528, 51.

Chapter 10

Improving the environmental performance of horticultural production: fruits and vegetables

Cornelius Adewale and David Granatstein, Washington State University, USA

1 Introduction

'Horticultural crops' are defined in different ways. An online dictionary considers horticulture as 'the science and art of growing fruits, vegetables, flowers, or ornamental plants' (*Merriam-Webster*, 2018). Other groups take a somewhat broader view and include other crops such as nuts, olives, medicinal and aromatic plants, root crops such as cassava and potato, and beverage crops such as coffee, tea and cacao. In a sense, horticulture refers to most crops other than field crops (grains, beans, oilseeds) or forages under cultivation. Some similar crops are also collected via wild harvest (e.g. berries, herbs, ornamental material) and would be considered horticulture, but would likely have a very different environmental impact than cultivated horticulture and might require different assessment metrics and methods. Given the diversity of crops, this chapter focuses on fruits and vegetables as they are increasingly considered essential to a healthy diet (USDA, 2010). The authors made this decision based on the resources available to address the topic, and therefore non-food crops (e.g. flowers, pharmaceuticals) or beverage crops are largely not included.

Life Cycle Assessment (LCA) as a process to evaluate the environmental impact of agricultural production has been applied to a number of horticultural crops. However, in reviewing the published literature, these studies tend to concentrate on two crops - apple for fruit and tomato for vegetable. These

http://dx.doi.org/10.19103/AS.2018.0044.13

two crops will be drawn upon for much of the discussion in Section 2 on key findings from LCA studies.

Generalizing again, there is a major distinction between most fruit and vegetable crops. Most fruits are perennials and grow on woody plants, while most vegetables are annuals and come from herbaceous plants. Thus, one might assume that the minimal soil cultivation in perennials could reduce the system impacts from fruits while the ability to rotate crops and reduce pest problems might lower impacts from vegetable production. There are numerous contrasts like this between fruits and vegetables to consider in an LCA, and even more diversity of issues among individual crops. Most horticultural crops have a defined climatic range, some broader than others, and climatic conditions can greatly affect production systems and the resulting LCA. Within a similar climate, growers have various practices and production systems to choose from in growing a crop, such as direct seed or transplanting, chemical pest control or biological control, and integrated production or organic production. Even for farms growing the same crop in the same locale with the same overall system, there will be differences in management that can alter the outcome of LCA. It may be that the variability within horticultural production is greater than for other categories of crops, and thus identifying widely applicable conclusions from LCA studies to help inform improved environmental performance could be challenging for these crops.

2 Key findings from LCA of horticultural crops

Findings from published LCA studies have been divided into the categories discussed below based on what is in the literature, as well as what are more universal issues and concerns across all the horticultural crops. These categories are by no means all-inclusive.

A recent meta-analysis of LCA results for fresh food products (Clune et al., 2017) provides an excellent summary of the range of global warming potential (GWP) values for different crops and livestock. This tends to be the most widely reported impact category for food LCAs. Field-grown vegetables have the lowest GWP, followed by field-grown fruits. Examples for individual fruit and vegetable crops are provided in Fig. 1. Fruits tend to have less variability among studies for the same crop than vegetables. Since GWP generally reflects higher use of fossil fuels which increases the environmental burden in a number of other impact categories, the data suggest that fruits and vegetables may be lower impact crops overall. The exception might be the greater use of crop protection materials and practices that would increase other impact categories.

Differences in the food composition (calories, protein, water content etc.) are not generally accounted for in these types of comparisons. A functional unit of mass is the most common choice in LCA studies. But a kg of fresh tomatoes may need to be considered differently than a kg of dry wheat. Horticultural

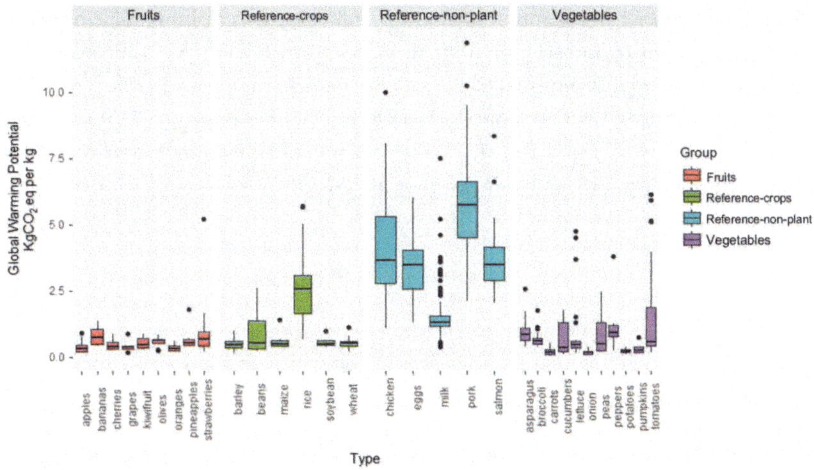

Figure 1 Examples of global warming potential for various fruits, vegetables and reference foods. Adapted from Clune et al. (2017).

crops tend not to be eaten so much for their caloric or protein contribution to diet as much as their vitamin, mineral, phytonutrient and fibre contributions. If environmental impact per unit of some of these other food characteristics were used, it could change the conclusion about relative footprint. But in order to be useful to consumers in terms of their decision-making, such information might have to be integrated into an evaluation of an entire diet.

2.1 Production system comparisons

A common question explored through LCA studies is whether switching to a different overall management system, such as integrated production (IP) or organic production, can lead to meaningful environmental improvements. These comparisons are made with 'conventional', which typically represents a much wider range of management than either of the two alternatives (which have published guidelines), and thus the choice of the conventional reference point can greatly influence the outcome. This accounts for some of the contrasting conclusions that come from comparison studies.

Alternative management systems for fruits and vegetables are commonly aimed at solving problems related to pest and disease management and use of synthetic fertilizers and pesticides that may have environmental or human health impacts. IP of apples led to lower impacts for seven categories, similar impacts for freshwater eutrophication and land use, and higher impacts for water resource depletion and mineral depletion than the conventional farms, while organic farms had higher impacts (in some cases ten-fold higher) for all categories except freshwater toxicity (Goossens et al., 2017). Many studies have found organic yields much lower (e.g. half) and this dramatically increases the impact when reported per unit mass compared with per unit area. This is illustrated in Fig. 2 where LCA was calculated for apple in four different European countries, and organic had lower impact for energy, GWP and most toxicities on an area basis but this reversed for all but aquatic and terrestrial toxicity when using a mass basis (Kägi et al., 2008). The effect of reporting unit

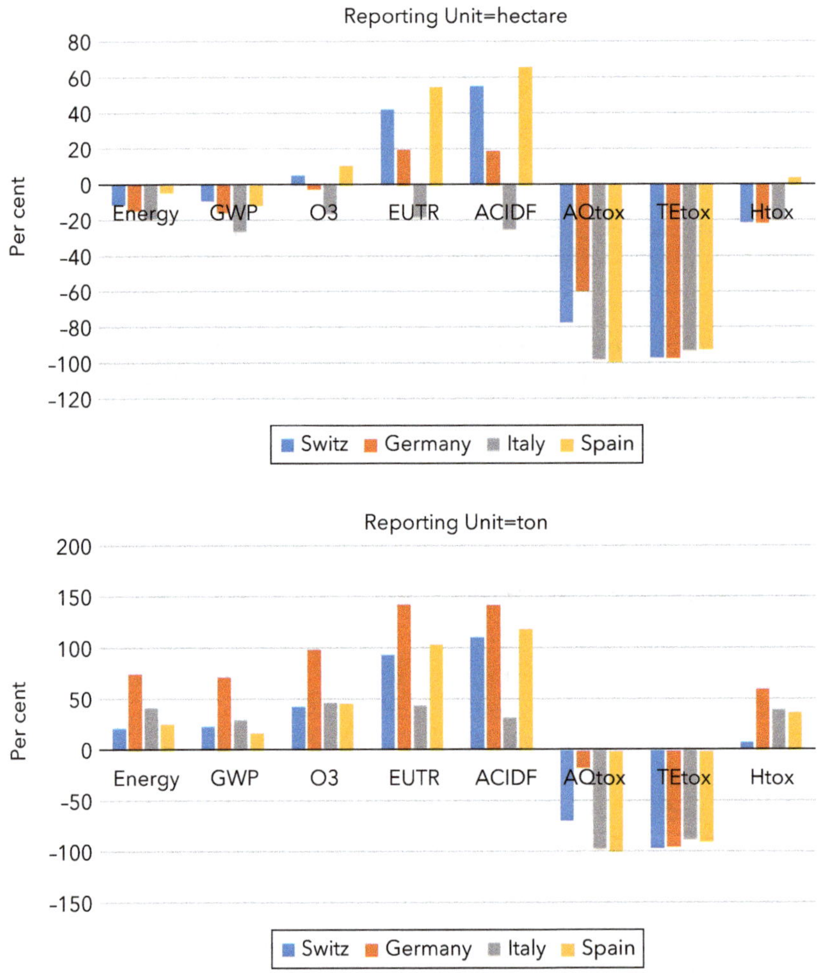

Figure 2 Per cent difference in LCA impact for organic apple versus integrated production (negative value means organic has lower impact than integrated) in four countries using two reporting units. GWP = global warming potential; O3 = ozone formation; EUTR = eutrophication; ACIDF = acidification; AQtox = aquatic toxicity; TEtox = terrestrial toxicity; Htox = human toxicity. All toxicity were based on CML model (kg DCB). Adapted from Kägi et al. (2008).

was similar for conventional versus organic vegetable system impacts from four different studies (De Backer, 2009; Foteinis and Chatzisymeon, 2016; Williams et al., 2006; Oliquino-Abasolo and Zamora, 2016). Organic production tends to have lower toxicity-related impacts regardless of the reporting unit; aquatic toxicity was 10^4 lower for organic apples in Nova Scotia, Canada (Keyes et al., 2015). Organic tomato production in China and the United Kingdom had lower energy use and GWP than conventional (He et al., 2016; Williams et al., 2006) but higher impacts for some other categories.

Manure and composts for fertilizer may lower certain impacts such as energy and resource depletion, and eutrophication, but can have higher acidification potential due to ammonia release (Alaphilippe et al., 2014; Martínez-Blanco et al., 2011a). The studies reviewed did not specify whether any indirect land use impacts involved in generating the manure were included, which would increase the impact from manure. Organic systems had lower fertilizer-related energy impacts on an area basis but higher on a mass basis due to lower yields (Kägi et al., 2008; Keyes et al., 2015).

Changing to a disease-resistant cultivar had more influence on LCA impacts than changing from conventional to organic management, illustrating the role genetics can play in reducing impact (Alaphilippe et al., 2014). Lower pesticide impacts with organic were counterbalanced by higher machinery-related impacts.

Environmental impacts were not greatly different between conventional and organic for apples grown in northern Italy using a mass basis (Longo et al., 2017). For Nova Scotia, Canada, in a cool, wet environment, results for organic were mixed, using a mass functional unit (Keyes et al., 2015). Organic impacts were much higher for energy use (+60%) and terrestrial acidification (+106%), while dramatically lower for human toxicity, metal depletion and aquatic ecotoxicity. Similar variable results were found for oranges in Spain (Ribal et al., 2017) and pears in China (Liu et al., 2010), illustrating how relative system performance can vary with location, impact category, functional unit and farm.

Using LCA to compare production systems has limitations, as shown in the examples above. One system seldom performs better across all impact categories, and if only a limited number of categories are considered (e.g. GWP being the most common), then any conclusions must clearly state that they pertain to that category only and not be extrapolated to overall system performance, as is often done in the popular media. Impacts can be converted to a common unit, such as monetary value, that can be combined across categories to give an overall assessment. The crop × location interaction is important and drawing global conclusions on a system's performance based on one location is misleading. Similarly, variability among farms lends conclusions more robust from studies using multiple farms versus single farm comparisons of two systems. In the case of organic system comparisons or others where yield might differ dramatically, the conclusion could vary depending on the reporting unit used.

Thus, using LCA for system comparisons of horticultural crops may be less useful in reaching broad conclusions with potential policy implications than for understanding the high impact aspects of a specific system and/or farm and seeking to reduce those by using practices from other systems or farms with lower impact. 'Conventional' growers can adopt lower impact pest management practices from organic, for example, if the practices meet other criteria such as cost and efficacy. A disease-resistant cultivar with market acceptance can benefit all production systems. Horticultural producers are already blending practices from various systems to create their own 'integrated' systems that boost production and income while reducing impacts on workers and the environment. LCA can be helpful in this process but cannot readily resolve trade-offs that occur when changing management systems.

2.2 Location effects

Geographic location can greatly influence the success of horticultural crop production and also its environmental footprint. Location played a role in toxicity impacts for apple in the study presented in Fig. 2, with France and Spain having substantially higher impacts using either reporting unit. Southern France production is challenged by a longer, warmer growing season with greater levels of pest and disease control required, which would impact the toxicity categories. For the same crop (apple) using the same general management system (IP), Switzerland appears to have the lowest overall impact using an area reporting unit, while Italy has the lowest impacts with a mass reporting unit (Kägi et al., 2008). Comparison of pesticide impacts for apple grown in New York State and Washington State, USA, under conventional, integrated and organic management regimes suggested higher impacts for organic in New York and lower impacts in Washington due to the absence of treatment for apple scab (*Venturia inaequalis*) in the latter (Kovach et al., 1992; Reganold et al., 2001). While these analyses utilized a method that has been extensively critiqued (Peterson and Schleier, 2014), organic scab control in a humid environment greatly increases pesticide impacts due to frequent application of less effective materials, whereas the absence of this disease in Washington leads to the lowest pesticide impacts in the organic system.

The water and land use impacts of tomato under greenhouse production were affected by location, being significantly higher in Sydney than in New South Wales Tablelands or Bundaberg Queensland in Australia (Page et al., 2014). The water scarcity footprint, freshwater ecotoxicity, marine ecotoxicity, terrestrial ecotoxicity and land use footprint were higher by 978%, 200%, 100%, 200% and 33%, respectively, in Sydney than in Tablelands using the same production system and a mass functional unit. Carbon footprint and marine eutrophication were somewhat higher for tomato produced in Tablelands. The carbon footprint for Sydney tomatoes was six times greater than Bundaberg. Additional transport energy needed for Bundaberg tomatoes was not included.

Strawberry production was compared among the four main regions in the United States, which have contrasting climates (Fig. 3). California, with the driest climate, had the lowest values for most impact categories. North Carolina values were consistently twice as high or more than for California. Florida production used 18 times more kg of pesticides than Oregon. The authors also found California to have the most kg of fruit produced per kg of N and P fertilizer used, and per MJ of energy, with the better resource efficiency translating to lower impacts (Tabatabaie and Murthy, 2016).

Several studies have looked at whether the environmental burden might be lowered by growing fruits or vegetables in a climate more amenable to their production and shipping them to other countries with less favourable climates. Saunders et al. (2006) found that producing apples in New Zealand reduced energy and GWP by 68% compared with growing them in the United Kingdom. When comparing New Zealand apples shipped to the United Kingdom with apples grown and stored there, these impacts were still 41% and 32% lower, respectively, for the New Zealand product. The situation for onions was less clear, with higher energy and GWP for onions grown in New Zealand, more energy for storing onions than for shipping but no difference

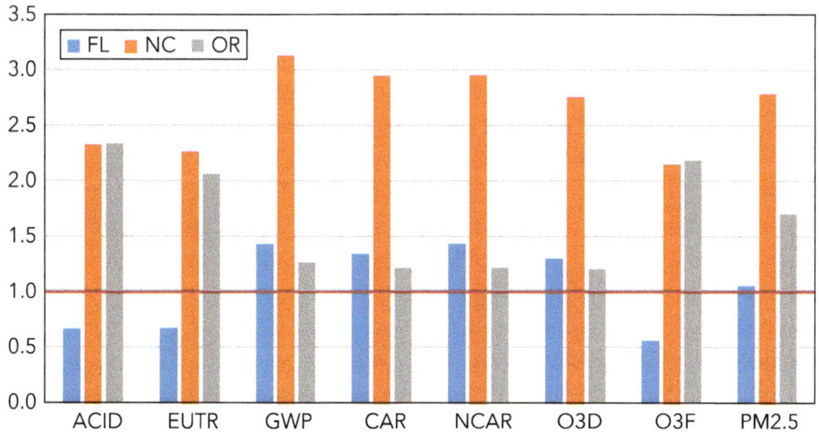

Figure 3 Relative effects of location on environmental impacts of strawberry production, USA. Values are ratio of Florida (FL), North Carolina (NC) and Oregon (OR) to California for each impact category. ACID = acidification potential; EUTR = eutrophication potential; GWP = global warming potential; CAR = carcinogenics; NCAR = non-carcinogenics; O3D = ozone depletion potential; O3F = photochemical ozone formation; PM2.5 = respiratory effects, PM2.5. Adapted from Tabatabaie and Murthy (2016).

in GWP and a smaller total energy footprint (−23%) but higher GWP (+8%) for New Zealand onions delivered to the United Kingdom. This points out the complexity of making broad generalizations about this strategy for reducing environmental impact. Payen et al. (2015) also observed that ranking, in terms of environmental impacts, between imported and local tomatoes in France was different depending on the impact category. Tomato produced in France had greater impacts for total energy consumption, global warming and eutrophication compared with imported tomato, even after including transport to France for the Moroccan tomato. The trade-off is that freshwater use impact for off-season tomatoes produced in Morocco under unheated greenhouses and then exported to France was greater by 267% than those produced in France. Without conversion to a common unit to sum across impact categories, no conclusion about the overall footprint of a product can be reached from this analysis. The expansion of 'protected cultivation' for fruits and vegetables worldwide represents an attempt to ameliorate certain climatic factors, but this comes at an environmental cost, as discussed in Section 2.8.

While it is clear that production of certain horticultural crops in a specific location can be less challenging than in other locations, this does not always lead to lower impacts. Certainly more favourable locations will reduce key impacts. For example, dry climates often reduce incidence of disease on fruit crops, lowering pesticide use and associated burden as well as the burden of application. The trade-off is that irrigation will typically be required and water scarcity may be exacerbated.

The transport stage does not always disadvantage products shipped from longer distances in terms of impacts if the production region itself has considerably lower impact potential (Brodt et al., 2013; Webb et al., 2013), but the choice of transport (e.g. rail versus truck) can determine the lowest impact. Axiomatic preference for local produce without adequate analysis of the whole supply chain can lead to erroneous conclusions depending on the relative yields, need for storage and other factors that would lead to a lower total GWP. However, other environmental impact categories need to be considered when analysing whole supply chain of fruits and vegetable for decision-making. A comprehensive review of location effects on various impact categories is an essential step to be considered to improve the environmental footprint of horticultural crops. Such analysis will reveal the trade-offs that usually occur between different environmental impacts as influenced by specific location.

2.3 Supply chain stage hot spots

LCA studies are not uniform in their setting of system boundaries. Most horticulture studies start with the 'cradle' and include upstream impacts for the procurement and manufacture of various production inputs (e.g. machinery, fertilizer, pesticides). All include the farm production stage, but many continue to storage and packing, and to transport to either a distribution centre or retail store. Some also include various end-of-life waste impacts.

Results across fruit studies are not consistent, as the relative importance of transport, packaging and storage will depend on the magnitude of impacts at the farm stage and the storage and transport requirements for that crop and its production region. Also, the study methods may not be the same. An LCA of apples from multiple regions in Italy found the transport stage contributed the most to GWP, ozone formation and acidification (40–50%), while the production stage contributed the most to eutrophication (Sessa et al., 2014). For apples in northern Italy, transport was 32% of energy use, while farm production was 15% (Longo et al., 2017). Packaging can have a large footprint, impacting ecotoxicity (e.g. cardboard manufacture; Keyes et al., 2015) and energy use (e.g. 71% of post-harvest energy; Longo et al., 2017). The farm stage dominated GWP and energy use for pears in China (Liu et al., 2010), blueberry and raspberry in Italy (Peano et al., 2015), and strawberry in Australia (Gunady et al., 2012).

For tomato products in Italy, the farm production and packaging stages resulted in the highest impact in all the considered categories due to their high consumption of fossil fuels and electricity (Del Borghi et al., 2014). Using organic fertilizers, adoption of crop rotation and use of efficient irrigation systems were important improvement options for the farm production stage. Specifically, crop rotation can lead to lower environmental impact per produced unit of output especially if active growth period is expanded and the amount of tillage in the rotation is reduced. Likewise, legume crops, which capture atmospheric nitrogen and 'fix' it into forms available to plant, can be included in the crop rotation to meet the needs of nitrogen-demanding crops. Crop rotations could also be planned in a way that reduce pest pressure, improve soil condition and improve plant nutrition. Studies have shown that unlike conventional inorganic fertilizers, organic fertilizer application could increase ecosystem respiration without obvious effects on GHG emissions (Zhang et al., 2016). These are benefits that could ultimately reduce the environmental impacts associated with the farm production stage. Reduction of packaging material mass and switching to different packaging materials would reduce impacts for the packaging stage. Reduction of packaging materials thickness, change of container size or shape, and use of a different top are some of the improvement options reported in past studies (Del Borghi et al., 2014). Direct delivery of raw loose salad reduced impacts in 14 of 15 categories studied due to packaging elimination and shorter supply chain (Tasca et al., 2017), an option for some crops.

For processed food products, the farm production stage may contribute less impact than post-harvest stages depending on the level of energy and materials used in post-harvest stages (Del Borghi et al., 2014; Palma et al., 2015; Tasca et al., 2017).

Not all LCA horticulture studies separate out the different stages of the supply chain, but many do. By comparing impacts in the different stages, one can evaluate where the biggest improvement might be made and target change at that stage. It may or may not be at the farm stage, and this is important to know both for business decisions and policy.

2.4 Yield effect

The effect of yield was already discussed in Section 2.1 with regard to conventional and organic comparisons and the effect of the reporting unit. Per unit area, yield has a much smaller influence on impact than per unit mass. In a crop such as apples, where tractor activities during harvest contribute to a large portion of impact, increasing yield will increase impact per area for those categories linked to fossil fuel use. For other aspects, higher yields will spread impact over more yield units and lead to lower impacts if the higher yields do not require additional inputs that raise impacts. Those impact categories that are primarily driven by inputs or management that act as a 'fixed cost' (same level of input needed regardless of yield) would be insensitive to yield level, while those that act more like 'variable cost' would be.

The yield effect as a source of variation for environmental impact is even more pronounced for vegetables production because different modalities in terms of cropping system characteristics are more common. As noted by Perrin et al. (2014), different production systems involve issues like the protection status (greenhouse versus open field), greenhouse type, greenhouse management, the growing media (substrate versus soil), location (temperate, Mediterranean or tropical), fertilization management (types and application rates), the production orientation (conventional, integrated or organic) and the product cultivar. All these issues can affect yield, and hence the environmental impacts. Improving yield via optimization of different production systems and choices could be vital for environmental impact reduction technique.

2.5 Farm-to-farm variability

For the same crop grown in the same region under the same general management system (e.g. IP), there can still be large farm-to-farm variation in management that leads to different environmental burdens. This was illustrated in a study of 12 fruit farms in eastern Switzerland all using IP (Mouron et al., 2006). The group was very homogeneous for climate, technology, educational level and markets; thus it provided the ability to look at the influence of management. Eight impact categories were measured, and principal component analysis identified that three categories (energy use, aquatic ecotoxicity and aquatic eutrophication) accounted for 90% of the variance and were linked to all the other categories. Impacts varied among farms by a factor of 2–3 on an area basis, and 4–7 on a total receipts (sales) basis. Reducing energy use led to lower impacts on ozone formation, GWP, terrestrial eutrophication and acidification, for example. Thus, looking for ways to reduce diesel use was a priority strategy, with proposed steps such as optimizing machinery and engine size, or better organization of harvest to reduce tractor hours. Milà i Canals et al. (2006) found similar variation among apple farms in New Zealand, pointing to opportunities to reduce impacts while achieving the same yields.

A larger survey of over 200 orange growers in the Valencia region of Spain enabled a more complex statistical analysis of variation among farms and included contrasts between conventional and organic production (Ribal et al., 2017). Farms were not normally distributed for impacts; a few farms seemed to cause the highest impacts, which were related to higher N fertilizer use and use of more toxic pesticides on conventional farms. There were no effects of farm size on the results. Cluster analysis revealed a group of conventional farms with higher impacts (C1) versus the remaining conventional farms (C2) which had impacts similar to the group of organic farms (O2). C2 farms had GWP and mineral resource depletion impacts that were 67% and 78% lower, respectively, than C1 farms. C1 farms used about double the N fertilizer per ha per year. C2 conventional farms had 25% lower diesel consumption than C1. There were no significant yield differences between C1 and C2, but C2 yields were significantly higher than O2. Thus, it appears possible to make major reductions in environmental impact in this region by focusing on more efficient machinery use, better N management and use of less toxic pest controls.

An environmental analysis of an agricultural district in southern Italy revealed vegetable production as the most variable and sensitive parameter that significantly affected the environmental impacts of the district (Cellura et al., 2012). There was large farm-to-farm variability for consumption of fertilizers, water usage, pesticide application and other agricultural inputs for vegetable production that contributed to the environmental burden. The manufacture of the greenhouse structures also accounted for a significant portion of the environmental impacts. The situation was similar for greenhouse tomato production in Colombia, where the greenhouse infrastructure accounted for most of the impacts except for freshwater ecotoxicity, marine ecotoxicity and GWP (where the greenhouse was the second largest contributor) (Bojacá et al., 2014). Despite the similar greenhouse structures, there was high yield variability among the growers that also affected their LCA burdens. Most impact categories had high variability (coefficient of variability ranging from 133% to 241%). Given the significance of greenhouse infrastructure, sourcing construction materials with the lowest impacts could be helpful for improving life cycle performance of greenhouses and the environmental impact of vegetable production in general.

Variability in life cycle impacts in these multi-farm studies is illustrated in Table 1 and can primarily be ascribed to individual farm management. This suggests that if LCA is used to help identify the higher-performing farms in a population in terms of minimizing environmental impacts and understand the management drivers for that better performance, that knowledge can be extended to other farmers to improve performance at a regional or industry-wide level. Three of the studies used the coefficient of variability (C.V.) and one used a minimum and maximum range to indicate variability. Variability is not the same for different impact categories, for different bases for reporting results (e.g. mass, area), or among studies. Thus, more emphasis is needed in LCA on understanding variability where results may be extrapolated beyond the enterprise(s) under study.

2.6 Energy use

Tree fruit operations use large amounts of fossil energy in their field operations, primarily in the form of diesel fuel for machinery. While there can be multiple trips through the orchard for crop protection sprays (30 or more is not uncommon), as well as trips for mowing, it appears that harvest is a consistent contributor to high energy use and the resulting impacts from it. Other contributors to energy use are related to manufacture of inputs such an N fertilizers, hail protection and other plastic covers, materials for support structures (posts, wire, fittings), and the plastics for irrigation systems. All these are bounded at the farm gate, which leaves further energy use downstream in terms of storing, sorting, packing and packaging, and transport to end users.

For tree fruit, organic systems generally have higher energy use per kg fruit than integrated or conventional, but less on a per ha basis (Kägi et al., 2008; Longo et al., 2017; Ribal et al., 2017). Most of the organic energy was related to machinery use, whereas with conventional most was related to fertilizer manufacture.

For tomato production, energy use in the post-harvest stage tends to be more significant. Hot spots are diesel use during production, natural gas and electricity during processing, and manufacture of the glass jars used for packaging sauce. Fuel use contributed to over half the energy use for cultivation in numerous scenarios (Ntinas et al., 2017), with energy for irrigation, water and fertilizers as additional hot spots for open-field production and energy for heating and structure materials for greenhouse production.

Plastics used for irrigation tubing, covers and netting can add large energy impacts (Tabatabaie and Murthy, 2016). No mention was made of plastic

Table 1 Examples of farm-to-farm variability in life cycle impacts

Crop (Citation)	Details	Results presented per:	Impact category	C.V. (%)	Rank
Tomato (Bojacá et al., 2014)	Colombia; unheated greenhouses, soil based; 117 farms	Mass	GWP	133	Lower
			OD	241	High
Orange (Ribal et al., 2017)	Valencia region, Spain; 145 organic farms, 122 conventional farms	Mass	GWP Org	111	Lower
			GWP Conv	256	High
			ADP Org	231	
			ADP Conv	132	
		Area	GWP Org	52	Lower
			GWP Conv	55	
			ADP Org	202	High
			ADP Conv	87	
Apple (Mouron et al., 2006)	Switzerland; homogeneous farms, farmer education; IP production; ~45 orchards on 12 farms, 4 year of data	Area	GWP	22	
			AqEutr	3	Lower
			TerrEutr	73	High
			Terr Ecotox	56	
		Receipts ($)	GWP	31	
			AqEutr	26	Lower
			TerrEutr	65	
			Terr Ecotox	95	High

Crop (Citation)	Details		GWP ($kg\ CO_2e/1000\ kg$)		Water consumption ($m^3/1000\ kg$)	
		Mass	Min	Max	Min	Max
Greenhouse vegetables (melon, pepper, zucchini) (Cellura et al., 2012)	Southeast Sicily; vegetables in greenhouses; ~60 farms	Cherry tomato	1219	1315	61.9	73.6
		Tomato	824	1019	70.6	103.8
		Melon	1138	2362	91.3	238.0
		Pepper	934	1829	71.5	240.7
		Zucchini	1415	1781	142.4	176.8

GWP = global warming potential; OD = ozone depletion potential; ADP = abiotic depletion, elements; AqEutr = aquatic eutrophication; TerrEutr = terrestrial eutrophication; TerrEcotox = terrestrial ecotoxicity; Org = organic; and Conv = conventional.

recycling and how it might affect the production stage footprint. Bio-based plastics are not currently used for these materials to any extent due to lower durability, but in the future they could help lower footprint depending on the indirect land impacts.

For addressing fuel use, there are several possible strategies. First, renewable energy sources such as biogas or electricity generated from solar energy could be used instead of fossil diesel to reduce impact (Adewale et al., 2016). Greenhouses could be designed to capture more solar energy and reduce other energy inputs for heating needs (Ntinas et al., 2017). Power requirement can be reduced through efficiency measures, such as more efficient irrigation management or equipment operation in fruit and vegetable production. Other improvement options for fruit and vegetable production include installation of air circulation systems in greenhouses, converting organic residue to biogas for electricity generation and use of materials with lower impact based on LCA evaluation. For instance, use of wood versus steel or longer-lived glass versus plastic for greenhouse construction.

Engineering innovation may help provide some novel solutions. For tree fruit, custom equipment has been developed for more efficient handling of fruit bins in the field (e.g. 'Bin Bandit', AutomatedAg, Inc., Moses Lake, WA, USA). The energy cost of a tractor moving one bin at a time, and often moving it multiple times, contributes to the large harvest footprint. Development of electric-powered equipment specifically for bin handling could also be helpful, particularly if the power could come from renewable sources. Research on solid set canopy delivery systems for crop protectant application (Niemann and Whiting, 2016) could lead to elimination of many of the tractor trips for spraying while also achieving more timely and effective pest control.

2.7 Water use efficiency

Water impacts were not reported in most of the studies, despite many horticultural crops being produced in regions that require irrigation due to low growing season precipitation. This climatic regime can reduce the management needed to control certain diseases and pests, reducing impacts related to crop protection while increasing those related to irrigation (blue water use, energy, piping systems). Studies used different measures of water use impact, including water resource depletion (WRD) (ILCD), water footprint, stress-weighted water footprint, virtual water content or simply volume of irrigation water. As with other impact categories, yield played an important role in terms of water impact per mass of fruit produced (Mordini et al., 2009). While water stress may lower yields, there is no yield advantage in supplying more than is needed and thus water acts as a 'fixed input'. A focus on better crop stress and soil moisture monitoring is a logical target for many producers.

There were no specific fruit or vegetable studies that compared the overall impacts of one irrigation system versus another, but drip irrigation was the most common system used and was considered the most water efficient. Comparisons were made between two growing regions for the same crop and how water use was affected. For example, strawberry grown in Spain used 130 m^3 water per ton fruit compared with 110 m^3 in the United Kingdom, a cooler, wetter climate (Williams et al., 2008). Organic systems had contrasting results when comparing water use to conventional, depending on crop and location (Chatzisymeon et al., 2017; Goossens et al., 2017). Greenhouse production of strawberries in Iran required nearly twice the water per ton fruit as those grown in open field (Khoshnevisan et al., 2013), whereas the reverse was true for tomato in Australia (Page et al., 2012). Use of protective netting over fruit orchards has led to a 20–30% reduction in irrigation water (L. Kalcsits, *pers. comm.*). In-row mulching with wood chips or other materials can also reduce irrigation water use by 20–25% (Granatstein and Mullinix, 2008).

Lopes et al. (2018) suggest ways to reduce the environmental impacts of irrigated agriculture in the Brazilian semi-arid region that could help lower water footprint in many places. They attributed the 32% loss of water used for irrigation to lack of knowledge of the water capacity of the soil, lack of enforcement of water use and the low cost of electricity for the producer. Recommended practice changes included reduced daily irrigation periods, analysis of the quality of the water used for irrigation, use of a tensiometer to monitor soil moisture and use of a timer device to turn off the irrigation system. The knowledge and tools exist to improve water use efficiency in horticultural production. Grower education and appropriate incentives are needed to achieve widespread implementation.

2.8 Other notable findings

There is a large degree of variability in the production of horticultural crops - among crops; between contrasting regions; between different production systems; among farms with the same crop in the same area. LCA studies have pointed out examples of groups of farms with high yields or economic returns but much lower than average environmental impacts, suggesting the knowledge, practices and tools for improvement exist and can be incorporated into commercial production (Mouron et al., 2006; Bojacá et al., 2014; Ribal et al., 2017). Continued education on best practices and public incentives for them are still needed. Producers of certain crops in a given region are trying to incentivize their peers by creating self-assessment programmes or ecolabels that monitor environmental performance within a designated group of farmers (e.g. Lodi Winegrape Commission, 2017) and encourage continuous improvement.

With the changing climate and more pressure to grow more fresh produce locally, the use of protected cultivation systems is increasing worldwide. High-value horticultural crops are the most likely to be able to economically justify these investments. Such systems can also reduce or eliminate certain crop protection costs such as insect and disease control, or bird damage. LCA was used to evaluate whether shifting to a fully enclosed system in apples might lead to reduced pest management that would result in a net lower impact (Boschiero et al., 2018). Enclosed net systems increased carbon footprint by 102–831 kg CO_2e ha^{-1} $year^{-1}$ more than the typical top-only hail netting system commonly used in Italy. If all insect pest and disease controls could be eliminated, this would reduce footprint by 305 (Gala) or 342 (Braeburn) kg CO_2e ha^{-1} $year^{-1}$, considerably less than the additional impact of the enclosures that require more plastic material. No evidence is yet available that all standard crop protection could be eliminated. While the trade-off is not positive for carbon footprint, it may be for other LCA impact categories such as human toxicity and ecotoxicity, but those were not modelled in this study.

The situation is similar for greenhouse production, primarily used for horticultural crops. While greenhouse production can boost yields per unit area and reduce pressure on the land base, as well as solve crop protection problems in many cases, the footprint from the materials involved and energy used can represent a large trade-off. Greenhouse strawberry production in Iran had much higher ecotoxicity impacts than open-field berries, while the latter had lower environmental impacts in all categories examined except acidification and eutrophication (Khoshnevisan et al., 2013).

Likewise, greenhouse tomato production in Mediterranean areas had higher impact for eutrophication and global warming potential (Martínez-Blanco et al., 2011b). Specifically, Dias et al. (2017) found heating with fossil fuels contributing between 50% and 85% of the total impact for ozone depletion, global warming, smog, acidification and respiratory effects for greenhouse tomato production in Ontario, Canada. Given the high impacts of greenhouse infrastructure, use of recycled and longer-lived materials, reducing material inputs and use of materials with lower impacts based on LCA evaluation are possible improvement options that could be adopted.

2.9 Summary of LCA findings

Results from LCAs of horticultural crops are quite varied, and no broad conclusion can be reached. However, a few key points are evident. Energy use contributes to environmental impact in all systems and should be a target for improved efficiency and lower impact energy sources. Machinery use (and the fossil fuel consumed) is a particularly important contributor in fruit systems. Crop protection structures, particularly greenhouses for vegetables, have a large effect on impacts, particularly from the materials used for the structure. Climatic differences among horticultural crop production regions can lead to dramatically different environmental burdens. In many cases, this involves trade-offs, for example, between lower impacts from crop protection and higher water-related impacts in drier regions. More needs to be done to be able to analyse impacts across category and reach conclusions about overall impact of the systems.

Production systems vary with regard to the environmental impacts from each part of the production process. As a result, no one production system could be generalized as the system which most effectively reduced environmental impact. Rather, individual farm management practices that help improve yield and lower environmental impacts should be encouraged. LCA can be a useful tool in identifying such practices, understanding the management drivers and extending such knowledge to other farmers for improved environmental performance at a regional or industry-wide level.

3 Unique challenges of LCA for horticultural crops

Many horticultural crops are perennials (e.g. fruits, beverage crops) for which the various phases of the life of the planting need to be considered in an LCA (Cerutti et al., 2014). Often, an LCA is done for a crop using data of a single year. In addition to typical year-to-year variation in yields and inputs due to weather, some fruit crops can experience alternate bearing which annual crops do not. A minimum of four years of data would average over two cycles of alternate bearing which would improve the estimate of average yield.

The point in the life of the perennial planting can affect the inputs and results from an LCA. If one only models a year during the highly productive phase, the impacts may be understated for a mass-based functional unit because years of no and low production during establishment and declining production with age are not included, which would lower the 'average' yield over the entire life of the planting. Similarly, on an area basis, there are major differences in inputs (and thus impacts) at different phases. For example, orchard establishment requires large inputs of durable materials (trees, trellis, irrigation system) whereas a full production year mostly requires fertilizers, pesticides, fuel for equipment and water. How these are included and amortized over the life of the planting will dramatically affect the impact calculations. Inclusion of the nursery phase, which can be very intensive, is another question for perennial crop LCAs. Impacts may be underestimated by as much as 30% if the nursery phase is excluded. The full production phase typically accounts for 60–70% of most impact categories (Cerutti et al., 2014). Removal of the planting as soon as productivity begins to decline would help increase average yield but would spread the fixed impacts from establishment over fewer years, perhaps offsetting the yield effect.

Accounting for carbon storage in agroecosystems is a general problem for LCA due to the complexities involved. Perennial tree crops add another dimension, as they serve as a carbon sink for a number of years. At the end of life, trees are often burned in the open field, but they can also be converted to mulch for use in a new orchard or used for firewood or biomass energy feedstock. Each of these end uses would affect the carbon calculation, and none of the studies reviewed addressed this aspect. One can speculate that open-field burning would have the highest impact, but achieving beneficial use for the trees also has a cost, such as the energy-intensive grinding operation needed to create mulch material.

As perennials, tree crop systems provide a relatively undisturbed ecosystem structure over time. This can contribute to a variety of services including habitat and biodiversity, land protection and watershed function, and aesthetic appeal, none of which were evaluated in studies reviewed.

Horticultural LCAs most often use impact categories for GWP, eutrophication and acidification. GWP is strongly related to fossil energy use, while the latter two are more linked to fertilizers. Many horticultural crops are grown in climates with more sun and less precipitation, making the assessment of water impacts relatively more important. The quality requirements for fresh fruits and vegetables typically require higher levels of crop protection, which call for better evaluation of human and ecosystem toxicity (Fantke, 2018).

Most horticultural crops, especially for fresh sales, have a number of quality criteria they must meet in order to go to market. These include size, shape, colour, firmness, blemishes and Brix. Achieving different levels of product quality undoubtedly comes with different environmental burdens, but these are rarely reflected in LCA (Schau and Fet, 2008). Fruits and vegetables are typically not considered primary calorie or protein sources in the diet. Assessing their impact on an energy basis would be less appropriate than for a cereal crop and focusing on their other dietary constituents could be more meaningful. The concept of nutrient density is more widely discussed today and a nutrient balance concept (Fern et al., 2015) has been developed that provides quantitative comparisons of foods and diets that may be suitable as an LCA metric for dietary health impacts of fruits and vegetables. Horticultural crops are mostly sold on a fresh weight basis, and a mass functional unit is most commonly used. However, dry matter content can vary according to cultivar, year, management system and location. It would be helpful to know whether use of a wet versus dry mass functional unit would materially change the outcome of an LCA comparison study.

Storage of fresh horticultural crops is mostly with refrigeration, which ideally is maintained to and at the point of sale. This adds significant environmental burden during post-harvest that other types of crops may not face, both for the energy involved and the more demanding storage structures. Therefore, putting farm production impacts in the context of the entire supply chain may be particularly important for horticultural crops.

4 Case study: OFoot Project, Washington State, USA

The case study of a small-scale vegetable farm, farm A, located near Ridgefield, Washington, USA, revealed a number of practical improvement options applicable for reduced environmental impact of vegetable production. In 2011, the farmer began participating in a multi-year collaboration with Washington State University Organic Farming Footprints (OFoot) Project. The OFoot Project developed a scientific tool to estimate the carbon footprint of organic farms. OFoot uses two models to estimate whole-farm GHG emissions and environmental impact of a farm. The first part is a carbon calculator that uses 'cradle-to-gate' analysis to estimate the GHG emissions of the equipment, infrastructure and consumables used on a farm. The second part is a process-based crop growth and field simulation model, which estimates the GHG emission attributable to soil and crop dynamics resulting from management practices. This biophysical process-based model, part of OFoot, uses a modified version of CropSyst (Stöckle et al., 2003) to simulate crop growth, field management and soil dynamics based on local soil and weather data. The outputs of OFoot include average change of soil organic carbon, N_2O emissions from the soil, N leached from the soil, N applied to the soil in fertilizer and soil amendments, applied irrigation and total farm CF.

Farm A was one of five OFoot 'focus' farms. The farm practices a 9-year rotation with eight cash crops and one growing season of a cover crop, with an additional short-term cover crop each winter. The farm has a climate typical of the maritime region, with cool, wet winters, and moderately warm, dry summers. In addition to being certified organic, practising diversified rotations and using frequent cover crops for nutrient management, the farm was selected as part of OFoot 'focus' farms because (1) it was representative of small-scale vegetable production in the Pacific Northwest with diverse annual and perennial crops (Ostrom and Donovan, 2016), and (2) detailed records required to carry out the partial LCA of the farm were available from the producer.

The OFoot carbon footprint analysis of the farm based on 2012 production data identified that fuel and tillage, electricity use and amendments/fertilizers represent a combined total of 75% of the farm's carbon footprint (Adewale et al., 2016). The 2012 baseline analysis of the CF of the farm also revealed the CF of onions produced on the farm to be 0.079 kg CO_2-eq kg^{-1} which is comparable to those reported in other studies at the time (Saunders et al., 2006; Lagerberg-Fogelberg and Carlsson-Kanyama, 2006; O'Halloran et al., 2008; Lindenthal et al., 2010). Following some of the recommendations for reduced CF based on results of the 2012 baseline analysis, the farm has been able to reduce the whole-farm CF (area basis) by 61% over a period of five years (Fig. 4) while increasing the vegetable biomass output by 3.3%.

Below is the summary of some of the practical steps implemented by the farmer for reduced footprint.

1 Soil amendments and soil emissions: Introduce an entire year fallow period between all market crops every 3–4 years. Devote the fallow area to growing nitrogen fixing cover crops, hence reducing the amount of imported fertilizers/soil amendments used in the farm. Local nitrogen capture through cover cropping helps avoid additional footprint associated with transport and production of imported fertilizers.
2 Timing: Terminate cover crops at proper growth stage and time which helps to maximize nutrient contributions from nitrogen fixing legumes.
3 Equipment and fuel: Replace the rototiller with less soil-disturbing secondary tillage tool, reduce frequency of rotovator use, and purchase a walk-behind tractor. Combination of these steps reduced the gasoline use by 39% and diesel use by 77% over the 5-year period.
4 Greenhouse management: Reduce the use of peat for greenhouse operations.
5 Infrastructure: Eliminate the use of synthetic weed barriers on the farm.
6 Irrigation and electricity: The farm installed an 8.64 kW photovoltaic system. This system annually offsets over 45% of the entire farm's electricity usage. The photovoltaic system was installed on about 50 m^2 of unused land adjacent to the farm, thus minimizing the land use impact.

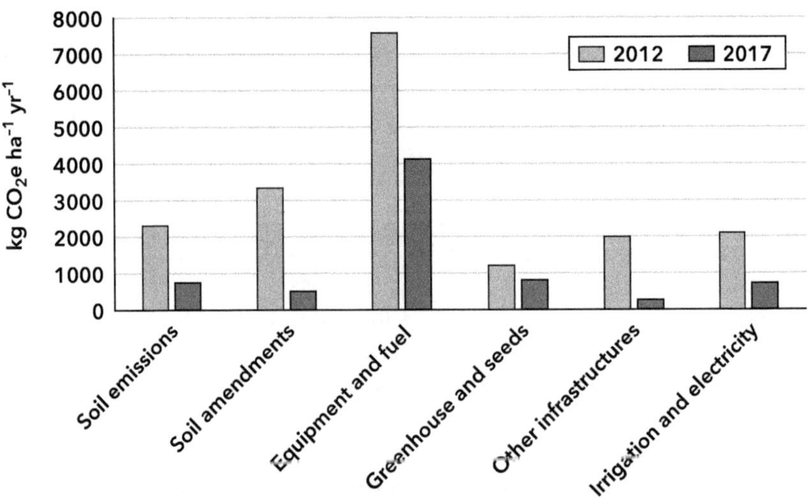

Figure 4 Carbon footprint of a certified organic vegetable farm in Washington State based on 2012 and 2017 production year data.

Through energy conservation, elimination of redundant labour and reduction of food waste, the farmer continues to incrementally refine the farm model and improve their environmental performance.

5 Future trends and conclusion

Horticultural crops can involve far more complex production systems than those for grains, pulses, oilseeds and forages. Accurately modelling these systems for LCA is challenging.

A clear future direction for horticulture is protected cultivation, especially with a changing climate and a continual onslaught of new invasive pests. The economic case may be very strong for these protected systems, while the attendant environment burden is likely to be high. The materials used in these systems generally have substantial footprints: concrete, steel, galvanized wire, plastic sheeting or netting. Redesigning systems such as orchards for protected cultivation is already occurring (Dorigoni, 2016), but with no conscious thought to reducing the footprint. One strategy worth examining is the ability to re-establish a new planting with no disturbance of the existing structure such that the durable components (concrete and steel) might be used for two or three orchard cycles versus one. Currently bioplastics cannot substitute for the longer-lived petroleum-based materials, and further research on bio-based materials is needed. Also, bio-based materials can have greater life cycle carbon emissions when indirect land use impacts are accounted for. Even increased ability to recycle or re-use end-of-life agricultural plastics could reduce the footprint.

Vertical farming is another expanding direction for horticulture that addresses the transport issue and would also provide relief from climate change and environmental stress. Often, existing buildings in urban areas are repurposed, but require large electricity inputs for artificial lighting as no natural light may be available. In other cases, new buildings are designed and built specifically for this purpose, typically with glass for light. How do these options compare with field-grown crops in terms of impact? LCA is well-suited to help answer that question.

Many horticultural crops are labour intensive and require a large workforce in the field. Social questions about farm labour are typically ignored by LCA practitioners and therefore outside the scope of published studies, but they certainly could be included. Examining the environmental impacts of the labour force may be a more comfortable step. For example, one could analyse the energy footprint of ten workers driving 50 km round trip each day to a 20-ha orchard in the United States for two months to pick apples. How does this energy impact compare to energy use for the orchard production itself? If significant, a response such as arranging vanpools for workers from the towns to the orchard might help measurably lower impact.

Further development is needed in evaluating environmental efficiency versus sustainability. High-input, high-yield systems always have lower impacts when reported on a mass basis than low-input systems that may have lower impacts when reported on an area basis (Cerutti et al., 2015). Improving energy efficiency, for example, would reduce the impact of a given system for energy and related impact categories. But if all the energy is from fossil fuel, then impact categories that reflect on this 'sustainability' aspect are needed as well. This argues for using more impact categories in horticultural LCAs that can help reconcile contradictory results between efficiency and sustainability.

6 Where to look for further information

The large majority of peer-reviewed journal articles appear in two journals: *International Journal of Life Cycle Assessment* and *Journal of Cleaner Production*. Other journals with articles include *Agriculture, Ecosystems, Environment*; *Environmental Science and Technology*; *Agricultural Systems*; *Environmental Management*. The 2015 book by Notarnicola et al., *Life Cycle Assessment in the Agri-food Sector* is an excellent resource. Proceedings from the various international conferences on LCA in the Agri-food Sector have many good articles. The American Center for Life Cycle Assessment (ACLCA) organizes an annual conference and hosts webinars (aclca.org).

7 References

Adewale, C., Higgins, S., Granatstein, D., Stöckle, C. O., Carlson, B. R., Zaher, U. E. and Carpenter-Boggs, L. 2016. Identifying hotspots in the carbon footprint of a small scale organic vegetable farm. *Agricultural Systems* 149:112–21.

Alaphilippe, A., Simon, S. and Hayer, F. (2014). Using life cycle analysis to analyse the environmental performance of organic and non-organic apple orchards (pp. 221-38). In: Bellon, S. and Penverne, S. (Eds), *Organic Farming: Prototype for Sustainable Agricultures*. Springer Science, Dordrecht, DE.

Bojacá, C. R., Wyckhuys, K. and Schrevens, E. (2014). Life cycle assessment of Colombian greenhouse tomato production based on farmer-level survey data. *Journal of Cleaner Production* 69:26–33.

Boschiero, M., Casera, C. and Kelderer, M. (2018). Carbon footprint of innovative plastic covers used as insect and pest control system in organic apple orchards. *Proceedings of 18th International Conference on Organic Fruit-Growing*, 19–21 February 2018. University of Hohenheim, Germany, pp. 71–7. http://www.ecofruit.net/proceedings.html

Brodt, S., Kramer, K. J., Kendall, A. and Feenstra, G. (2013). Comparing environmental impacts of regional and national-scale food supply chains: A case study of processed tomatoes. *Food Policy* 42:106–14. doi:10.1016/j.foodpol.2013.07.004.

Cellura, M., Ardente, F. and Longo, S. (2012). From the LCA of food products to the environmental assessment of protected crops districts: A case-study in the south of Italy. *Journal of Environmental Management* 93:194–208. doi:10.1016/j.jenvman.2011.08.019.

Cerutti, A. K., Beccaro, G. L., Bruun, S., Bosco, S., Donno, D., Notarnicola, B. and Bounous, G. (2014). Life cycle assessment application in the fruit sector: State of the art and recommendations for environmental declarations of fruit products. *Journal of Cleaner Production* 73:125–35.

Cerutti, A. K., Beccaro, G. L., Bosco, S., De Luca, A. I., Falcone, G., Fiore, A., Iofrida, N. and Giudice, A. L. (2015). Life cycle assessment in the fruit sector. Chapter 6 (pp. 333–88). In: Notarnicola, B., Salomone, R., Petti, L., Renzulli, P., Roma, R. and Cerutti, A. (Eds), *Life Cycle Assessment in the Agri-food Sector*. Springer International Publishing, Switzerland.

Chatzisymeon, E., Foteinis, S. and Borthwick, A. G. L. (2017). Life cycle assessment of the environmental performance of conventional and organic methods of open field pepper cultivation system. *International Journal of Life Cycle Assess*ment 22:896–908. doi:10.1007/s11367-016-1204-8.

Clune, S., Crossin, E. and Verghese, K. (2017). Systematic review of greenhouse gas emissions for different fresh food categories. *Journal of Cleaner Production* 140:766–83.

De Backer, E. 2009. Assessing the ecological soundness of organic and conventional agriculture by means of life cycle assessment (LCA): A case study of leek production. *British Food Journal* 111:1028–61. https://www.emeraldinsight.com/doi/abs/10.1108/00070700910992916

Del Borghi, A., Gallo, M., Strazza, C. and Del Borghi, M. (2014). An evaluation of environmental sustainability in the food industry through Life Cycle Assessment: The case study of tomato products supply chain. *Journal of Cleaner Production* 78:121–30. doi:10.1016/j.jclepro.2014.04.083.

Dias, G. M., Aver, N. W., Khosla, S., VanAcker, R., Young, S. B., Whitney, S. and Hendricks, P. 2017. Life cycle perspectives on the sustainability of Ontario greenhouse tomato production: Benchmarking and improvement opportunities. *Journal of Cleaner Production* 140:831–9.

Dorigoni, A. (2016). Innovative fruit tree architecture as a nexus to improve sustainability in orchards. *Acta Horticulturae* 1137:1–10.

Fantke, P. (2018). Modeling the environmental impacts of pesticides in agriculture. Chapter 4. In: Weidema, B. (Ed.), *Assessing the Environmental Impact of Agriculture*. Burleigh Dodds Science Publishing, Cambridge, UK (*in press*).

Fern, E. B., Watzke, H., Barclay, D. V., Roulin, A. and Drewnowski, A. (2015). The nutrient balance concept: A new quality metric for composite meals and diets. *PLoS ONE* 10:e0130491. doi:10.1371/journal.pone.0130491.

Foteinis, S. and Chatzisymeon, E. (2016). Life cycle assessment of organic versus conventional agriculture: A case study of lettuce cultivation in Greece. *Journal of Cleaner Production* 112:2462–71. doi:10.1016/j.jclepro.2015.09.075.

Goossens, Y., Annaert, B., De Tavernier, J., Mathijs, E., Keulemans, W. and Geeraerd, A. (2017). Life cycle assessment (LCA) for apple orchard production systems including low and high productive years in conventional, integrated and organic farms. *Agricultural Systems* 153:81–93.

Granatstein, D. and Mullinix, K. (2008). Mulching options for Northwest organic and conventional orchards. *HortScience* 43:45–50.

Gunady, M., Biswas, W., Solah, V. A. and James, A. P. (2012). Evaluating the global warming potential of the fresh produce supply chain for strawberries, romaine/cos lettuces

(*Lactuca sativa*), and button mushrooms (*Agaricus bisporus*) in Western Australia using life cycle assessment (LCA). *Journal of Cleaner Production* 28:81–7.

He, X., Qiao, Y., Liu, Y., Dendler, L., Yin, C. and Martin, F. (2016). Environmental impact assessment of organic and conventional tomato production in urban greenhouses of Beijing city, China. *Journal of Cleaner Production* 134:251–8. doi:10.1016/j.jclepro.2015.12.004.

Kägi, T., Casado, D., Delval, P., Gaillard, G., Otto, S. and Strassemeyer, J. (2008), Life cycle assessment of integrated and organic apple production systems in Europe. *Proceedings of the 6th International Conference on Life Cycle Assessment in the Agri-Food Sector*, Zurich, CH, 12–14 November 2008, pp. 95–6. http://www.agroscope.admin.ch/aktuell/02720/02722/03985/index.html

Khoshnevisan, B., Rafiee, S. and Mousazadeh, H. (2013). Environmental impact assessment of open field and greenhouse strawberry production. *European Journal of Agronomy* 50:29–37.

Kovach, J., Petzoldt, C., Degni, J. and Tette, J. (1992). A method to measure the environmental impact of pesticides. New York Food and Life Sciences Bulletin No. 139. NYS Agricultural Experiment Station, Geneva, NY.

Keyes, S., Tyedmers, P. and Beazley, K. (2015). Evaluating the environmental impacts of conventional and organic apple production in Nova Scotia, Canada, through life cycle assessment. *Journal of Cleaner Production* 104:40–51.

Lagerberg-Fogelberg, C. and Carlsson-Kanyama, A. (2006). Environmental assessment of foods: An LCA inspired approach. Environmental Information in Food Supply System. FOI.

Lindenthal, T., Markut, T., Hörtenhuber, S. and Rudolph, G. (2010). Greenhouse gas emissions of organic and conventional foodstuffs in Austria. LCAfood, pp. 319–24.

Liu, Y., Langer, V., Høgh-Jensen, H. and Egelyn, H. (2010). Life cycle assessment of fossil energy use and greenhouse gas emissions in Chinese pear production. *Journal of Cleaner Production* 18:1423–30.

Lodi Winegrape Commission (2017). Lodi rules for sustainable winegrowing. http://www.lodigrowers.com/wp-content/uploads/2013/04/2018-LODI-RULES-Binder-standards-3rd-Ed.-with-intro.pdf

Longo, S., Mistretta, M., Guarino, F. and Cellura, M. (2017). Life cycle assessment of organic and conventional apple supply chains in the North of Italy. *Journal of Cleaner Production* 140:654–63.

Lopes, J., Medeiros, D. L. and Kiperstok, A. (2018). Combining cleaner production and life cycle assessment for reducing the environmental impacts of irrigated carrot production in Brazilian semi-arid region. *Journal of Cleaner Production* 170:924–39. doi:10.1016/j.jclepro.2017.08.198.

Martínez-Blanco, J., Anton, A., Rieradevall, J., Castellari, M. and Munoz, P. (2011a). Comparing nutritional value and yield as functional units in the environmental assessment of horticultural production with organic or mineral fertilization. *International Journal of Life Cycle Assessment* 16:12–26. doi:10.1007/s11367-010-0238-6.

Martínez-Blanco, J., Munoz, P., Anton, A. and Rieradevall, J. (2011b). Assessment of tomato Mediterranean production in open-field and standard multi-tunnel greenhouse, with compost or mineral fertilizers, from an agricultural and environmental standpoint. *Journal of Cleaner Production* 19:985–97. doi:10.1016/j.jclepro.2010.11.018.

Merriam-Webster (2018). Merriam-Webster On-line dictionary. www.merriam-webster.com/dictionary/horticulture (accessed 21 March 2018).

Milà i Canals, Burnip, G. M. and Cowell, S. J. (2006). Evaluation of the environmental impacts of apple production using Life Cycle Assessment (LCA): Case study in New Zealand. *Agriculture, Ecosystems & Environment* 114:226–38.

Mordini, A., Nemecek, T. and Gaillard, G. (2009). Carbon and water footprint of oranges and strawberries: A literature review. Agroscope Reckenholz-Tänikon Research Station ART, Zurich, CH. http://www.agroscope.ch/

Mouron, P., Nemecek, T., Scholz, R. W. and Weber, O. (2006). Management influence on environmental impacts in an apple production system on Swiss fruit farms: Combining life cycle assessment with statistical risk assessment. *Agriculture, Ecosystems & Environment* 114:311–22.

Niemann, S. M. and Whiting, M. D. (2016). Spray coverage in apple and cherry orchards using a solid set canopy delivery system. *Acta Horticulturae* 1130:647–54. doi:10.17660/ActaHortic.2016.1130.96.

Ntinas, G. K., Neumair, M., Tsadilas, C. D. and Meyer, J. (2017). Carbon footprint and cumulative energy demand of greenhouse and open-field tomato cultivation systems under Southern and Central European climatic conditions. *Journal of Cleaner Production* 142:3617–26. doi:10.1016/j.jclepro.2016.10.106.

O'Halloran, N. J, Fisher, P. D. and Rab, M. A. (2008). Vegetable industry carbon footprint scoping study preliminary estimation of the carbon footprint of the Australian vegetable industry. In Discussion Paper (Vol. 4).

Oliquino-Abasolo, A. and Zamora, O. B. (2016). Agro-environmental sustainability of conventional and organic vegetable production systems in Tayabas, Quezon, Philippines. *Journal of Environmental Science and Management* 19:58–71.

Ostrom, M. and Donovan, C. (2016). Profile of small farms in Washington State agriculture; 2015 update. Washington State Univ. Extension Publ. TB39, Pullman, WA, 10pp. http://pubs.cahnrs.wsu.edu/publications/wp-content/uploads/sites/2/publications/tb39.pdf

Page, G., Ridoutt, B. and Bellotti, B. (2012). Carbon and water footprint tradeoffs in fresh tomato production. *Journal of Cleaner Production* 32:219–26. doi:10.1016/j.jclepro.2012.03.036.

Page, G., Ridoutt, B. and Bellotti, B. (2014). Location and technology options to reduce environmental impacts from agriculture. *Journal of Cleaner Production* 81:130–6. doi:10.1016/j.jclepro.2014.06.055.

Palma, G., Padilla, M., Saheb, M., Tatar, Y., Tugulay, A. and Kellou, I. (2015). LCA of tomato sauce made in France with French and with Turkish tomato paste. *Acta Horticulturae* 1081:81–7.

Payen, S., Basset-Mens, C. and Perret, S. (2015). LCA of local and imported tomato: An energy and water trade-off. *Journal of Cleaner Production* 87:139–48. doi:10.1016/j.jclepro.2014.10.007.

Peano, C., Baudino, C., Tecco, N. and Girgenti, V. (2015). Green marketing tools for fruit growers associated groups: Application of the Life Cycle Assessment (LCA) for strawberries and berry fruits ecobranding in northern Italy. *Journal of Cleaner Production* 104:59–67.

Perrin, A., Basset-Mens, C. and Gabrielle, B. (2014). Life cycle assessment of vegetable products: A review focusing on cropping systems diversity and the estimation of field emissions. *International Journal of Life Cycle Assessment* 19:1247–63. doi:10.1007/s11367-014-0724-3.

Peterson, R. K. D., Schleier III, J. J. (2014) A probabilistic analysis reveals fundamental limitations with the environmental impact quotient and similar systems for rating pesticide risks. *Peer Journal* 2:e364. doi:10.7717/peerj.364.

Reganold, J. P., Glover, J. D. Andrews, P. K. and Hinman, H. R. (2001). Sustainability of three apple production systems. *Nature* 410:926–30.

Ribal, J., Ramírez-Sanz, C., Estruch, V., Clemente, G. and Sanjuán, N. (2017). Organic versus conventional citrus. Impact assessment and variability analysis in the Comunitat Valenciana (Spain). *International Journal of Life Cycle Assessment* 22:571–86.

Saunders, C., Barber, A. and Taylor, G. (2006). Food Miles - Comparative Energy/ Emissions Performance of New Zealand's Agriculture Industry. Research Report No. 285. Lincoln University, New Zealand.

Schau, E. M. and Fet, A. M. (2008). LCA studies of food products as background for environmental product declarations. *International Journal of Life Cycle Assessment* 13:255–64.

Sessa, F., Marino, M., Montanaro, G., Dal Piaz, A., Zanotelli, D., Mazzetto, F. and Tagliavini, M. (2014). Life cycle assessment of apples at a country level: The case study of Italy. *Proceedings of the 9th Internationl Conference on Life Cycle Assessment in the Agri-Food Sector*, San Francisco.

Stöckle, C. O., Donatelli, M. and Nelson, R. (2003). CropSyst, a cropping systems simulation model. *European Journal of Agronomy* 18:289–307.

Tabatabaie, S. and Murthy, G. (2016). Cradle to farm gate life cycle assessment of strawberry production in the United States. *Journal of Cleaner Production* 127:548–54.

Tasca, A. L., Nessi, S. and Rigamonti, L. (2017). Environmental sustainability of agri-food supply chains: An LCA comparison between two alternative forms of production and distribution of endive in northern Italy. *Journal of Cleaner Production* 140:725–41. doi:10.1016/j.jclepro.2016.06.170.

USDA. (2010). Dietary Guidelines for Americans, 2010. 7th Edition. US Department of Agriculture and US Department of Health and Human Services, Washington DC. https://www.cnpp.usda.gov/sites/default/files/dietary_guidelines_for_americans/PolicyDoc.pdf (accessed 21 March 2018).

Webb, J., Williams, A. G., Hope, E., Evans, D. and Moorhouse, E. (2013). Do foods imported into the UK have a greater environmental impact than the same foods produced within the UK? *International Journal of Life Cycle Assessment* 18:1325–43. doi:10.1007/s11367-013-0576-2.

Williams, A. G., Audsley, E. and Sandars, D. L. (2006). Energy and environmental burdens of organic and non-organic agriculture and horticulture. In: Atkinson, C., Ball, B., Davies, D. H. K., Rees, R., Russell, G., Stockdale, E. A., Watson, C. A., Walker, R. and Younie, D. (Eds), Aspects of Applied Biology, 79, What Will Organic Farming Deliver? COR 2006. Association of Applied Biologists, pp. 19–23.

Williams, A., Pell, E., Webb, J., Moorhouse, E. and Audsley, E. (2008). Strawberry and tomato production for the UK compared between the UK and Spain. *Proceedings of the 6th International Conference on LCA in the Agri-food Sector*, Zurich, CH, pp. 254–62.

Zhang, M., Li, B. and Xiong, Z. Q. (2016). Effects of organic fertilizer on net global warming potential under an intensively managed vegetable field in southeastern China: A three-year field study. *Atmospheric Environment* 145:92–103. doi:10.1016/j.atmosenv.2016.09.024.

Chapter 11

Improving the environmental performance of pig and poultry production

Ben Putman, Martin Christie and Greg Thoma, University of Arkansas, USA

1 Introduction

Currently, pork and chicken account for nearly three-fourths of global meat consumption. As of 2012, pigs and meat-producing poultry (broilers) produced 36% and 35% of the world's meat supply, respectively (FAO, 2018a). By 2026, production is projected to reach 131.6 and 127.5 million metric tons of poultry and pig meat, respectively (FAO and OECD, 2018). With increasing concern for the environmental impacts associated with human activity, and due to the magnitude of the pork and poultry sectors, there is mounting pressure to meet the growing demand for these animal-based proteins with fewer resources and lower emissions.

Environmental impacts occur throughout the pork and chicken supply chains, but a relatively small fraction is attributable to the activities after the farm gate; for example, a life-cycle assessment (LCA) of global pig and poultry supply chains published in 2013 estimated that the greenhouse gas (GHG) emissions from the entire post-farm gate supply chain amounted to 6% for pigs and 7% for broilers (Macleod et al., 2013). The majority of GHG emissions were attributed to the production of feed, which for pig and poultry production is often the single largest contributor to many environmental impact categories (Leinonen et al., 2012; Pelletier, 2008; Prudêncio da Silva et al., 2014).

The contribution of feed production to the environmental impacts of pork and chicken supply chains is universal, although the magnitude and nature of

http://dx.doi.org/10.19103/AS.2018.0044.15

those impacts vary regionally. Production occurring in Asia is heavily reliant on rice-based feeds, which have a higher carbon footprint than other grains due to the methane emissions from flooded rice paddies. In Europe and Latin America, soybean meal from South America provides a majority of the protein in pig diets and is a significant source of GHG emissions due to the carbon emissions associated with land-use change (Meul et al., 2012; Van Der Werf et al., 2005).

The emission intensity of pork and poultry operations is largely determined by how efficiently the animals can convert feed to body weight; this conversion efficiency is commonly known as the feed conversion ratio (FCR). This metric is particularly useful for pigs and poultry because (1) most of these animals are not grazed, so feed consumption is easily quantified and controlled and (2) most of the environmental impacts are linked with feed production. Due to physiological differences, broilers tend to have lower FCRs than pigs; however, both species tend to have lower FCRs as the scale of production increases (Macleod et al., 2013). Small-scale, extensive production, common in developing nations, generally have higher FCRs because of differences in genetics, feed quality and management strategies. The type of production system also affects the contribution of the breeding flock/herd because extensive production systems tend to have higher mortality and lower fertility rates.

Manure management is another significant source of emissions in both pig and poultry production. The excretion rate of volatile solids and nitrogen to manure and, therefore, the methane and nitrous oxide formation potential depends on the digestibility of feed and the FCR of the animal: low digestibility and high FCR tend to increase these emissions. This situation is most common in small-scale, extensive production, which tends to have higher manure-related emissions relative to more intensive alternatives. The conversion of volatile solids into methane depends on the type of manure management system and the ambient temperature. Systems that encourage anaerobic conditions produce more methane, and this effect is amplified at higher temperatures.

2 Improvement options

The focus of this section is solely on those improvement options which are available to the livestock producer. While gains can be made by more efficiently producing the crops fed to pigs and chickens, these management options are addressed in other chapters. This section provides an overview of management options under the direct control of pig and poultry production sectors and includes the types and sources of feed; manure and waste management; animal genetics; and herd/flock health.

A common methodology for evaluating alternative approaches to achieve improved environmental performance is LCA. Broadly, there are two system models used by LCA practitioners and understanding the differences in these models is important in assessing the available literature and opportunities for improved performance.

2.1 LCA system models

This topic is covered in more detail elsewhere in this book, but a brief introduction is provided here for context. The two system models are known as attributional and consequential. An attributional system takes an engineering-like view of the production system, considering that the resource flows in the processes which are directly linked to the physical supply chain. A consequential system model has an economic view of the system where the system considered is 'expanded' to include effects on other markets that are affected by changes in the studied system. Thus, attributional models are static and backward-looking while consequential models are change-oriented and therefore more appropriate for the evaluation of improvement opportunities (i.e. for evaluating the effect of a change to the system).

2.2 Feed rations

The environmental impacts associated with pig and poultry production are largely due to/attributed to feed production and consumption. In livestock systems, feed-use efficiency by the animal is critical because improving efficiency means both less feed needs to be produced and less manure per unit of live weight is produced. One approach to reduce feed-related impacts is to use feed additives. Typical feed additives include synthetic amino acids (AA) and enzymes, among others (Apple et al., 2013, 2017). The use of synthetic AAs ensures that all essential AA are available and can also reduce the required amount of crude proteins such as soybean or rapeseed meal in both pig and poultry rations by improved balancing of some essential AA which may be deficient in plant proteins (Brotzge et al., 2014; Kebreab et al., 2016). Studies have shown that the use of synthetic AAs can lower the potential acidification, eutrophication, terrestrial ecotoxicity, climate change impact and cumulative energy demand of pig and poultry rations; however, the magnitude of these improvements varies depending on the feed formulation (Mosnier et al., 2011; Ogino et al., 2013). The effect is twofold: not only is the impact of the feed ration lower, but the resultant excreta is lower in nitrogen compared to rations without synthetic AAs. Thus, the potential environmental impacts from nitrogen-containing emissions from excreta are reduced (Garcia-Launay et al., 2014; Kebreab et al., 2016). An evaluation of the environmental impacts associated with protein sources in poultry feed rations in the United Kingdom found that slight reductions could be achieved when combined with synthetic AAs. However, this conclusion was dependent upon the method of accounting for land-use change. The reported improvements were not statistically significant when indirect land-use change was accounted (Leinonen et al., 2013).

The incorporation of enzymes into feed rations has been investigated as a means of improving nutrient utilization and feed conversion in pigs and poultry (Nielsen et al., 2008). One enzyme, phytase, is used in swine rations to increase phosphorous uptake in growing pigs (Jongbloed et al., 1992; Liebert et al., 2005; Nasir et al., 2014; Nelson et al., 1971; Nielsen and Wenzel, 2007). This enzyme reduces the amount of inorganic phosphate supplementation and consequently the amount of phosphorous excreted to the environment; both effects reduce environmental impacts of production. Phytase has been shown to reduce the GWP by 17%, acidification by 110% and eutrophication by 700% (Kebreab et al., 2016; Nielsen and Wenzel, 2007).

Other enzymes, such as proteases, have been added to broiler feeds to increase feed digestibility; Oxenbøll et al. (2011) reported reduction in the eutrophication, acidification potential and climate change impacts of broiler production by 3%, 5% and 2%, respectively. The use of alternative feed additives has shown some success as well. For example, the addition of bee pollen to broiler feed has been linked as significantly higher weight gains, improved haematological parameters and reduced mortality (Klaric et al., 2018).

Studies have evaluated different feed processing technologies to improve utilization and conversion efficiency. Pelleting provides some observable benefit ($P < 0.05$) to the daily gain and feed-conversion efficiency for broilers, but particle size does not have have a statistically significant effect (Chewning et al., 2012; Engberg et al., 2002; Lv et al., 2015; Rezaeipour and Gazani, 2014). In swine, reducing particle size has been shown to improve feed efficiency. However, there are trade-offs to be considered with additional processing. Finely ground feed requires more energy to produce and if the particle size is too small pigs are more likely to develop gastric ulcers (Healy et al., 1994; Wondra et al., 1995; Mavromichalis et al., 2000). There are no LCA studies evaluating the benefits of these technologies but improving feed conversion is always an opportunity for improved environmental performance.

The use of antimicrobial growth promoters (AGPs) has been banned in the EU since 2006, and as of 1 January 2017, the use of medically important (for human use) antimicrobials in US livestock feed is no longer allowed for growth promotion purposes. Despite this, AGPs are still used in some countries of Asia (Collignon and Voss, 2015; Krishnasamy et al., 2015; Van Boeckel et al., 2015) in an effort to maintain animal health and, thus, high productivity. However, some studies suggest that alternative, non antimicrobial feed additives may also be effective at maintaining animal health and productivity. Pettigrew (2006), for example, suggests that spray-dried plasma and dietary zinc can protect against enteric infection in piglets, providing similar benefits as AGPs. Gong et al. (2014) report on the use of herbs as alternatives to antimicrobials, suggesting potential avenues for further research in maintaining herd health without AGPs.

The use of crop by-products in pig and poultry rations has been evaluated with regard to a potential reduction in the environmental impacts associated with producing those animals (Mackenzie et al., 2016; van Zanten et al., 2015, 2016, 2018). The accounting approach used can affect the results. van Zanten et al. (2018) showed different results for attributional compared to consequential LCA when soybean meal was replaced by waste-fed larvae; the attributional method showed reductions in impact while the opposite was found for the consequential modelling approach. In general, the consequential model, provided appropriate care is taken in identifying the marginal substitution, is more robust in determining the benefit/cost of by-product use in animal feed because the attributional modelling approach relies on a normative division of impacts between the main and by-product, while the consequential modelling approach attempts to account for additional changes in affected systems (Sonnemann and Vigon, 2011; Weidema et al., 2018).

2.3 Animal performance

Improvement options for animal performance are commonly categorized in two ways: improvements in the animals themselves (e.g. genetic advances) or improvements for the herd/flock (e.g. operational improvements that increase productivity). Livestock genetics is continually used to produce animals with desirable characteristics for both the consumer and the producer (Paxton et al., 2010). Several companies including Cobb-Vantress, Aviagen and Hubbard focus their research efforts on developing new genetic lines of meat-producing chickens.

Geneticists have done much work to identify and manipulate specific genes that control the physiological traits that are of economic interest (Dunn et al., 2004). Such genetic advances, typically accomplished through genetic selection, in pigs and chickens can improve feed conversion and decrease associated environmental impacts (Varley, 2009). Consider the following example: in the United States, from 1957 to 2005, FCR in poultry production decreased by 50%, and 85–90% of the improved performance was estimated to be attributed to genetics (Zuidhof et al., 2014).

Despite the long history of using genetic selection, there are still opportunities for improvement. Research has shown that the most promising areas for future improvement will involve enhancing the digestive efficiency of pigs and reducing the maintenance requirements in breeding poultry, although the latter could be an option for pigs as well (Kyriazakis, 2011). There are several avenues by which maintenance requirements can be reduced. For example, broiler genes that control neck feather orientation and density have been identified and selected for expression such that the rate of heat dissipation is increased. Heat dissipation is crucial for maintaining body temperature which directly impacts flock FCR (Yunis and Cahaner, 1999). New broiler strains are developed with the intent of increasing average daily gain, increasing the ratio of desirable to non-desirable meat cuts and decreasing the FCR. Optimizing the human-edible meat produced from individual animals leads to environmental benefits. Pigs produced in the United States have seen significant improvements in the amount of edible meat per animal. The hot carcass weight (the weight after the head, hide, intestinal tract and internal organs are removed) of pigs produced in the United States has grown from 63% of the market weight in 1959 to 75% in 2009, which is a significant driver behind the reductions in impacts experienced over the same time period (Boyd et al., 2012).

While genetics is a significant factor for pig and poultry performance, housing conditions must be optimized and maintained to achieve their full genetic potential. Failure to maintain an optimal environment results in lower average daily gains, higher FCR, and increased mortality for meat-producing livestock (Averós et al., 2010; Li et al., 2017; Najafi et al., 2015; Pradère, 2014; Purswell et al., 2012; Renaudeau et al., 2011). Such declines in performance lead to longer growing periods, increased resource use, feed consumption and thus larger environmental footprint.

During brooding (approximately the first 2 weeks of a broiler's life), it is essential that temperature, air quality, feed and water management are a priority to ensure proper development of the thermo-regulation systems (Henrique, 2015). Failure to maintain optimal conditions can inhibit weight gain, increase mortality and negatively affect feed conversion (May and Lott, 2000; Moraes et al., 2002). The decreased performance can continue even after conditions are improved (Reece and Lott, 1980). In addition to environmental conditions, proper stocking density and bedding depth should be maintained for optimal health and productivity (Dozier et al., 2005; Ekstrand et al., 1997; Proudfoot et al., 1979). For example, Dozier et al. (2005) reported a reduction in cumulative weight gain and increased footpad lesions with increasing stocking density. Other animal welfare issues have been thought to carry performance trade-offs; however, some studies have shown that these changes may, in fact, lead to a slight improvement in environmental performance. Specifically, Leinonen et al. (2014) reported that more energy-efficient colony cage systems for broilers and layers had about 3% reduction in GWP and 8% reduction in primary energy use than baseline cage systems. In addition, Bandekar et al. (2019) reported that switching from individual gestation stalls to group pens in sow barns resulted in about 1% reduction in GWP and energy consumption for a functional unit of 1 kg live weight of finished pig at slaughter. In this case, improvements arose from a small increase in the number of live piglets weaned per sow.

Swine and poultry production in less-developed countries represent an important avenue for smallholders to increase their prosperity. Some of the major opportunities for improvement in these systems include improving animal genetics, health and nutrition. Access to processing systems will also increase the marketability for smallholder products. Recent advances in genomics may enable producers in these regions to more rapidly advance productivity by enabling them to bypass the traditional selective breeding technology (Larkins, 2009).

2.4 Manure and waste management

Emissions from manure in poultry and pig production are determined by the rate of volatile solids and nitrogen excretion, which, as previously mentioned, is influenced by FCR and the composition of diets. Feed rations with low digestibility coupled with high FCRs generally result in higher rates of volatile solids and nitrogen excretion; however, it is the way in which this manure is managed that determines the rate at which these compounds are subsequently converted into impactful emissions such as CH_4, N_2O and NH_3. Higher temperatures and anaerobic conditions generally increase the conversion rate of volatile solids to CH_4.

Manure is typically managed in poultry operations by applying bedding (rice hulls, wood chips etc.) to the floors and allowing it to collect excreta from the flock. This combination of bedding, manure, urine, feed and so on is known as litter. Litter is typically removed from poultry barns every year or two and either transported directly to fields for application or stored until removal. A study of waste management techniques on five Georgia farms found that litter stored on an impervious surface and covered decreased the risk of elevating nitrogen and phosphorus levels in nearby groundwater (Myers et al., 1999). Emissions from housing for both pigs and poultry can also be mitigated with air treatment technologies such as scrubbers or biofilters (Van der Heyden et al., 2015).

Litter is a well-documented fertilizer and soil amendment (Sommers and Sutton, 1980; Stephenson et al., 1990) due to its nitrogen (N) and phosphorus (P) content; however, both have the potential for ecosystem damage. If the litter is mismanaged, groundwater and surface water can become contaminated with P and N due to run-off or leaching (Sharpley, 1997; Sharpley et al., 2007). Following best practices for land application of poultry litter will help minimize environmental disturbances. Some best practices include the following: keep the litter as dry as possible, do not apply in excess of required soil phosphorous, avoid applications when the ground is wet or precipitation is expected and keep records of applications (Chapman, 1996; Christianson and Harmel, 2015). A 20-year study on phosphorous run-off and leaching found that the addition of alum to poultry litter resulted in an 83% and 80% reduction in dissolved P and total P, respectively, in leachate samples (Anderson et al., 2018). In a 7-year study of litter application, it was observed that soils treated with litter-containing alum had approximately three times less soluble P than soils treated with just litter (Moore and Edwards, 2007). The use of alum should be considered as an approach for acute management of P loss from soils; however, alternate solutions for nutrient recovery and recycling have greater potential for long-term sustainability (Goulding et al., 2008; Jarvie et al., 2015; Schoumans et al., 2010; Suh and Yee, 2011).

A significant amount of research has been conducted on nutrient recovery from animal manure. Some of the technologies include anaerobic digestion (García-González et al., 2016), gasification (Cotana et al., 2014; Sharara et al., 2016), solid separation coupled with composting and/or digestion (Brockmann et al., 2014; Vanotti et al., 2009), struvite precipitation by both chemical (Kataki et al., 2016; Liu et al., 2011b; Sommer et al., 2004; Taddeo et al., 2018; Yetilmezsoy et al., 2017; Zhang et al., 2018) and novel electrochemical approaches using magnesium electrodes (Huang et al., 2016; Lin et al., 2018; Liu et al., 2011a; Ren et al., 2017). These and other emerging technologies hold promise for both mitigation of environmental impacts and enhancing nutrient-use efficiency of the animal food system.

Swine manure is also used as a fertilizer, and land application is an effective way to utilize manure nutrients for crop production. The benefits are accounted through displacement of equivalent amounts of synthetic fertilizers (including production and application). The LEAP Partnership has published a draft methodology for determining fertilizer equivalence for manure-nutrient sources (FAO, 2018b) that is based on the estimation of the quantity and plant-use efficiency of nutrients in manures compared to synthetic fertilizers. Adoption of this standard, after final review, is important given that Brockmann et al. (2014) have demonstrated that different methods of calculating the fertilizer equivalent displaced by swine manure can result in (sometimes) opposite conclusions for some impact categories, emphasizing the importance of careful and critical interpretation of LCA literature.

Direct land application of swine manure slurry has been associated with lower potential impacts than other aerobic treatment options including biological separation and constructed wetlands (Bayo et al., 2012; Lopez-Ridaura et al., 2009). The most common choice for pretreatment of slurry involves the redistribution of its nutrients like slurry separation, which concentrates nutrients into forms that are easier to transport. The impacts of slurry pretreatment have been shown to be equal to or less than direct land application; although, the differences between impact reductions associated with implementing these technologies can be minor, and the best option will likely depend on other factors, that is local policies, regulations and cost (Brockmann et al., 2014; ten Hoeve et al., 2014). Research suggests that less energy-intensive methods of solid–liquid separation could be an effective pretreatment option for reducing impacts (De Vries et al., 2013).

Dalgaard (2007) evaluated a range of solid separation alternatives and anaerobic digestion for the management of nutrients and GHG emissions associated with swine manure management. An advantage of solid separation is that the N/P ratio can be partially controlled, depending on efficiency, leading to an enriched P content in the solid fraction thus reducing adjacent land requirements for disposition of the liquid slurry. The solids fraction can also be transported a greater distance due to reduced volume. Accounting for field emissions and fertilizer replacement yielded between 10% and 48% reduction in GHG emissions depending on the separation efficiency.

In addition, the way in which manure is applied affects the associated environmental impacts. Deep injection or immediate ploughing of the slurry is one of the most effective methods for land-applying slurry to minimize eutrophication and acidification potential (Prapaspongsa et al., 2010; Webb et al., 2010) as well as essentially eliminating ammonia volatilization. A meta-analysis reports that ammonia release is reduced, while nitrous oxide emissions are increased (Hou et al., 2015), but not to a level that offsets overall benefits.

Using swine manure as a source of bioenergy through anaerobic digestion is a promising strategy for reducing environmental impacts (De Vries et al., 2012; Prapaspongsa et al., 2010). It has been shown to be more effective at reducing fossil fuel consumption and GHG emissions than other options, such as reducing feed consumption or manure storage time, but some combination of approaches is likely to produce the greatest improvement (Nguyen et al., 2010, 2012). Williams et al. (2016) evaluated the use of turkey litter as a fuel for electricity generation as an alternative to its use as a fertilizer and soil amendment. They report a 14%, 55% and 70% reduction in fossil energy demand, eutrophication and acidification potential, respectively, after accounting for the loss of fertilizer value of the litter. The benefit derives primarily from reduced ammonia emissions from storage and land application. They report only 3% reduction in GHG emissions due to the loss of soil carbon when the litter is not land applied.

In addition to manure/litter from pig and poultry operations, other materials are resulting from production that must be properly managed to minimize their impact on the environment. It is necessary to properly dispose of mortalities and condemnations not suitable for food/feed. In poultry operations, discarding mortalities into burial pits is a commonly used method of disposal, but this practice poses threats to local water quality. Treatment of burial pits and carcasses with hydrated lime has been shown to decrease the population and mobility of pathogenic *E. coli* (Avery et al., 2009; Sánchez et al., 2008).

3 Case studies

One way to determine the most effective improvement options for livestock production is to analyse the changes that have occurred in the recent past to determine which had the greatest influence. These types of retrospective analyses have been conducted for a variety of agricultural products across several countries (Bai et al., 2014; Boyd et al., 2012; Capper et al., 2009; Putman et al., 2017, 2018). Retrospective analyses covering pig and poultry production in the United States are presented here as case studies for developing an understanding of how direct and indirect changes can affect the environmental performance of pig and poultry production.

Bai et al. (2014) report a retrospective study of swine production in China between 1960 and 2010 and report significant increases in N and P losses associated with the transition from backyard to landless systems – 30 and 95 times larger in 2010 than in 1960. This is primarily associated with the lack of available land area for the disposition of manure. The authors project continued increases in emissions through 2030 for business as usual; however, they estimate that reductions of 64% and 95% for N and P losses, respectively, are achievable through the adoption of a combination of modern technologies for housing, herd and manure management, ration formulation and precision feeding.

Putman et al. (2017) evaluated the environmental performance of poultry production over a comparable period (1965–2010) and found that improvements in bird performance accounted for 12.6–91.7% of the decline in environmental impacts depending on the impact category. The system boundaries of this assessment included grandparent and parent breeding generations in addition to broiler production. Environmental impacts were assessed based on 1000 kg of live weight poultry, which included the spent hens from the breeding stock that contributed to the human food supply chain. The major differences between poultry production in 1965 and 2010 were changes in feed rations, declining mortality and increasing fertility for breeding birds in 2010, and a heavier market weight in 2010 that was achieved in fewer days and with better FCR. The study also considered changes in background systems outside the control of the poultry operator. These background system changes included crop yields, tractor efficiency, transport efficiency and the electricity grid mix.

Despite these declines in the intensity metrics of environmental impacts of poultry production, the sector-level impact of the US poultry industry increased over the assessment period, driven by significant growth in poultry consumption. Improvements in background systems, particularly crop production, were the primary drivers behind the decline in land occupation and fossil energy consumption. The decreases in eutrophication, acidification and GWPs were largely attributed to advancements in bird performance. The study did not delineate the specific causal factors responsible for those improvements, but general industry shifts, for example adoption of fully enclosed housing, were the most likely drivers of change. The shift in ration ingredients from 1965 to 2010 was the predominant factor in the reduction of water use associated with poultry production as a result of a shift away from more water-intensive wheat in favour of corn and soy; however, this shift also created additional eutrophication and acidification potential in 2010, but those increased impacts were outweighed by other improvements.

Putman et al. (2018) conducted a similar retrospective analysis of pig production. The differences in pig production systems in the United States in 1960 and 2015 resulted in decreased impacts per kg of live weight ranging from 7–8% (GWP, energy use) to 75% (land use). The system boundaries for this assessment included the breeding generation (sow barn) in addition to the grow-finish pigs. The decline in land occupation was driven by increasing crop yields.

This study evaluated the impacts of production in 5-year intervals, providing insights into the effects of the changes occurring during the period of the assessment. Results from the intervening years suggest that the modernization of pig farming that occurred in the United States during the 1970s and 1980s initially increased impacts associated with production, driven by additional energy used for heating and ventilation. As these housing advances became more common, the national swine herd realized improvements in health and productivity, ultimately resulting in a decline in impacts across all categories assessed.

These retrospective assessments highlight the importance of feed in the pig and poultry industries and show that some benefits accrued to these sectors are driven by external factors in the general economy. Obviously, continual improvement in the general economy benefits all sectors, and market signals sent through purchasing choices can help improve long-term sustainability performance.

4 Future trends

The growing population's demands for inexpensive, high-quality agricultural products require agriculture industry professionals to increase production efficiencies throughout a product's supply chain. Doing so requires a comprehensive understanding of the full life cycle, including raw material extractions and the processing and distribution of intermediate products. Modern computing power and modelling techniques make computer-based analysis an essential part of sustainable agriculture research. Physical models of pork and poultry production are used to provide simulated animal performance, growing efficiencies, resource expenditure and estimates of supply chain environmental impacts. Data management tools improve data organization and accessibility, both of which may improve collaborative research efforts. Machine learning and artificial intelligence technology are continually improving and have an important place in analysing complex agricultural supply chains.

Models, such as the PPEC model and the Poultry Calculator, are used to simulate the effects of management practices, regional weather and facility construction on the environmental impacts of swine and poultry production (UA CoE[1]). Physical models provide an opportunity to perform many virtual experiments in a short period. Running simulations with different inputs is much quicker and less expensive than designing and executing traditional experiments. Validation of these models poses the greatest challenge in their development. Validation requires the developer to find data from an actual facility to compare simulated results. While these models cannot simulate exactly what occurs on a farm, they are directionally correct and can be used to inform decision-making or guide future research investigations.

Data-driven decision-making in agriculture must be adopted as the demand for food continually increases. Decision support systems are tools that can be used to identify improvement opportunities at any point in the supply chain. Systems such as FarmBeats provide robust, low-cost, cloud-enabled data collection systems that can be used to identify and quantify sources of variation in production parameters. FarmBeats can be used to improve farm productivity by varying farm inputs, such as irrigation and fertilizer, according to the need at specific locations on the farm (Vasisht et al., 2017). The major challenges of implementing these tools include costs of farm-wide data collection equipment, internet connectivity challenges in rural areas and the learning curve related to operating new software (Lindblom et al., 2017). Data collection and analysis tools should be used to move towards decision-making that will optimize economic and environmental trade-offs.

Collaborations of sustainability researchers with computer scientists, horticulture specialists and engineers will spur innovation. Computer scientists have developed several machine learning algorithms for yield prediction, crop quality control, animal welfare management, water management and soil management (Liakos et al., 2018). Soil sensors have been used to identify and implement the best irrigation strategies based on soil type and local weather (Goldstein et al., 2018). Unmanned aerial vehicles (UAVs) equipped with decision-making software can monitor plant health and treat certain infestations (Alsalam et al., 2017).

1 https://resilientfood.uark.edu/research/.

The future of data science and automation in agriculture becomes clearer every day. Machine learning and decision-making algorithms analyse data to find patterns orders of magnitude faster than a human brain. As the demand for agricultural products increases, it is crucial that the industry adopts technologies that will lead to improved efficiencies.

5 Conclusion

Swine and poultry are the fastest-growing meat products consumed globally and continued improvement in their productivity, and environmental performance, will be critical to meet the burgeoning demand for animal proteins sustainably. The interplay of genetics, environment and management of these production systems are major determining factors in the overall environmental sustainability of the production systems in both developed and developing economies. Management focussed on maintaining and improving optimal animal health and well-being by providing suitable shelter and minimizing disease pressures may provide productivity gains. Feed sourcing and precision nutrition, using specialty feed ingredients such as synthetic AA, also show promise in improving productivity and environmental sustainability.

Considering the very large contribution from feed production to the environmental impacts of pig and poultry systems, optimizing feed efficiency seems the most effective way to continue the trend in declining impacts observed over the past 50 years. Research suggests that while some of the genetic traits affecting feed utilization in pigs and poultry have already been maximized, others, such as those regulating maintenance energy costs, still have the potential for improvement. Furthermore, other factors including the nutrient profile of the feed, the way in which it is processed and the use of feed additives in precision feeding provide additional opportunities for greater feed utilization (Hill et al., 2014; Kyriazakis, 2011; Patience et al., 2015).

6 Where to look for further information

The FAO publication '*Greenhouse gas emissions from pig and chicken supply chains*' provides an excellent introduction to the global structure of pig and poultry production and identifies key elements of production systems relevant for understanding GHG emissions and mitigation activities (Macleod et al., 2013). Putman et al. (2017, 2018) and Bai (2014) have provided a retrospective view of pig and poultry production systems and identified significant drivers for past improvement in environmental performance - many of these same drivers are anticipated to remain relevant concerning continued mitigation opportunities in the sectors.

The Pig Site (www.thepigsite.com) and The Poultry Site (www.thepoultrysite.com) are industry-centred information portals with news and resources relevant to global production covering a wide range of issues from antimicrobial resistance to feeding strategies. The Food and Agriculture Organization of the United Nations is a good source of information regarding swine and poultry production, and food systems in general. There is valuable statistical information regarding production and consumption in the FAOSTAT database (http://www.fao.org/faostat/en/). The authors' research site has posted calculators for swine and poultry production that enable users to estimate environmental footprints for production facilities (www.resilientfood.uark.edu). Finally, the biannual LCA Food Conference has high-quality, current presentations of ongoing research from the global community (https://www6.inra.fr/lcafoodconferencearchives/).

7 References

Alsalam, B. H. Y., Morton, K., Campbell, D. and Gonzalez, F. 2017. Autonomous UAV with vision based on-board decision making for remote sensing and precision agriculture. In: *2017 IEEE Aerospace Conference*. IEEE, pp. 1–12. doi:10.1109/AERO.2017.7943593.

Anderson, K. R., Moore, P. A., Miller, D. M., DeLaune, P. B., Edwards, D. R., Kleinman, P. J. A. and Cade-Menun, B. J. 2018. Phosphorus leaching from soil cores from a twenty-year study evaluating alum treatment of poultry litter. *J. Environ. Qual.* 47(3), 530–7. doi:10.2134/jeq2017.11.0447.

Apple, J. K., Yancey, W. S., Hollenbeck, J. J., Johnson, T. M., Bass, B. E., Tsai, T., Maxwell, C. V., Hanigan, M. D., Radcliffe, J. S., Richert, B. T., et al. 2013. Effects of amino acid supplementation of reduced crude protein (RCP) diets on LM quality of growing-finishing swine. *J. Anim. Sci.* 9(Suppl. 2), 102.

Apple, J. K., Maxwell, C. V., Bass, B. E., Yancey, J. W. S., Payne, R. L. and Thomson, J. 2017. Effects of reducing dietary crude protein levels and replacement with crystalline amino acids on growth performance, carcass composition, and fresh pork quality of finishing pigs fed ractopamine hydrochloride. *J. Anim. Sci.* 95(11), 4971–85. doi:10.2527/jas2017.1818.

Averós, X., Brossard, L., Dourmad, J.-Y., de Greef, K. H., Edge, H. L., Edwards, S. A. and Meunier-Salaün, M.-C. 2010. A meta-analysis of the combined effect of housing and environmental enrichment characteristics on the behaviour and performance of pigs. *Appl. Anim. Behav. Sci.* 127(3–4), 73–85. doi:10.1016/J.APPLANIM.2010.09.010.

Avery, L. M., Williams, A. P., Killham, K. and Jones, D. L. 2009. Heat and lime-treatment as effective control methods for *E. coli* O157: H7 in organic wastes. *Bioresour. Technol.* 100(10), 2692–8. doi:10.1016/J.BIORTECH.2008.12.044.

Bai, Z. H., Ma, L., Qin, W., Chen, Q., Oenema, O. and Zhang, F. S. 2014. Changes in pig production in China and their effects on nitrogen and phosphorus use and losses. *Environ. Sci. Technol.* 48(21), 12742–9. doi:10.1021/es502160v.

Bandekar, P. A., Leh, M., Bautista, R., Matlock, M. D., Thoma, G. and Ulrich, R. 2019. Life cycle assessment of alternative swine management practices. *J. Anim. Sci.* 97(1), 472–84. doi:10.1093/jas/sky425.

Bayo, J., Gómez-López, M. D., Faz, A. and Caballero, A. 2012. Environmental assessment of pig slurry management after local characterization and normalization. *J. Clean. Prod.* 32, 227–35. doi:10.1016/j.jclepro.2012.04.003.

Boyd, G., Cady, R., Wittig, L., Bryan, G., Anderson, D., Sutton, A. L. and Holden, P. 2012. A 50-year comparison of the carbon footprint and resource use of the U.S. swine herd: 1959–2009. Available at: https://www.pork.org/research/a-50-year-comparison-of-the-carbon-footprint-and-resource-use-of-the-us-swine-herd-1959-2009/ (accessed on 3 April 2019).

Brockmann, D., Hanhoun, M., Négri, O. and Hélias, A. 2014. Environmental assessment of nutrient recycling from biological pig slurry treatment - impact of fertilizer substitution and field emissions. *Bioresour. Technol.* 163, 270–9. doi:10.1016/J.BIORTECH.2014.04.032.

Brotzge, S. D., Chiba, L. I., Adhikari, C. K., Stein, H. H., Rodning, S. P. and Welles, E. G. 2014. Complete replacement of soybean meal in pig diets with hydrolyzed feather meal with blood by amino acid supplementation based on standardized ileal amino acid digestibility. *Livest. Sci.* 163, 85–93. doi:10.1016/j.livsci.2014.02.001.

Capper, J. L., Cady, R. A. and Bauman, D. E. 2009. The environmental impact of dairy production: 1944 compared with 2007. *J. Anim. Sci.* 87(6), 2160–7. doi:10.2527/jas.2009-1781.

Chapman, S. L. 1996. Soil and solid poultry waste nutrient management and water quality. *Poult. Sci.* 75(7), 862–6. doi:10.3382/ps.0750862.

Chewning, C. G., Stark, C. R. and Brake, J. 2012. Effects of particle size and feed form on broiler performance. *J. Appl. Poult. Res.* 21(4), 830–7. doi:10.3382/japr.2012-00553.

Christianson, L. E. and Harmel, R. D. 2015. 4R water quality impacts: an assessment and synthesis of forty years of drainage nitrogen losses. *J. Environ. Qual.* 44(6), 1852–60. doi:10.2134/jeq2015.03.0170.

Collignon, P. and Voss, A. 2015. China, what antibiotics and what volumes are used in food production animals? *Antimicrob. Resist. Infect. Contr.* 4, 16. doi:10.1186/s13756-015-0056-5.

Cotana, F., Messineo, A., Petrozzi, A., Coccia, V., Cavalaglio, G. and Aquino, A. 2014. Comparison of ORC turbine and Stirling engine to produce electricity from gasified poultry waste. *Sustainability* 6(9), 5714–29. doi:10.3390/su6095714.

Dalgaard, R. L. 2007. The environmental impact of pork production from a life cycle perspective. Dep. Agroecol. Environ. Aarhus University.

De Vries, J. W., Vinken, T. M. W. J., Hamelin, L. and de Boer, I. J. M. 2012. Comparing environmental consequences of anaerobic mono- and co-digestion of pig manure to produce bio-energy - a life cycle perspective. *Bioresour. Technol.* 125, 239–48. doi:10.1016/j.biortech.2012.08.124.

De Vries, J. W., Aarnink, A. J. A., Groot Koerkamp, P. W. G. and de Boer, I. J. M. 2013. Life cycle assessment of segregating fattening pig urine and feces compared to conventional liquid manure management. *Environmental Science and Technology* 47, 1589–97. doi:10.1021/es302951a.

Dozier, W. A., Thaxton, J. P., Branton, S. L., Morgan, G. W., Miles, D. M., Roush, W. B., Lott, B. D. and Vizzier-Thaxton, Y. 2005. Stocking density effects on growth performance and processing yields of heavy broilers. *Poult. Sci.* 84(8), 1332–8. doi:10.1093/ps/84.8.1332.

Dunn, I. C., Miao, Y. W., Morris, A., Romanov, M. N., Wilson, P. W. and Waddington, D. 2004. A study of association between genetic markers in candidate genes and reproductive traits in one generation of a commercial broiler breeder hen population. *Heredity* 92(2), 128–34. doi:10.1038/sj.hdy.6800396.

Ekstrand, C., Algers, B. and Svedberg, J. 1997. Rearing conditions and foot-pad dermatitis in Swedish broiler chickens. *Prev. Vet. Med.* 31(3–4), 167–74. doi:10.1016/S0167-5877(96)01145-2.

Engberg, R. M., Hedemann, M. S. and Jensen, B. B. 2002. The influence of grinding and pelleting of feed on the microbial composition and activity in the digestive tract of broiler chickens. *Br. Poult. Sci.* 43(4), 569–79. doi:10.1080/0007166022000004480.

FAO. 2018a. Meat and poultry meat products. WWW Document. FAO's Animal Production and Health Division.

FAO. 2018b. *Nutrient Flows and Associated Environmental Impacts in Livestock Supply Chains: Guidelines for Assessment* (vol. 1). Food and Agricultural Organization of the United Nations, Rome, Italy.

FAO and OECD. 2018. *OECD – FAO Agricultural Outlook 2018–2020*. OECD Publishing, Paris and Food and Agriculture Organization of the United Nations, Rome, Rome. doi:10.1017/CBO9781107415324.004.

García-González, M. C., Riaño, B., Teresa, M., Herrero, E., Ward, A. J., Provolo, G., Moscatelli, G., Piccinini, S., Bonmatí, A., Bernal, M. P., et al. 2016. Treatment of swine manure: case studies in European's N-surplus areas. *Sci. Agric. (Piracicaba, Braz.)* 73(5), 444–54. doi:10.1590/0103-9016-2015-0057.

Garcia-Launay, F., van der Werf, H. M. G., Nguyen, T. T. H., Le Tutour, L. and Dourmad, J. Y. 2014. Evaluation of the environmental implications of the incorporation of feed-use amino acids in pig production using Life Cycle Assessment. *Livest. Sci.* 161, 158–75. doi:10.1016/j.livsci.2013.11.027.

Goldstein, A., Fink, L., Meitin, A., Bohadana, S., Lutenberg, O. and Ravid, G. 2018. Applying machine learning on sensor data for irrigation recommendations: revealing the agronomist's tacit knowledge. *Precis. Agric.* 19(3), 421–44. doi:10.1007/s11119-017-9527-4.

Gong, J., Yin, F., Hou, Y. and Yin, Y. 2014. Review: Chinese herbs as alternatives to antibiotics in feed for swine and poultry production: potential and challenges in application. *Can. J. Anim. Sci.* 94(2), 223–41. doi:10.4141/cjas2013-144.

Goulding, K., Jarvis, S. and Whitmore, A. 2008. Optimizing nutrient management for farm systems. *Philos. Trans. R. Soc. Lond., B, Biol. Sci.* 363(1491), 667–80. doi:10.1098/rstb.2007.2177.

Healy, B. J., Hancock, J. D., Kennedy, G. A., Bramel-Cox, P. J., Behnke, K. C. and Hines, R. H. 1994. Optimum particle size of corn and soft sorghum for nursery pigs. *J. Anim. Sci.* 72:2227–36.

Henrique, F. 2015. How to give chicks the best start in the brooding period. WWW Document. Cobb-Vantress Academy.

Hill, W. G., Wolc, A., O'Sullivan, N. P. and Avendaño, S. 2014. Breeding for sustainability: maintaining and enhancing multi-trait genetic improvement. In: Burton, E., Gatcliffe, J., O'Neill, H. M. and Scholey, D. (Eds), *Sustainable Poultry Production in Europe*. CABI Publishing, Wallingford, pp. 193–213. doi:10.1079/9781780645308.0193.

Hou, Y., Velthof, G. L. and Oenema, O. 2015. Mitigation of ammonia, nitrous oxide and methane emissions from manure management chains: a meta-analysis and integrated assessment. *Glob. Chang. Biol.* 21(3), 1293–312. doi:10.1111/gcb.12767.

Huang, H., Zhang, P., Zhang, Z., Liu, J., Xiao, J. and Gao, F. 2016. Simultaneous removal of ammonia nitrogen and recovery of phosphate from swine wastewater by struvite electrochemical precipitation and recycling technology. *J. Clean. Prod.* 127, 302–10. doi:10.1016/j.jclepro.2016.04.002.

Jarvie, H. P., Sharpley, A. N., Flaten, D., Kleinman, P. J., Jenkins, A. and Simmons, T. 2015. The pivotal role of phosphorus in a resilient water-energy-food security nexus. *J. Environ. Qual.* 44(4), 1049–62. doi:10.2134/jeq2015.01.0030.

Jongbloed, A. W., Mroz, Z. and Kemme, P. A. 1992. The effect of supplementary Aspergillus niger phytase in diets for pigs on concentration and apparent digestibility of dry matter, total phosphorus, and phytic acid in different sections of the alimentary tract. *J. Anim. Sci.* 70(4), 1159–68. doi:10.2527/1992.7041159x.

Kataki, S., West, H., Clarke, M. and Baruah, D. C. 2016. Phosphorus recovery as struvite from farm, municipal and industrial waste: feedstock suitability, methods and pre-treatments. *Waste Manag.* 49, 437–54. doi:10.1016/j.wasman.2016.01.003.

Kebreab, E., Liedke, A., Caro, D., Deimling, S., Binder, M. and Finkbeiner, M. 2016. Environmental impact of using specialty feed ingredients in swine and poultry production: A life cycle assessment. *J. Anim. Sci.* 94(6), 2664–81. doi:10.2527/jas.2015-9036.

Klaric, I., Domacinovic, M., Seric, V., Miskulin, I., Pavić, M. and Paradinović, K. 2018. Effects of bee pollen and propolis on performance, mortality, and some hematological blood parameters in broiler chickens. *Slov. Vet. Res.* 55(1). doi:10.26873/SVR-385-2018.

Krishnasamy, V., Otte, J. and Silbergeld, E. 2015. Antimicrobial use in Chinese swine and broiler poultry production. *Antimicrob. Resist. Infect. Contr.* 4, 17. doi:10.1186/s13756-015-0050-y.

Kyriazakis, I. 2011. Opportunities to improve nutrient efficiency in pigs and poultry through breeding. *Animal* 5(6), 821–32. doi:10.1017/S1751731110002545.

Larkins, B. 2009. *Emerging Technologies to Benefit Farmers in sub-Saharan Africa and South Asia*. National Academies Press, Washington DC. doi:10.17226/12455.

Leinonen, I., Williams, A. G., Wiseman, J., Guy, J. and Kyriazakis, I. 2012. Predicting the environmental impacts of chicken systems in the United Kingdom through a life cycle assessment: broiler production systems. *Poult. Sci.* 91(1), 8–25. doi:10.3382/ps.2011-01634.

Leinonen, I., Williams, A. G., Waller, A. H. and Kyriazakis, I. 2013. Comparing the environmental impacts of alternative protein crops in poultry diets: the consequences of uncertainty. *Agric. Syst.* 121, 33–42. doi:10.1016/j.agsy.2013.06.008.

Leinonen, I., Williams, A. G. and Kyriazakis, I. 2014. The effects of welfare-enhancing system changes on the environmental impacts of broiler and egg production. *Poult. Sci.* 93(2), 256–66. doi:10.3382/ps.2013-03252.

Li, M. M., Seelenbinder, K. M., Ponder, M. A., Deng, L., Rhoads, R. P., Pelzer, K. D., Radcliffe, J. S., Maxwell, C. V., Ogejo, J. A., White, R. R., et al. 2017. Effects of dirty housing and a Typhimurium DT104 challenge on pig growth performance, diet utilization efficiency, and gas emissions from stored manure. *J. Anim. Sci.* 95(3), 1264–76. doi:10.2527/jas.2016.0863.

Liakos, K. G., Busato, P., Moshou, D., Pearson, S. and Bochtis, D. 2018. Machine learning in agriculture: a review. *Sensors* 18(8), 2674. doi:10.3390/s18082674.

Liebert, F., Htoo, J. K. and Sünder, A. 2005. Performance and nutrient utilization of laying hens fed low-phosphorus corn-soybean and wheat-soybean diets supplemented with microbial phytase. *Poult. Sci.* 84(10), 1576–83. doi:10.1093/ps/84.10.1576.

Lin, X., Han, Z., Yu, H., Ye, Z., Zhu, S. and Zhu, J. 2018. Struvite precipitation from biogas digestion slurry using a two-chamber electrolysis cell with a magnesium anode. *J. Clean. Prod.* 174, 1598–607. doi:10.1016/J.JCLEPRO.2017.10.224.

Lindblom, J., Lundström, C., Ljung, M. and Jonsson, A. 2017. Promoting sustainable intensification in precision agriculture: review of decision support systems development and strategies. *Precis. Agric.* 18(3), 309–31. doi:10.1007/s11119-016-9491-4.

Liu, Y. H., Kumar, S., Kwag, J. H., Kim, J. H., Kim, J. D. and Ra, C. S. 2011a. Recycle of electrolytically dissolved struvite as an alternative to enhance phosphate and nitrogen recovery from swine wastewater. *J. Hazard. Mater.* 195, 175–81. doi:10.1016/j.jhazmat.2011.08.022.

Liu, Y. H., Kwag, J. H., Kim, J. H. and Ra, C. S. 2011b. Recovery of nitrogen and phosphorus by struvite crystallization from swine wastewater. *Desalination* 277(1–3), 364–9. doi:10.1016/j.desal.2011.04.056.

Lopez-Ridaura, S., van der Werf, Hv, Paillat, J. M. and Le Bris, B. 2009. Environmental evaluation of transfer and treatment of excess pig slurry by life cycle assessment. *J. Environ. Manage.* 90(2), 1296–304. doi:10.1016/j.jenvman.2008.07.008.

Lv, M., Yan, L., Wang, Z., An, S., Wu, M. and Lv, Z. 2015. Effects of feed form and feed particle size on growth performance, carcass characteristics and digestive tract development of broilers. *Anim. Nutr.* 1(3), 252–6. doi:10.1016/j.aninu.2015.06.001.

Mackenzie, S. G., Leinonen, I., Ferguson, N. and Kyriazakis, I. 2016. Can the environmental impact of pig systems be reduced by utilising co-products as feed? *J. Clean. Prod.* 115, 172–81. doi:10.1016/j.jclepro.2015.12.074.

Macleod, M., Gerber, P. J., Opio, C., Falcucci, A., Tempio, G., Henderson, B., Mottet, A. and Steinfeld, H. 2013. *Greenhouse Gas Emissions from Pig and Chicken Supply Chains*. Food and Agriculture Organization of the United Nations, Rome, Italy. Available at: http://www.fao.org/3/i3460e/i3460e.pdf.

Mavromichalis, I., Hanck, J. D., Senne, B. W., Gugle, T. L., Kennedy, G. A., Hines, R. H., et al. 2000. Enzyme supplementation and particle size of wheat in diets for finishing pigs. *J. Anim. Sci.* 78:3086–95.

May, J. D. and Lott, B. D. 2000. The effect of environmental temperature on growth and feed conversion of broilers to 21 days of age. *Poult. Sci.* 79(5), 669–71. doi:10.1093/ps/79.5.669.

Meul, M., Ginneberge, C., Van Middelaar, C. E., de Boer, I. J. M., Fremaut, D. and Haesaert, G. 2012. Carbon footprint of five pig diets using three land use change accounting methods. *Livest. Sci.* 149(3), 215–23. doi:10.1016/j.livsci.2012.07.012.

Moore, P. A. and Edwards, D. R. 2007. Long-term effects of poultry litter, alum-treated litter, and ammonium nitrate on phosphorus availability in soils. *J. Environ. Qual.* 36(1), 163–74. doi:10.2134/jeq2006.0009.

Moraes, V., Malheiros, R., Furlan, R., Bruno, L., Malheiros, E. and Macari, M. 2002. Effect of environmental temperature during the first week of brooding period on broiler chick body weight, viscera and bone development. *Rev. Bras. Cienc. Avic.* 4(1), 1–8. doi:10.1590/S1516-635X2002000100003.

Mosnier, E., van der Werf, H. M. G., Boissy, J. and Dourmad, J. Y. 2011. Evaluation of the environmental implications of the incorporation of feed-use amino acids in the manufacturing of pig and broiler feeds using Life Cycle Assessment. *Animal* 5(12), 1972–83. doi:10.1017/S1751731111001078.

Myers, L. M., Bush, P. B., Segars, W. I. and Radcliffe, D. E. 1999. Impact of poultry mortality pits on farm groundwater quality. In: *Georgia Water Resources Conference*, pp. 234–9.

Najafi, P., Zulkifli, I., Amat Jajuli, N. A., Farjam, A. S., Ramiah, S. K., Amir, A. A., O'Reily, E. and Eckersall, D. 2015. Environmental temperature and stocking density effects on acute phase proteins, heat shock protein 70, circulating corticosterone and performance in broiler chickens. *Int. J. Biometeorol.* 59(11), 1577–83. doi:10.1007/s00484-015-0964-3.

Nasir, Z., Broz, J. and Zijlstra, R. T. 2014. Supplementation of a wheat-based diet low in phosphorus with microbial 6-phytase expressed in Aspergillus oryzae increases digestibility and plasma phosphorus but not performance in lactating sows. *Anim. Feed Sci. Technol.* 198, 263–70. doi:10.1016/j.anifeedsci.2014.10.008.

Nelson, T. S., Shieh, T. R., Wodzinski, R. J. and Ware, J. H. 1971. Effect of supplemental phytase on the utilization of phytate phosphorus by chicks. *J. Nutr.* 101(10), 1289–93. doi:10.1093/jn/101.10.1289.

Nguyen, T. L. T., Hermansen, J. E. and Mogensen, L. 2010. Fossil energy and GHG saving potentials of pig farming in the EU. *Energy Policy* 38(5), 2561–71. doi:10.1016/j.enpol.2009.12.051.

Nguyen, T. L. T., Hermansen, J. E. and Mogensen, L. 2012. Environmental costs of meat production: the case of typical EU pork production. *J. Clean. Prod.* 28, 168–76. doi:10.1016/j.jclepro.2011.08.018.

Nielsen, P. H. and Wenzel, H. 2007. Environmental assessment of Ronozyme® P5000 CT phytase as an alternative to inorganic phosphate supplementation to pig feed used in intensive pig production. *Int. J. Life Cycle Assess.* 12(7), 514–20. doi:10.1065/lca2006.08.265.2.

Nielsen, P. H., Dalgaard, R. L., Korsbak, A. and Pettersson, D. 2008. Environmental assessment of digestibility improvement factors applied in animal production: Use of Ronozym®e WX CT xylanase in Danish pig production. *Int. J. Life Cycle Assess.* 13(1), 49–56. doi:10.1065/lca2007.07.352.

Ogino, A., Osada, T., Takada, R., Takagi, T., Tsujimoto, S., Tonoue, T., Matsui, D., Katsumata, M., Yamashita, T. and Tanaka, Y. 2013. Life cycle assessment of Japanese pig farming using low-protein diet supplemented with amino acids. *Soil Sci. Plant Nutr.* 59(1), 107–18. doi:10.1080/00380768.2012.730476.

Oxenbøll, K. M., Pontoppidan, K. and Fru-Nji, F. 2011. Use of a protease in poultry feed offers promising environmental benefits. *Int. J. Poult. Sci.* 10(11), 842–8. doi:10.3923/ijps.2011.842.848.

Patience, J. F., Rossoni-Serão, M. C. and Gutiérrez, N. A. 2015. A review of feed efficiency in swine: biology and application. *J. Anim. Sci. Biotechnol.* 6(1), 33. doi:10.1186/s40104-015-0031-2.

Paxton, H., Anthony, N. B., Corr, S. A. and Hutchinson, J. R. 2010. The effects of selective breeding on the architectural properties of the pelvic limb in broiler chickens: A comparative study across modern and ancestral populations. *J. Anat.* 217(2), 153–66. doi:10.1111/j.1469-7580.2010.01251.x.

Pelletier, N. 2008. Environmental performance in the U.S. broiler poultry sector: Life cycle energy use and greenhouse gas, ozone depleting, acidifying and eutrophying emissions. *Agric. Syst.* 98(2), 67–73. doi:10.1016/j.agsy.2008.03.007.

Pettigrew, J. E. 2006. Reduced use of antibiotic growth promoters in diets fed to weanling pigs: dietary tools, part 1. *Anim. Biotechnol.* 17(2), 207–15. doi:10.1080/10495390600956946.

Pradère, J. P. 2014. Improving animal health and livestock productivity to reduce poverty. *Rev. Sci. Tech.* 33(3), 735–44, 723.

Prapaspongsa, T., Christensen, P., Schmidt, J. H. and Thrane, M. 2010. LCA of comprehensive pig manure management incorporating integrated technology systems. *J. Clean. Prod.* 18(14), 1413–22. doi:10.1016/j.jclepro.2010.05.015.

Proudfoot, F. G., Hulan, H. W. and Ramey, D. R. 1979. The effect of four stocking densities on broiler carcass grade, the incidence of breast blisters, and other performance traits. *Poult. Sci.* 58(4), 791-3. doi:10.3382/ps.0580791.

Prudêncio da Silva, V., van der Werf, H. M. G., Soares, S. R. and Corson, M. S. 2014. Environmental impacts of French and Brazilian broiler chicken production scenarios: an LCA approach. *J. Environ. Manage.* 133, 222–31. doi:10.1016/j.jenvman.2013.12.011.

Purswell, J., Dozier, W. A., Olanrewaju, H., Davis, J. D., Xin, H. and Gates, R. S. 2012. Effect of temperature-humidity index on live performance in broiler chickens grown from 49 to 63 days of age. In: *2012 IX International Livestock Environment Symposium (ILES IX)*, pp. 1–9. doi:10.13031/2013.41619.

Putman, B., Thoma, G., Burek, J. and Matlock, M. 2017. A retrospective analysis of the United States poultry industry: 1965 compared with 2010. *Agric. Syst.* 157, 107–17. doi:10.1016/j.agsy.2017.07.008.

Putman, B., Hickman, J., Bandekar, P., Matlock, M. D. and Thoma, G. J. 2018. A retrospective assessment of U.S. pork production : 1960 to 2015. Final Report. Fayetteville, AR.

Reece, F. N. and Lott, B. D. 1980. The effect of ammonia and carbon dioxide during brooding on the performance of broiler chickens. *Poult. Sci.* 59(7), 1654–61.

Ren, S., Li, M., Sun, J., Bian, Y., Zuo, K., Zhang, X., Liang, P. and Huang, X. 2017. A novel electrochemical reactor for nitrogen and phosphorus recovery from domestic wastewater. *Front. Environ. Sci. Eng.* 11(4), 17. doi:10.1007/s11783-017-0983-x.

Renaudeau, D., Gourdine, J. L. and St-Pierre, N. R. 2011. A meta-analysis of the effects of high ambient temperature on growth performance of growing-finishing pigs. *J. Anim. Sci.* 89(7), 2220–30. doi:10.2527/jas.2010-3329.

Rezaeipour, V. and Gazani, S. 2014. Effects of feed form and feed particle size with dietary L- threonine supplementation on performance, carcass characteristics and blood biochemical parameters of broiler chickens. *J. Anim. Sci. Technol.* 56, 20. doi:10.1186/2055-0391-56-20.

Sánchez, M., González, J. L., Gutiérrez, M. A. D., Guimaraes, A. C. and Gracia, L. M. N. 2008. Treatment of animal carcasses in poultry farms using sealed ditches. *Bioresour. Technol.* 99(15), 7369–76. doi:10.1016/J.BIORTECH.2008.01.042.

Schoumans, O. F., Rulkens, W. H., Oenema, O. and Ehlert, P. A. I. 2010. Phosphorus recovery from animal manure. Alterra Report 2158. Available at: http://library.wur.nl/WebQuery/wurpubs/404368.

Sharara, M. A., Sadaka, S. S., Costello, T. A., VanDevender, K., Carrier, J., Popp, M., Thoma, G. J. and Djioleu, A. 2016. Combustion kinetics of swine manure and algal solids. *J. Therm. Anal. Calorim.* 123(1), 687–96. doi:10.1007/s10973-015-4970-9.

Sharpley, A. N. 1997. Rainfall frequency and nitrogen and phosphorus runoff from soil amended with poultry litter. *J. Environ. Qual.* 26(4), 1127. doi:10.2134/jeq1997.00472425002600040026x.

Sharpley, A. N., Herron, S. and Daniel, T. 2007. Overcoming the challenges of phosphorus based management in poultry farming. *J. Soil Water Conserv.* 62, 375–89.

Sommer, S. G., Schjoerring, J. K. and Denmead, O. T. 2004. Ammonia emission from mineral fertilizers and fertilized crops. *Adv. Agron.* 82, 557–622. doi:10.1016/S0065-2113(03)82008-4.

Sommers, L. E. and Sutton, A. L. 1980. Use of waste materials as sources of phosphorus. In: Khasawneh, F. E., Sample, E. C. and Kamprath, E. J. (Eds), *The Role of Phosphorus in Agriculture*. ASA, CSSA, SSSA, Madison, WI, pp. 515–44.

Sonnemann, G. W. and Vigon, B. (Eds). 2011. *Global Guidance Principles for Life Cycle Assessment Databases, UNEP/SETAC Life Cycle Initiative*. UNEP/SETAC Life Cycle Initiative, Shonan, Japan.

Stephenson, A. H., McCaskey, T. A. and Ruffin, B. G. 1990. A survey of broiler litter composition and potential value as a nutrient resource. *Biol. Wastes* 34(1), 1–9. doi:10.1016/0269-7483(90)90139-J.

Suh, S. and Yee, S. 2011. Phosphorus use-efficiency of agriculture and food system in the U.S. *Chemosphere* 84(6), 806–13. doi:10.1016/j.chemosphere.2011.01.051.

Taddeo, R., Honkanen, M., Kolppo, K. and Lepistö, R. 2018. Nutrient management via struvite precipitation and recovery from various agroindustrial wastewaters: process feasibility and struvite quality. *J. Environ. Manage.* 212, 433–9. doi:10.1016/j.jenvman.2018.02.027.

ten Hoeve, M., Hutchings, N., Peters, G. M., Svanström, M., Jensen, L. S. and Bruun, S. 2014. Life cycle assessment of pig slurry treatment technologies for nutrient redistribution in Denmark. *Journal of Environmental Management* 132, 60–70. doi:10.1016/j.jenvman.2013.10.023.

Van Boeckel, T. P., Brower, C., Gilbert, M., Grenfell, B. T., Levin, S. A., Robinson, T. P., Teillant, A. and Laxminarayan, R. 2015. Global trends in antimicrobial use in food animals. *Proc. Natl. Acad. Sci. U.S.A.* 112(18), 5649–54. doi:10.1073/pnas.1503141112.

Van der Heyden, C., Demeyer, P. and Volcke, E. I. P. 2015. Mitigating emissions from pig and poultry housing facilities through air scrubbers and biofilters: state-of-the-art and perspectives. *Biosyst. Eng.* 134, 74–93. doi:10.1016/j.biosystemseng.2015.04.002.

Van Der Werf, H. M. G., Petit, J. and Sanders, J. 2005. The environmental impacts of the production of concentrated feed: the case of pig feed in Bretagne. *Agric. Syst.* 83(2), 153–77. doi:10.1016/j.agsy.2004.03.005.

van Zanten, H. H. E., Bikker, P., Mollenhorst, H., Meerburg, B. G. and de Boer, I. J. M. 2015. Environmental impact of replacing soybean meal with rapeseed meal in diets of finishing pigs. *Animal* 9(11), 1866–74. doi:10.1017/S1751731115001469.

van Zanten, H. H. E., Bikker, P., Meerburg, B. G. and de Boer, I. J. M. 2016. Environmental consequences of using alternative protein sources in pig diets. In: *10th International Conference on Life Cycle Assessment of Food*. Dublin.

van Zanten, H. H. E., Bikker, P., Meerburg, B. G. and de Boer, I. J. M. 2018. Attributional versus consequential life cycle assessment and feed optimization: alternative protein sources in pig diets. *Int. J. Life Cycle Assess.* 23(1), 1–11. doi:10.1007/s11367-017-1299-6.

Vanotti, M. B., Szogi, A. A., Millner, P. D. and Loughrin, J. H. 2009. Development of a second-generation environmentally superior technology for treatment of swine manure in the USA. *Bioresour. Technol.* 100(22), 5406–16. doi:10.1016/j.biortech.2009.02.019.

Varley, M. 2009. Taking control of feed conversion ratio. WWW Document. Pig Progress.

Vasisht, D., Kapetanovic, Z., Won, J., Jin, X., Chandra, R., Kapoor, A., Sinha, S. N., Sudarshan, M. and Stratman, S. 2017. FarmBeats : an IoT platform for data-driven agriculture. In: *Proceedings of the 14th USENIX Symposium on Networked Systems Design and Implementation*. USENIX, Boston, MA. doi:10.1093/abbs/gmw048.

Webb, J., Pain, B., Bittman, S. and Morgan, J. 2010. The impacts of manure application methods on emissions of ammonia, nitrous oxide and on crop response-A review. *Agric. Ecosyst. Environ.* 137(1–2), 39–46. doi:10.1016/j.agee.2010.01.001.

Weidema, B. P., Pizzol, M., Schmidt, J. H. and Thoma, G. J. 2018. Attributional or consequential Life Cycle Assessment: A matter of social responsibility. *J. Clean. Prod.* 174, 305–14. doi:10.1016/j.jclepro.2017.10.340.

Williams, A. G., Leinonen, I. and Kyriazakis, I. 2016. Environmental benefits of using turkey litter as a fuel instead of a fertiliser. *J. Clean. Prod.* 113, 167–75. doi:10.1016/j.jclepro.2015.11.044.

Wondra, K. J., Hancock, J. D., Behnke, K. C., Hines, R. H. and Stark, C. R. 1995. Effect of particle size and pelleting on growth performance, nutrient digestibility and stomach morphology. *J. Anim. Sci.* 73:757–63.

Yetilmezsoy, K., Ilhan, F., Kocak, E. and Akbin, H. M. 2017. Feasibility of struvite recovery process for fertilizer industry: A study of financial and economic analysis. *J. Clean. Prod.* 152, 88–102. doi:10.1016/j.jclepro.2017.03.106.

Yunis, R. and Cahaner, A. 1999. The effects of the naked neck (Na) and frizzle (F) genes on growth and meat yield of broilers and their interactions with ambient temperatures and potential growth rate. *Poult. Sci.* 78(10), 1347–52. doi:10.1093/ps/78.10.1347.

Zhang, T., Wang, Q., Deng, Y. and Jiang, R. 2018. Recovery of phosphorus from swine manure by ultrasound/H2O2 digestion, struvite crystallization, and ferric oxide hydrate/biochar adsorption. *Front. Chem.* 6, 464. doi:10.3389/fchem.2018.00464.

Zuidhof, M. J., Schneider, B. L., Carney, V. L., Korver, D. R. and Robinson, F. E. 2014. Growth, efficiency, and yield of commercial broilers from 1957, 1978, and 2005. *Poult. Sci.* 93(12), 2970–82. doi:10.3382/ps.2014-04291.

Chapter 12

The environmental impact of valorising agricultural by-products

Jerke de Vries and Rik Eweg, Van Hall Larenstein University of Applied Sciences, The Netherlands

1 Introduction

Agricultural by-products are an important source of biomass containing nutrients such as nitrogen (N), phosphorus (P) and potassium (K) and carbon (C) and other compounds such as cellulose, hemicellulose and fibres. Agricultural by-products include a varied range of residual biomass such as crop residues (primary residue), residues from processing industry (secondary residue) and manure. Where crop residues and manure are often cycled back to the soil and crop to maintain soil quality and crop growth, residues from processing industries are often used as animal feed or to a lesser extent for human food. By-products can also be used to produce bio-based materials and products. This use, generally called valorisation, is becoming more relevant in light of a circular and bio-based economy that should reduce the need for depletable fossil resources and their related environmental impact (Tuck et al. 2012). Hence, the demand for biomass is deemed to increase up to approximately 24 billion tonnes (bn t) dry matter (DM) in 2050 worldwide (Piotrowski et al. 2015). Agricultural by-products, mostly crop residue, make up approximately 12% of that at 2.8 bn t DM (Lal 2005) and are available in limited quantities as their production is dependent on the production of main products such as grains and oils.

http://dx.doi.org/10.19103/AS.2018.0044.16
Published by Burleigh Dodds Science Publishing Limited, 2019.

By-products can be valorised and converted through different treatment pathways such as fermentation, pyrolysis and anaerobic digestion (Fig. 1). These treatment pathways deliver different bio-based materials and products for a range of markets varying in size and economic value. Markets include fuels, chemicals for pharmaceutical and bulk chemical applications, animal feed and materials. Another 'market' is the often original use of by-products for building soil organic matter (SOM) and fertilising crops when returned to the soil either with or without addition of manure (Fig. 1). Although various treatment pathways are possible, conversion to bioenergy such as bioethanol, electricity and heat have been the most considered treatment options in the past decade. Only a few studies have focused on a combination of multiple treatment pathways such as bio-refining and bio-cascading (Fig. 2). Biorefinery as an integrated treatment pathway is comprised of a combination of technologies and is becoming more relevant in light of providing multiple bio-based materials from the same source of biomass. Bio-cascading is the more general term for valorising different components of biomass.

As valorisation can take place through different treatment pathways and by-products are available in limited quantities, competition for by-products between treatment pathways will easily occur and will affect the environmental consequences. Changing the use of by-products from its original application, such as animal litter or feedstock to another valorisation route will induce the need of a substitute for its original use. For example, when co-digesting beet pulp with animal manure for energy production, the beet pulp cannot be utilised for animal feed (De Vries et al. 2012a). Henceforth, a substitute for the animal feed will need to be introduced, for example barley. Consequently,

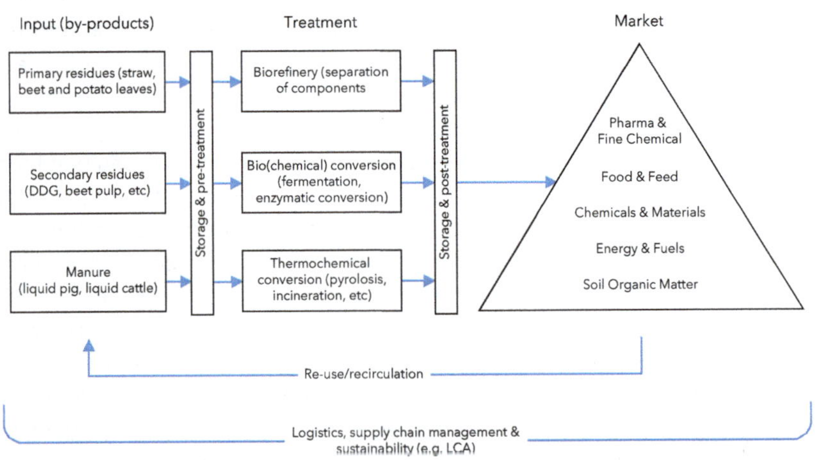

Figure 1 Valorisation of by-products through different treatment pathways, markets, re-use and recirculation, logistics, supply chain management and sustainability.

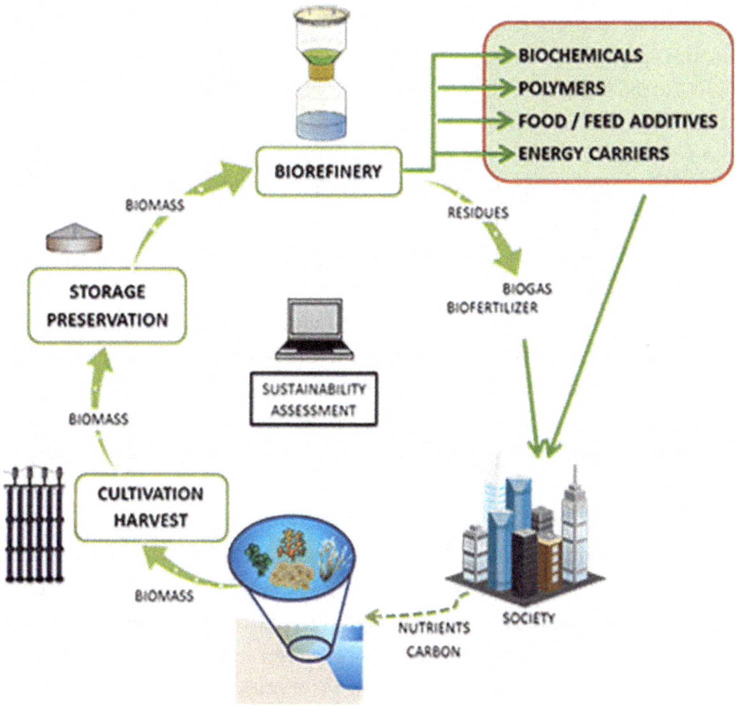

Figure 2 Concept of the bio-refinery in a circular society perspective. Courtesy of Gröndahl (2013).

the environmental impact of producing the barley needs to be included. Consequences related to the production of barley include land use and (indirect) land use changes (LULUC). LULUC can significantly affect the soil carbon (C) balance and sequestration and therewith the greenhouse gas (GHG) balance (Wiloso et al. 2016). Such changes can easily negate any environmental advantages obtained from valorising by-products (Tonini et al. 2015). These environmental trade-offs are also called pollution swapping (De Vries et al. 2015a). Pollution swapping can occur between different stages inside or outside the production system and also between different environmental impacts. To summarise, sustainable valorisation of by-products strongly depends on how competition between the original and aimed uses are included and therewith the related environmental consequences (e.g. consider Plevin et al. (2013) and Hedegaard et al. (2008)).

Important environmental impacts related to the valorisation of agricultural by-products include GHG emissions such as carbon dioxide (CO_2), methane (CH_4) and nitrous oxide (N_2O), LULUC and fossil energy depletion (Tilman et al. 2002). Water consumption may also be of importance for industrial processing of by-products (Lynch et al. 2016). Other impacts are eutrophication and acidification potentials which are more related to land use and production of the main crops. Impact on LULUC becomes relevant in light of changing uses of biomass or increasing demands (Parajuli et al. 2015). Where land use change is related to direct changes in cultivation and use of biomass, for example C sequestration differences when using straw for bioenergy, indirect land use change (iLUC) is related to shifting of production to other areas and biomes in the world, for example land expansion causing deforestation (Plevin et al. 2010). C emitted and sequestered in the soil is directly linked to the C and GHG balance and is critical for maintaining soil quality and therewith crop productivity (Wiloso et al. 2016). Soil quality and crop productivity in turn are critical for economic and social stability of the production system (Lal 2004; Pawelzik et al. 2013). Because of this central role, soil as the basic production substrate for all biomass will have to be the central pivotal point around which decisions for valorising biomass and agricultural by-products will have to take place. One question that will have to be answered is: 'What must we feed the soil, or more specifically, the soil biota when biomass is used for other purposes?'

Consequential life cycle assessment (LCA) is one of the most comprehensive tools available to assess the environmental consequences and trade-offs of valorising agricultural by-products. This does imply, however, that system expansion is applied wherever changes in the original use and applications of biomass and its end products occurs.

This chapter serves to provide a literature overview of opportunities to reduce the environmental impact of valorising agricultural by-products through different treatment pathways. In the following headings, we provide an overview and discussion on:

- the main treatment pathways for agricultural by-products,
- as an example the available and collectable by-products in Northwestern Europe (NW EU),
- environmental impacts including GHG reduction potential and soil C sequestration related to the valorisation of by-products through the main treatment pathways,
- future developments for reducing environmental impact of valorising by-products.

2 Main treatment pathways for valorising agricultural by-products

The main treatment technologies for valorising by-products roughly include mechanical treatment, for example pressing and breaking, biochemical treatment, for example fermentation and enzymatic conversion and (thermo)

Published by Burleigh Dodds Science Publishing Limited, 2019.

Table 1 Bio-refinery platforms based on converting biomass feed stocks to energy and bio-based products (IEA-Bioenergy 2018)

Route	Feed stock	Main treatment technologies	Intermediate and end products
1	Grass silage and food residues	Mechanical and fermentation	Bio-plastics, insulation material, fertiliser and electricity
2	Wood chips	Mechanical and (thermo)chemical	Pulp, paper, turpentine, tall oil, bark, electricity and heat
3	Starch	Mechanical, enzymatic and fermentation	Bioethanol and feed
4	Wood chips	Mechanical, enzymatic, fermentation and thermochemical	Bioethanol, electricity, heat and phenols
5	Oil seed crops	Mechanical and chemical	Bio-diesel, glycerine and feed
6	Oil-based residues	Mechanical and chemical	Bio-diesel, glycerine, bio-oil and fertiliser
7	Wood chips	Mechanical and (thermo)chemical	Biofuels, electricity, heat and waxes
8	Straw	Mechanical and (thermo)chemical	Biofuels and methanol

chemical treatment, for example pyrolysis and incineration (Fig. 1). Bio-refineries use a combination of technologies to convert inputs from biomass, for example the (hemi)cellulose, sugars, and starch and deliver outputs in the forms of for example energy, chemical feed stocks, biofuels and animal feed. Bio-refineries are considered as an alternative to conventional oil refineries (De Jong and Jungmeier 2015). The IEA Bioenergy consortium has defined eight bio-refinery platforms based on converting biomass feed stocks to energy and various intermediate and end products (Table 1). Only one of them (number 8) includes agricultural by-products as feedstock.

Primary residues such as straw contain lignocellulose that can be converted to for example biofuel or chemical feedstock. Secondary residues like beet pulp and distillers dried grains (DDG) are currently used to feed cattle, pigs and poultry. However, they also contain valuable components, such as hemicellulose, pectin and protein that can be used as either food ingredients, for chemicals and biofuels (Panagiotopoulos et al. 2010). Manure contains valuable nutrients and organic matter that can be utilised either as fertiliser or for nutrient extraction and energy or biofuel (De Azevedo et al. 2017).

3 Availability and collectability of by-products

3.1 Availability of agricultural by-products and their original use

Availability of the main agricultural by-products worldwide is estimated at 2.8 bn t DM for cereal crops, 3.1 bn t DM for the main cereal crops and legumes and 3.7 bn t DM for the main food crops (Lal 2005). In NW EU the available by-products are estimated at approximately 111 million tonnes (Mt) DM/year (Table 2). Primary residue from crop production represents approximately 64.7 Mt DM/year (or 59%), secondary residue from processing crops approximately 17.6 Mt DM/year (or 16%) and liquid manure approximately 28.2 Mt DM/year (or 26%). In the EU27 the availability of by-products was estimated at 258 Mt dry primary residue per year (Scarlat et al. 2010) and 597 Mt liquid pig and cattle manure (Foged et al. 2012).

Most of the by-products are produced in France (38%) and Germany (29%) followed by the United Kingdom (14%), Denmark (5%) and the Netherlands (4%). Of the primary residue, 42% is produced in France, 30% in Germany, 14% in the United Kingdom and 6% in Denmark. Approximately two-thirds of the secondary residue and half of the liquid manure are produced in France and Germany. Benelux covers approximately 14% of the liquid manure production.

Most of the primary residue is used for soil conditioning or SOM build-up by leaving it in the field (around 63%). The other part (approximately 37%) is used for litter in animal housing where afterwards it is returned to the soil for crop nutrition and SOM build-up (Helin et al. 2012). Less than 1% of straw is used for combined heat and power production. Secondary residues are often used for either animal feed, for example as a protein source, or for human food consumption. Some of the secondary residues, such as beet pulp, molasses and potato peels are also used for bioenergy production through anaerobic digestion. Manure is mostly used as fertiliser and for SOM build-up. About 11.5% of the manure in the EU is treated, of which 3.1% is treated by separation, 6.4% is anaerobically treated to produce bioenergy and another 2% is further treated by other technologies (Foged et al. 2012).

3.2 Collection of by-products

The collection of by-products includes the harvesting from the field or from the industrial application or animal housing system. The collection of primary residue from the field is dependent on many factors including yield, environmental conditions, available equipment, plant variety and crop rotation (Scarlat et al. 2010). Collection rates of residue vary between 30% and 50% depending on the need for SOM. Collection should occur approximately within a range of 50 km from the processing plant (Monforti et al. 2013) in order to limit the travel range and need for fuel. Collection rates of primary residue in Table 2 were based on the 'collectable' residue rates. Optimal collection rates may be higher depending on the local opportunity (Monforti et al. 2015). Similarly, for primary residues such as leaves, the collectable range will lie in the same order of magnitude but collection is less attractive due to its low availability. Leaves do contain a fair amount of protein (roughly 30% of the dry

Table 2 Availability of some of the main agricultural by-products in North Western Europe: Belgium (BE), Denmark (DK), Germany (DE), Ireland (IE), France (FR), Luxembourg (LU), the Netherlands (NL), Sweden (SE) and the United Kingdom (UK)

Agricultural by-products (Mt DM/year)	BE	DK	DE	IE	FR	LU	NL	SE	UK	Total	Original use	Source
Primary residue												
Crop residue (wheat, barley, rye, oat, maize, rice, rapeseed, sunflower)	0.97	3.45	18.2	0.73	25.6	0.06	0.61	2.03	8.37	60.0	Litter and SOM	Monforti et al. (2013) Collectable residue
Potato leaves	0.09	0.04	0.27	0.01	0.19	0	0.18	0.03	0.16	0.97	SOM	Eurostat (2018a)[a]
Beet leaves	0.19	0.12	1.20	0	1.39	0	0.24	0.10	0.35	3.61	SOM	Eurostat (2018a)[b]
Other root crops leaves	0.01	0.01	0.01	0.01	0.02	0	0	0	0.05	0.10	SOM	Eurostat (2018a)[c]
Total	1.26	3.62	19.7	0.75	27.2	0.06	1.03	2.16	8.93	**64.7**		
Secondary residue												
Beet pulp	0.96	0.18	0.05	0	0.33	0	0.02	0.16	0.43	2.19	Feed/energy	CEFS (2016)[d]
Molasses	0.11	0.07	0.71	0	0	0	0.18	0.05	0.10	1.22	Food/feed/energy	CEFS (2016)[d]
Distillers dried grains (DDG)	0.37	0	0.76	0.01	1.55	0	0.37	0.16	0.64	3.86	Feed	ePURE (2016)[e]
Soy bean meal	0	0	0.04	0	0.20	0	0	0	0	0.23	Feed/protein	Eurostat (2018a)[f]
Rapeseed meal	0.03	0.41	3.03	0.02	2.95	0.01	0.00	0.19	1.30	7.94	Feed/protein	Eurostat (2018a)[g]
Sunflower meal	0	0	0.03	0	0.95	0	0	0	0	0.98	Feed/protein	Eurostat (2018a)[h]
Brewers grain	0.09	0.03	0.42	0.03	0.09	0	0.11	0.02	0.19	0.98	Feed/protein	the Brewers of Europe (2016)[i]
Potato peels	0.02	0.01	0.07	0	0.05	0	0.04	0.01	0.04	0.24	Feed/protein/energy	Eurostat (2018a)[j]
Total	1.22	0.70	4.33	0.06	4.62	0.01	0.36	0.44	2.05	**17.6**		

(*Continued*)

Table 2 (*Continued*)

Agricultural by-products (Mt DM/year)	BE	DK	DE	IE	FR	LU	NL	SE	UK	Total	Original use	Source
Manure												
Liquid pig	0.50	0.99	2.15	0.12	1.18	0.01	0.97	0.12	0.37	6.39	Fertiliser/SOM	Foged et al. (2012)[k]
Liquid cattle	0.96	0.58	4.88	2.53	7.01	0.07	1.51	0.58	3.73	21.8	Fertiliser/SOM	Foged et al. (2012)[k]
Total	1.45	1.57	7.03	2.65	8.19	0.08	2.47	0.70	4.10	**28.2**		

Values 0 are <0.05. SOM = soil organic matter.
[a] Based on 2.25 t DM/ha (Tijmensen et al. 2002; Neeteson and Greenwood 1987).
[b] Based on 6.75 t DM/ha (Tijmensen et al. 2002).
[c] Assumed similar as potato leaves.
[d] 5-year average 2011–16.
[e] 2016 data converted assuming 0.65 kg DDG/L of ethanol and 90% DM (US Grains Council 2012).
[f] Average of 2013–18. Based on 77% extraction and 89% DM (ingredients101 2018).
[g] Average of 2013–18. Based on 66% extraction (Nova 2007) and 88.5% DM (KW 2018).
[h] Average of 2013–18. Based on 75% extraction and 89% DM (Le Clef and Kemper 2015).
[i] Based on 20 kg brewers grain/hL of beer (Mussatto et al. 2006) and 22% DM (Duynie 2018b).
[j] Average of 2013–18. Based on 38 kg peels/t of potatoes (Ahokas et al. 2014) and 16.5% DM (Duynie 2018a).
[k] Based on 8.23% DM for liquid pig manure and 7.45% DM for liquid cattle manure Menzi (2002) in De Vries et al. (2015a).

weight) that may be interesting for food and feed applications (Kiskini 2017). The collection of secondary residues and manure depends on the original use and storage. Secondary residues and manure are more easily collected than primary residues as they are stored already.

4 Environmental impact of valorising by-products

4.1 Primary residue

Table 3 provides an overview of the environmental impact of the main treatment pathways for agricultural by-products. Studies were often focused on crop residues such as corn stover and wheat or rice straw for biofuel production purposes and manure as they are most abundant, for example De Azevedo et al. (2017), Morales et al. (2015) and Prapaspongsa et al. (2010). Biofuel production from valorising wheat straw and corn stover reduces GHG emissions between 262 and 903 kg CO_2-equivalents (eq)/t DM/year compared to its fossil reference. However, effects of changing soil C sequestration as a result of removing crop residues are not included in all studies (e.g. the 903 kg CO_2-eq/t DM in Table 3). These studies, therefore, overestimate the GHG reduction potentials. In Section 5.2 we further elaborate on this issue and the impact on the GHG reduction estimates. Assuming a maximum reduction potential of 903 kg CO_2-eq/t DM for primary residue, GHGs can be reduced by up to 54.2 Mt CO_2-eq on a NW EU scale (Table 4). This represents about 2.2% of the total GHG emissions and 22% of the GHG emissions from agriculture (for the given countries in Table 4: 2447 Mt CO_2-eq total and 249 Mt CO_2-eq agricultural emissions in 2015, Eurostat (2018b)). When including changes in C sequestration, the estimated GHG reduction potential is 37 Mt CO_2-eq. Fossil fuel depletion (FFD) varies between 1120 and 2219 MJ/t DM/year (input of fossil-based and primary energy for conversion). Fossil fuel savings run up to 18 GJ/t DM/year (output of bioenergy) (Cherubini and Ulgiati 2010). On the contrary, the acidification potential varied between 1.64 and 1.95 kg SO_2-eq/t DM/year and eutrophication potential varied from 0.82 to 1.09 kg PO_4-eq/t DM/year. Acidification potentials were lower compared to the fossil reference (up to 2.43 kg SO_2-eq/t DM/year) and eutrophication potentials are generally higher compared to the fossil reference (up to 0.36 kg PO_4-eq/t DM/year). Higher eutrophication potentials were related to using fertiliser in the production of the main crop.

4.2 Secondary residue

Little information is available about the environmental impact of valorising secondary residues. Some work has been done on the valorisation of glycerol from rapeseed oil and potato juice to propionic acid and also anaerobic co-digestion of residues with manure (dealt with in Section 4.3).

Table 3 Overview of environmental impacts (SOM = soil organic matter, GHG = greenhouse gases, FFD = fossil fuel depletion, Resp inorg = respiratory inorganics, Acidif = acidification potential, Terr Eutr = terrestrial eutrophication potential and Aq Eutr = aquatic eutrophication potential) of the main treatment pathways for valorising agricultural by-products

							Environmental impact							
By-product	Treatment pathway	End product/ Market	Country	Original use	Substitute included (yes/no)?	Unit	GHG[y/n] (kg CO_2-eq)	GHG saved comp to fossil-based reference (kg CO_2-eq)	FFD (MJ)	Resp inorg (kg $PM_{2.5}$-eq)	Acidif (kg SO_2-eq)	Terr Eutr (m^2 UES)	Aq Eutr (kg NO_3-eq)	Source
Primary residue														
Corn stover	Bio-refinery	Bioethanol, heat, electricity and phenols	AU	SOM	Yes	per t DM per year	287[y]	333	1677[a]	n.d.	1.95	n.d.	1.09 kg PO_4-eq	Cherubini and Ulgiati (2010)
Wheat straw	Bio-refinery	Bioethanol, heat, electricity and phenols	AU	SOM	Yes	per t DM per year	273[y]	262	1887[a]	n.d.	1.64	n.d.	0.82 kg PO_4-eq	Cherubini and Ulgiati (2010)
Straw	Bio-refinery	Diesel and methanol		SOM	No	per t DM per year	93.2[n]	903	2219[a]	n.d.	n.d.	n.d.	n.d.	IEA-Bioenergy (2018)
Secondary residue														
Glycero and potato juice	Bio-refinery (fermentation)	Propionic acid	DK	Heat/ fertiliser	No	per t of rapeseed	~39[n]	~56	~1120[b]	n.d.	n.d.	n.d.	0.17 kg PO_4-eq	Ekman and Börjesson (2011)
Potato peels	Bio-reactor	Hydrogen, feed	NL	Feed	No	per t DM	−25[n]	25	−413	n.d.	n.d.	n.d.	n.d.	Djomo et al. (2008)

Manure														
Liquid pig	Anaerobic mono-digestion	Energy and Fuel	DK, NL	Fertiliser	Yes	per t manure	−16 to -40^{y}	16 to 147	−117	0.13	0.09	144	−0.13 to 0.93	Hamelin et al. (2011) and De Vries et al. (2012a)
Liquid pig	Anaerobic co-digestion	Energy and Fuel	DK, NL	Fertiliser and Feed	Yes	per t substrate	−89 to 105^{y}	89 to −105	−2398 to 1390	−0.13 to 0.36	−0.33 to 1.61	n.d.	−0.34 to 1.12	Hamelin et al. (2011) and De Vries et al. (2012a)
Liquid pig	Incineration	Electricity and Heat	DK	Fertiliser	Yes	per t manure	-11.8^{n}	119	n.d.	0.04	n.d.	64	0.58	Prapaspongsa et al. (2010)
Liquid pig	Gasification	Energy and Fuel	DK	Fertiliser	Yes	per t manure	-8.8^{n}	116	n.d.	0.05	n.d.	64	0.58	Prapaspongsa et al. (2010)
Liquid pig	Separation	SOM	DK, NL	Fertiliser	Yes	per t manure	37 to 88^{y}	16 to −3	348 to 181	0.76 kg PM_{10}-eq	4.8 to 5.9	n.d.	1.02 to 4.21	Ten Hoeve et al. (2014) and De Vries et al. (2012b)
Liquid cattle	Integrated management	Fertiliser and SOM	NW EU	Fertiliser	Yes	per t manure	-78.6^{n}	190	−805	0.05 kg PM_{10}-eq	0.79	n.d.	0.04	De Vries et al. (2015a)
Liquid pig	Integrated management	Fertiliser and SOM	NW EU	Fertiliser	Yes	per t manure	-45.9^{n}	472	−1043	0.27	2.44	n.d.	0.93	De Vries et al. (2015a)

n.d. = no data. Negative numbers under GHG emissions and other impacts represent a net reduction in environmental impact. Negative numbers under GHG saved represent net additions in GHG. $GHG^{y/n}$ means that either soil C sequestration is included or excluded/unknown. Substitute included (Yes/No) means that a substitute is considered for the change of the original use of the by-product.

[a] These figures include only the fossil fuel input for producing bioenergy.

[b] Primary energy consumption.

Table 4 Estimated maximum GHG reduction potential of valorising agricultural by-products through the main treatment pathways in NW EU based on Tables 2 and 3

Potential GHG reduction (Mt CO_2-eq)	BE	DK	DE	IE	FR	LU	NL	SE	UK	Total
Primary residue										
Crop residue (wheat, barley, rye, oat, maize, rice, rapeseed, sunflower)	0.87	3.11	16.4	0.66	23.1	0.05	0.55	1.84	7.56	**54.2**
Crop residue with changes in C sequestration[a]	0.59	2.11	11.1	0.44	15.6	0.04	0.37	1.24	5.11	36.7
Secondary residue										
Rapeseed meal	0	0.04	0.29	0	0.28	0	0	0.02	0.12	0.76
Potato peels	6.1E-04	2.6E-04	1.7E-03	5.9E-05	1.2E-03	2.8E-06	1.1E-03	1.3E-04	9.1E-04	0.01
Total	0	0.04	0.29	0	0.28	0	0	0.02	0.13	**0.77**
Manure										
Liquid pig	2.85	5.66	12.3	0.67	6.78	0.04	5.54	0.70	2.11	36.6
Liquid cattle	2.43	1.48	12.4	6.44	17.8	0.19	3.83	1.48	9.50	55.6
Total	5.28	7.14	24.7	7.12	24.6	0.23	9.37	2.18	11.6	**92.3**
										147

Belgium (BE), Denmark (DK), Germany (DE), Ireland (IE), France (FR), Luxembourg (LU), the Netherlands (NL), Sweden (SE) and the United Kingdom (UK).
Values 0 are <0.05.
[a] C-sequestration estimates included based on the calculation in Section 4.4.

GHG reduction potential of producing propionic acid was estimated at 56 kg CO_2-eq/t rapeseed. However, a substitute for the original use of the glycerol is not considered and therefore possibly overestimating the reduction potential. Glycerol is currently used for heat and power production because it is produced in excess of market demand. This means that when adding a substitute, additional GHG emissions will be emitted. Assuming natural gas as the substitute with 0.07 kg CO_2-eq/MJ and a lower heating value of glycerol of 16.5 MJ/kg, an additional amount of 45 kg CO_2-eq will be emitted per ton of rapeseed. This additional amount negates 80% of the saved GHGs (45/56 kg CO_2-eq). On a NW EU basis, this means a potential reduction in GHG emission of up to 0.76 Mt CO_2-eq (without a substitute) which represents only 0.03% of the GHG emissions from agriculture. With a substitute for heat production this would be approximately 0.15 Mt CO_2-eq or 0.01% of the GHG emissions from agriculture. Eutrophication potential was 0.17 kg PO_4-eq/t rapeseed.

Another study examined the production of hydrogen from potato steam peels instead of using it as animal feed (Djomo et al. 2008). Results of the study showed a potential to reduce GHGs up to a factor of two to three compared to a fossil reference. GHG emissions were −25 kg CO_2-eq/t DM potato peels and FFD was −303 MJ/t DM potato peels. Again, a substitute for animal feed when using the potato peels was not included. On a NW EU basis, this means a potential reduction in GHG emission of <0.01 Mt CO_2-eq which represents <0.01% of the GHG emissions from agriculture (Table 4).

4.3 Manure

There exist numerous environmental impact studies on manure treatment pathways, from simple separation to the production of energy and strategies for integrated management (Table 3). Strategies for integrated management encompass (technological) changes in all phases of the manure management system from storage through processing and (field) application. This includes acidification of manure, separation and synchronising of manure application with crop demand and also soil treatment (De Vries et al. 2015a). The GHG emission of valorising manure varies between a positive emission of 105 kg CO_2-eq/t manure to a reduction of 472 kg CO_2-eq/t manure (Table 3). On a NW EU basis, this means a potential reduction in GHG emissions of up to 92 Mt CO_2-eq which represents about 4% of the total GHG emissions and 37% of the GHG emissions from agriculture (Table 4). FFD varies between 348 and 1390 MJ/t manure. Fossil fuel savings run up to 1043 MJ/t manure through the production of energy (De Vries et al. 2015a). Acidification potential varied between −0.33 and 5.9 kg SO_2-eq/t manure and eutrophication potential varied from −0.34 to 4.21 kg NO_3-eq/t manure.

4.4 LULUC and soil carbon sequestration

LULUC and consistent removal of by-products pose adverse effects on soil C sequestration and soil quality in the short and long run (Lal 2005). LULUC-induced emission may run up to 4.1 t CO_2-eq per demanded hectare (ha) or 1.2–1.4 t CO_2-eq/t dry biomass (Tonini et al. 2015). As a part of LULUC, soil C sequestration varies between 50 and 1000 kg of C/ha/year corresponding to 183 and 3667 kg CO_2/ha/year (Goglio et al. 2015; Lal 2004). Sequestration is affected by the type of crops that are produced, their rotation sequence and the type of soil management that is applied, for example tillage or no tillage (Lal 2004). These parameters determine the amount of C that is turned into humus and stored in the soil for a longer period of time.

In the current literature, it is estimated that 40–50% of crop residues can be removed without diminishing the stable soil C pool (Monforti et al. 2013, 2015; Scarlat et al. 2010). In no-till farming, residue removal rates could be as high as 82% (Scarlat et al. 2010), but in other circumstances removal rates of 30–40% may already induce hazard to soil quality and the stable soil C pool (Lal 2005).

To estimate soil C sequestration, the C sequestration efficiency can be used. For example, assuming a conservative sequestration efficiency of 17% for wheat straw (Hua et al. 2014), a DM content of 90%, an organic matter content of 94% of DM and 50% of C in the organic matter, approximately 80 kg of C or 292 kg of CO_2/t of dry straw can be sequestered yearly. On a NW EU basis, this approximates to 4.8 Mt of C or 18 Mt of CO_2. This is roughly 32% of the GHG reduction potential of primary residue (18/54.2 Mt CO_2-eq, Table 4) and 12% of the total GHG reduction potential (18/147 Mt CO_2-eq, Table 4).

For pig and cattle manure, assuming a C sequestration efficiency of 11% and 6.5% organic matter, the sequestration is roughly 4 kg of C or 14 kg of CO_2/t wet manure yearly. On a NW EU basis, this approximates 1.4 Mt of C or 5.2 Mt of CO_2 being roughly 1% of the total GHG reduction potential (1.4/147 Mt CO_2-eq in Table 4).

Concluding, soil C sequestration strongly influences the GHG reduction potentials, especially for primary residue, and therefore needs to be assessed when scientific studies lack such inclusion. Using the C sequestration efficiency is a straightforward way to simply estimate the (reduced) C sequestration and its impact on the GHG reduction potential.

5 Future opportunities and perspectives

5.1 Availability and collectability of by-products

The availability and collectability of by-products is critical for valorisation opportunities. Here, the available primary residue was based on the 'collectable estimate' given in Monforti et al. (2015). They also provided an 'optimal estimate' that was 42.9% higher relative to the default collection. Optimal collection rates of primary residue for NW EU would be approximately 86 Mt DM/year with additional potential to reduce GHG emissions up to 77 Mt CO_2-eq from primary residue without lost C sequestration and 52 Mt CO_2-eq with lost C sequestration. Agricultural by-products, however, are not always optimally distributed meaning that collecting residues may not be economically attractive. To be economically feasible for exploitation, valorisation plants will have to be placed at strategic distances from where the residue is located (Golecha and Gan 2016).

Next to collectability, the availability of by-products will depend on their competing uses. When using by-products for different treatment pathways they compete either with the conventional raw materials or with different applications of the same by-product. This may increase their market values. In recent years, prices of by-products used for anaerobic co-digestion, such as energy crops and industrial by-products, increased up to €72/t due to the increased demand for anaerobic digestion (Velghe and Wierinck 2013). Moreover, when by-products were originally used as animal feed, the GHG reduction potential of anaerobic digestion was negated by LULUC, e.g. De Vries et al. (2012a). Since agricultural by-products are dependent on a main product, they remain limited in their availability. In other words, the amount of by-products will generally not increase if market demand rises. Therefore, a good balance between different treatment pathways will be needed to not overexploit the available by-products, resulting in skyrocketing prices and increased consequences for the environment. This will require not only good guidelines on a national and international level but also a level playing field for alternative uses of by-products such as bio-based chemicals and materials (Carus et al. 2011).

Availability of by-products may increase through intensification of agricultural production and using marginal lands. Crop production increased by 2.5% annually since the 1960s up to the late 1990s. Estimates now show that crop production may increase but this depends strongly on the genetic yield potential and environmental factors (Cassman 1999). The environmental consequences of intensifying agriculture has, however, been severely debated and remains controversial.

Asia and the United States will be the main producers of biomass and by-products in the future (Daioglou et al. 2016). Exchange of biomass between countries will not be very likely due to the degradability and bulkiness of the material but will need to be valorised locally as much as possible. Valorisation of using biomass will depend much on the local business opportunities.

5.2 Environmental assessment and reduction potential of valorising by-products

The total maximum GHG reduction potential of 147 Mt CO_2-eq (Table 4) represents about 6% of the total GHG emissions and 59% of GHG emission from agriculture in NW EU. Valorising all wheat straw on a worldwide scale (2.8 bn t DM/year) through bio-refinery with a GHG reduction potential of 611 kg CO_2-eq/t DM (903 kg CO_2-eq/t DM minus 32% C sequestration loss, Table 3 and Section 4.4.), this could be as much as 1.71 bn t CO_2-eq or <4% of the worldwide GHG emissions (approximately 45 bn tons CO_2-eq in 2014). Previous estimates do not yet include valorisation of all by-products as studies on environmental impact reduction are still limited (Tables 3 and 4). The main potential to reduce environmental impact, however, lies in the valorisation of animal manure (63% of the total reduction potential, or 92.3 Mt CO_2-eq of 147 Mt CO_2-eq). Manure management contributes up to 50% of the N_2O and around 20% of the CH_4 emissions in the livestock sector of the EU (Leip et al. 2015). Changes in manure management are often limited to reducing GHGs and affect other emissions in no or limited ways (Table 3). New strategies for integrated manure management show opportunities to further reduce environmental impact but have yet to be implemented in practice.

Few environmental assessments have been done for valorising primary residues such as beet and potato leaves and secondary residues coming from industry after processing. Beet and potato leaves contain protein that can be extracted and used for food products (Tamayo Tenorio et al. 2016). Secondary residues that have been researched for their human food potential, such as brewers' spent grain have shown to be potentially interesting as a protein source as well (Lynch et al. 2016). We estimate that reduction potentials of these pathways will be <1–2% of the total GHG emission in NW EU. However, changing the original use of secondary residues may well introduce trade-offs such as iLUC and therewith negate the GHG reduction potential. Although limited, it is still important to understand and assess the environmental impact of valorisation in a preliminary stage to support decision making.

5.3 Environmental trade-offs and pollution swapping in the chain

Environmental trade-offs and pollution swapping occurs when one environmental impact is reduced and another increases as a direct or indirect consequence of a taken action. Pollution swapping occurs on different scales from process level up to chain level and may include various environmental impacts. In manure management, for example, pollution swapping occurs when covering manure storages or when injecting manure instead of broadcast spreading (De Vries et al. 2015b). When covering manure storages and injecting manure, ammonia emissions are reduced but nitrous oxide emissions are increased. Swapping may also occur on chain level when more N is contained in the manure and is potentially lost during field application.

When valorising agricultural by-products, pollution swapping between environmental impacts also occurs as shown in Table 3. Where GHG emissions were reduced, other environmental impacts such as eutrophication potential or FFD were increased. This is common due to the needed substitute for the original use of the by-product when used for other treatment pathways. The substitute needed may well be an additional amount of crop and consequently an amount of land to produce the needed crop. Next to LULUC, inputs such as fertiliser and manure are needed for production and cause eutrophication and other land-related impacts (Kim and Dale 2005).

Another important environmental trade-off that occurs when by-products are valorised is the shift in C sequestration and the related GHG emissions as demonstrated in Section 4.4. Changes in the C balance as a result of LULUC and C sequestration are essential for estimating GHG reduction potentials when valorising by-products. Wiloso et al. (2016) provide a comprehensive and straightforward overview of how the C balance is related to LULUC and C sequestration. C sequestration, when lacking in environmental assessments, can easily be estimated based on the C sequestration efficiency.

A solution to avoid pollution swapping lies in the (re-)design of the complete valorisation chain. In such a process all requirements of the stakeholders in the chain need to be included. The environment can be included as a 'stakeholder' in order to establish requirements for environmental impact reduction and trade-offs. (Re-)design of a complete system, however, is a complex task and requires strong dedication of stakeholders and a good process manager (De Vries et al. 2015b).

Other environmental impacts not considered here but that are relevant when valorising by-products are water consumption, toxicity effects and human health impacts. When valorising by-products to biofuels, water consumption of processing varies between roughly 800 and 4000 L of water per litre of biofuel produced (Singh et al. 2011). When water availability is limited, the scarcity of water will be relevant. This may require the use of a water stress index (Pfister et al. 2009). The water stress index relates the water use to the local availability or scarcity and thus impact of using water.

Other factors in addition to environmental ones, such as social and economic factors, will have to be included when making decisions on how to valorise agricultural by-products. Examples include increased logistics in small rural areas due to the collection of residues and market prices. In this way a broader scope to sustainability of the treatment pathway will be ensured.

5.4 Economic viability of valorising by-products

The economic viability of valorising agricultural by-products will be determined by the cost of collecting the by-products and their market prices, the investment and operations cost of the treatment pathway and the market value and opportunities for the end products. Market prices of primary residues were estimated at between €16 and €72/t ($19 and $84) (Carriquiry et al. 2011) and for some secondary residues such as brewers' grain at €35/t (Lynch et al. 2016). Typically, manure has a negative value in NW EU meaning the farmers need to pay for removing the manure from their farms. Removal prices vary between €15 and €25/t for liquid pig manure in the Netherlands (NVV 2016).

Bio-refinery pathways are estimated to be viable with revenues between €5.5 and €220 M/year or €394 to €1410 /t of residue per year (IEA-Bioenergy 2018). For straw to diesel and methanol this was estimated to be €160 M/year or €1068/t/year. Other, less economically viable pathways include anaerobic digestion. Anaerobic digestion has shown little economic feasibility if not subsidised or if the digestate could not be sold (Astill and Shumway 2016; De Dobbelaere et al. 2015).

6 Summary and conclusion

Agricultural by-products consist of a wide range of biomass types including crop residue, residues from processing industries (secondary residue) and manure. The main treatment technologies for valorising agricultural by-products include mechanical treatment such as pressing and breaking, biochemical treatment such as fermentation and enzymatic conversion and (thermo)chemical conversion such as pyrolysis and incineration. Bio-refineries are comprised of a combination of technologies using multiple parts of the biomass to produce bio-based end products. Integrated approaches such as bio-refineries and integrated manure management strategies offer opportunities to valorise by-products while reducing environmental impact such as GHG emissions and FFD (Anon 2014).

On a worldwide scale, crop residue is available at approximately 2.8 bn t DM/year (Lal 2005). Assuming a default collection rate for crop residues (Monforti et al. 2015), approximately 111 Mt DM of agricultural by-products are available in NW EU on a yearly basis. Primary residues comprise 59%, secondary residues comprise 16% and liquid pig and cattle manure comprise 26% of the available by-products. Crop residues can be collected at a rate of approximately 40–50% without diminishing soil C stocks and reducing soil health. Optimal collection rates for crop residues could be 43% higher but depend on specific conditions.

The saved GHGs, compared to a fossil reference, when valorising primary residue were between 262 and 903 kg CO_2-eq/t DM (without lost C sequestration). Including a simple estimate of C sequestration, the maximum reduction potential was 32% lower. The maximum estimated GHG reduction potential for primary residue in NW EU was 54 Mt CO_2-eq or 2% of the total GHG emissions without lost C sequestration and 37 Mt CO_2-eq with lost C sequestration. On a worldwide scale this could be as much as 1.71 bn t CO_2-eq or <4% of the worldwide GHG emissions. GHG reductions were often accompanied by higher eutrophication potentials. The saved GHGs of valorising secondary residue were between 25 and 56 kg CO_2-eq/t DM. GHG reductions were <0.1% on NW EU scale. Valorisation of secondary residue was also highly susceptible to LULUC. The saved GHGs when valorising manure were between 105 and 472 kg CO_2-eq/t DM. The maximum GHG reduction potential for manure in NW EU was 92 Mt CO_2-eq or 4% of the total GHG emissions.

In total, the saved GHG emissions of valorising agricultural by-products in NW EU ran up to approximately 147 Mt CO_2-eq or 6% of the total and 59% of the agricultural GHG emissions. Altering manure management had the greatest potential to reduce GHG emissions (63% of the total estimated reduction potential). The opportunity to reduce GHG emissions is, however, easily diminished if LULUC is involved or soil C sequestration is lost when removing crop residue. A simple approach based on the C sequestration efficiency showed that a C sequestration potential of about 262 kg CO_2/t dry straw is lost when residue is removed. This represented about 32% of the GHG reduction potential when valorising primary residue. Removing crop residue from the field requires estimates of soil C sequestration. This is relevant especially when answering the question: 'What must we feed the soil, or more specifically, the soil biota when biomass is used for valorisation?' Not answering this question properly will lead to reduced sequestration and SOM and subsequently to reduced soil health and long-term productivity.

Other trade-offs besides LULUC and C sequestration include increased eutrophication potentials. Eutrophication is related to producing a substitute to replace the original use of the by-product. Trade-offs are important to take into account when establishing new valorisation chains and treatment pathways.

Future opportunities for reducing environmental impact of valorising agricultural by-products lie in increased availability of by-products through improved production systems such as conservation agriculture and no-till farming (Kim et al. 2009). The availability of crop residue in NW EU can run up to 86 Mt DM in optimal collection circumstances further reducing GHG emissions up to 77 Mt CO_2-eq (instead of 52 Mt CO_2-eq under collectable residue rate) without lost C sequestration or 54 Mt CO_2-eq with lost C sequestration. Information was limited on the environmental impact of valorising primary residue such as beet leaves and secondary residues from industries. The GHG reduction potential was estimated at <0.1% of the total GHG emission in NW EU but may easily be negated when LULUC is induced.

Finally, next to availability and environmental aspects, the viability of valorising agricultural by-products depends on the economics and revenues that can be generated from the end products. A level playing field for stimulating other bio-based products than fuel or energy is deemed necessary for best environmental results when valorising agricultural by-products.

7 Where to look for further information

The cited literature provides a good overview of the current assessments done and the information on the availability of agricultural by-products. The IEA-Bioenergy recently published information on valorisation of biomass to bioenergy. This includes information on the environmental impacts and value chains of different biorefinery treatment pathways (IEA-Bioenergy 2018), for example https://www.iea-bioenergy.task42-biorefineries.com/en/ieabiorefinery/Factsheets.htm.

Knowledge institutes such as the Nova Institute in Germany, Wageningen University and Research in the Netherlands and the Joint Research Centre of the European Union regularly publish data and information on the bioeconomy. They communicate through reports and newsletters available on their websites (Nova 2018a,b; WUR 2018; JRC 2018), for example:

- http://news.bio-based.eu/biomass-cascading-use-equals-best-lca/
- https://www.wur.nl/en/Research-Results/Research-Institutes/food-biobased-research/about/Biobased-Products-2.htm
- https://ec.europa.eu/jrc/en

General information on bioeconomy networks and research in the European Union can be found at (BBE 2018; EU 2018):

- https://www.biobasedeconomy.nl/
- https://ec.europa.eu/research/bioeconomy/index.cfm

Data on the production of crops, biofuels and industrial products can be found at different locations (Eurostat 2018a; FAO 2018; Oil World 2018), for example:

- https://ec.europa.eu/eurostat/data/database
- http://www.fao.org/statistics/en/
- https://www.oilworld.biz/

8 References

Ahokas, M., Välimaa, A. L., Lötjönen, T., Kankaala, A., Taskila, S. and Virtanen, E. 2014. Resource assessment for potato biorefinery: Side stream potential in Northern Ostrobothnia. *Agronomy Research*, 12(3), pp. 695–704.

Anon. 2014. Biomass: Cascading use equals best life cycle assessment – Bio-based News. *Biobased News*. Available at: http://news.bio-based.eu/biomass-cascading-use-equals-best-lca/ (accessed 23 March 2018).

Astill, G. M. and Shumway, C. R. 2016. Profits from pollutants: Economic feasibility of integrated anaerobic digester and nutrient management systems. *Journal of*

Environmental Management, 184, pp. 353–62. Available at: http://www.sciencedirect.com/science/article/pii/S0301479716307848.

BBE. 2018. BioBased economy. Available at: https://www.biobasedeconomy.nl/ (accessed 7 July 2018).

Carriquiry, M. A., Du, X. and Timilsina, G. R. 2011. Second generation biofuels: Economics and policies. *Energy Policy*, 39(7), pp. 4222–34. Available at: http://www.sciencedirect.com/science/article/pii/S0301421511003193.

Carus, M., Carrez, D., Kaeb, H., Ravenstijn, J. and Venus, J. 2011. Level playing field for bio-based chemistry and materials. Available at: http://greengran.nl/download/Policy paper on Bio-based Economy in the EU.pdf.

Cassman, K. G. 1999. Ecological intensification of cereal production systems: Yield potential, soil quality, and precision agriculture. *Proceedings of the National Academy of Sciences*, 96(11), pp. 5952–9. Available at: http://www.pnas.org/content/96/11/5952.abstract.

CEFS. 2016. *CEFS Sugar Statistics 2016*. CEFS, Brussels.

Cherubini, F. and Ulgiati, S. 2010. Crop residues as raw materials for biorefinery systems – A LCA case study. *Applied Energy*, 87(1), pp. 47–57. Available at: http://www.sciencedirect.com/science/article/pii/S0306261909003596.

Daioglou, V., Stehfest, E., Wicke, B., Faaij, A. and van Vuuren, D. P., 2016. Projections of the availability and cost of residues from agriculture and forestry. *GCB Bioenergy*, 8(2), pp. 456–70.

De Azevedo, A., Fornasier, F., da Silva Szarblewski, M., de Cassia de Souza Schneider, R., Hoeltz, M. and de Souza, D. 2017. Life cycle assessment of bioethanol production from cattle manure. *Journal of Cleaner Production*, 162, pp. 1021–30. Available at: https://www.sciencedirect.com/science/article/pii/S0959652617313100 (accessed 23 May 2018).

De Dobbelaere, A., De Keulenaere, B., De Mey, J., Lebuf, V., Meers, E., Ryckaert, B., Schollier, C. and Driessche, J. van 2015. Small scale anaerobic digestion: case studies in Western Europe. Available at: http://hdl.handle.net/1854/LU-7003618.

De Jong, E. and Jungmeier, G. 2015. Chapter 1 – Biorefinery concepts in comparison to petrochemical refineries A2 – Pandey, Ashok. In: R. Höfer, M. Taherzadeh, K. M. Nampoothiri and C. Larroche (Eds), *Industrial Biorefineries & White Biotechnology*. Elsevier, Amsterdam, pp. 3–33. Available at: https://www.sciencedirect.com/science/article/pii/B978044463453500001X.

De Vries, J. W., Vinken, T. M. W. J., Hamelin, L. and De Boer, I. J. M. 2012a. Comparing environmental consequences of anaerobic mono- and co-digestion of pig manure to produce bio-energy – A life cycle perspective. *Bioresource Technology*, 125, pp. 239–48. Available at: http://dx.doi.org/10.1016/j.biortech.2012.08.124.

De Vries, J. W., Groenestein, C. M. and De Boer, I. J. M. 2012b. Environmental consequences of processing manure to produce mineral fertilizer and bio-energy. *Journal of Environmental Management*, 102, pp. 173–83. Available at: http://dx.doi.org/10.1016/j.jenvman.2012.02.032.

De Vries, J. W., Groenestein, C. M., Schröder, J. J., Hoogmoed, W. B., Sukkel, W., Groot Koerkamp, P. W. G. and De Boer, I. J. M. 2015a. Integrated manure management to reduce environmental impact: II. Environmental impact assessment of strategies. *Agricultural Systems*, 138, pp. 88–99. Available at: http://dx.doi.org/10.1016/j.agsy.2015.05.006.

De Vries, J. W., Hoogmoed, W. B., Groenestein, C. M., Schröder, J. J., Sukkel, W., De Boer, I. J. M. and Groot Koerkamp, P. W. G. 2015b. Integrated manure management to reduce environmental impact: I. Structured design of strategies. *Agricultural Systems*, 139, pp. 29–37. Available at: http://www.sciencedirect.com/science/article/pii/S0308521X15000773.

Djomo, S. N., Humbert, S. and Dagnija, B. 2008. Life cycle assessment of hydrogen produced from potato steam peels. *International Journal of Hydrogen Energy*, 33(12), pp. 3067–72. Available at: http://www.sciencedirect.com/science/article/pii/S0360319908001444.

Duynie. 2018a. Aardappelpersvezels – co-producten Duynie. Available at: https://www.duynie.nl/producten/aardappelpersvezels/327 (accessed 21 May 2018).

Duynie. 2018b. Bierbostel – co-producten Duynie. Available at: https://www.duynie.nl/producten/bierbostel/319 (accessed 21 May 2018).

Ekman, A. and Börjesson, P. 2011. Environmental assessment of propionic acid produced in an agricultural biomass-based biorefinery system. *Journal of Cleaner Production*, 19(11), pp. 1257–65. Available at: https://www.sciencedirect.com/science/article/pii/S0959652611000849 (accessed 23 May 2018).

ePURE. 2016. European renewable ethanol key figures 2016. Available at: https://epure.org/media/1610/2016-industry-statistics.pdf.

EU. 2018. Bioeconomy – Research & Innovation – European Commission. Available at: https://ec.europa.eu/research/bioeconomy/index.cfm (accessed 7 July 2018).

Eurostat. 2018a. Eurostat data on agricultural production. Available at: http://ec.europa.eu/eurostat/data/database.

Eurostat. 2018b. Greenhouse gas emission statistics – emission inventories – statistics explained. Available at: http://ec.europa.eu/eurostat/statistics-explained/index.php/Greenhouse_gas_emission_statistics (accessed 6 July 2018).

FAO. 2018. Statistics | Food and Agriculture Organization of the United Nations. Available at: http://www.fao.org/statistics/en/ (accessed 7 July 2018).

Foged, H. L., Flotats, X., Bonmati, A., Palatsi, J., Magri, A. and Schelde, K. 2012. Inventory of manure processing activities in Europe. Technical Report No. I concerning 'Manure Processing Activities in Europe' to the European Commission, Directorate-General Environment (p. 138).

Goglio, P., Smith, W. N., Grant, B. B., Desjardins, R. L., McConkey, B. G., Campbell, C. A. and Nemecek, T. 2015. Accounting for soil carbon changes in agricultural life cycle assessment (LCA): A review. *Journal of Cleaner Production*, 104, pp. 23–39. Available at: http://www.sciencedirect.com/science/article/pii/S0959652615005879.

Golecha, R. and Gan, J. 2016. Biomass transport cost from field to conversion facility when biomass yield density and road network vary with transport radius. *Applied Energy*, 164, pp. 321–31. Available at: https://www.sciencedirect.com/science/article/pii/S0306261915015202 (accessed 16 June 2018).

Gröndahl, F. 2013. Seafarm application to Formas. Available at: http://www.seafarm.se/.

Hamelin, L., Wesnæs, M., Wenzel, H. and Petersen, B. M. 2011. Environmental consequences of future biogas technologies based on separated slurry. *Environmental Science and Technology*, 45(13), pp. 5869–77.

Hedegaard, K., Thyø, K. A. and Wenzel, H. 2008. Life cycle assessment of an advanced bioethanol technology in the perspective of constrained biomass availability. *Environmental Science & Technology*, 42(21), pp. 7992–9. Available at: doi:10.1021/es800358d.

Helin, T., Vesterinen, P., Ahola, H., Niemelä, K., Doublet, S., Couturier, C., Piotrowski, S., Carus, M., Tambuyser, B., Hasija, R., Singh, R. and Adholeya, A. 2012. BIOCOmmodity REfinery Deliverable D1. 1 : Availability of lignocellulosic biomass types of interest in the study regions.

Hua, K., Wang, D., Guo, X. and Guo, Z. 2014. Carbon sequestration efficiency of organic amendments in a long-term experiment on a vertisol in Huang-Huai-Hai Plain, China. *PLoS ONE*, 9(9), p. e108594. Available at: http://www.ncbi.nlm.nih.gov/pubmed/25265095 (accessed 16 June 2018).

IEA-Bioenergy. 2018. Factsheets biorefineries. Available at: https://www.iea-bioenergy.task42-biorefineries.com/en/ieabiorefinery/Factsheets.htm (accessed 11 June 2018).

ingredients101. 2018. Soybean meal. Available at: http://www.ingredients101.com/soybeanml.htm (accessed 21 May 2018).

JRC. 2018. EU Science Hub – European Commission. Available at: https://ec.europa.eu/jrc/en (accessed 7 July 2018).

Kim, S. and Dale, B. E. 2005. Life cycle assessment of various cropping systems utilized for producing biofuels: Bioethanol and biodiesel. *Biomass and Bioenergy*, 29(6), pp. 426–39. Available at: http://www.sciencedirect.com/science/article/pii/S0961953405000978.

Kim, H., Kim, S. and Dale, B. E. 2009. Biofuels, land use change, and greenhouse gas emissions: Some unexplored variables. *Environmental Science & Technology*, 43(3), pp. 961–7. Available at: https://doi.org/10.1021/es802681k.

Kiskini, A. 2017. Sugar beet leaves: From biorefinery to techno-functionality. Wageningen University. Available at: http://edepot.wur.nl/421994.

KW. 2018. Rapeseed meal. Available at: https://www.kwalternativefeeds.co.uk/products/view-products/rapeseed-meal/ (accessed 21 May 2018).

Lal, R. 2004. Soil carbon sequestration to mitigate climate change. *Geoderma*, 123(1), pp. 1–22. Available at: http://www.sciencedirect.com/science/article/pii/S0016706104000266.

Lal, R. 2005. World crop residues production and implications of its use as a biofuel. *Environment International*, 31(4), pp. 575–84. Available at: https://www.sciencedirect.com/science/article/pii/S0160412004001564 (accessed 8 June 2018).

Le Clef, E. and Kemper, T. 2015. 8 – Sunflower seed preparation and oil extraction BT – Sunflower. AOCS Press, pp. 187–226. Available at: https://www.sciencedirect.com/science/article/pii/B9781893997943500143.

Leip, A., Billen, G., Garnier, J., Grizzetti, B., Lassaletta, L., Reis, S., Simpson, D., Sutton, M. A., de Vries, W., Weiss, F. and Westhoek, H. 2015. Impacts of European livestock production: Nitrogen, sulphur, phosphorus and greenhouse gas emissions, land-use, water eutrophication and biodiversity. *Environmental Research Letters*, 10(11), p. 115004. Available at: http://stacks.iop.org/1748-9326/10/i=11/a=115004?key=crossref.b8ce885804d5c860e008c03ed18e7ab8 (accessed 16 March 2018).

Lynch, K. M., Steffen, E. J. and Arendt, E. K. 2016. Brewers' spent grain: A review with an emphasis on food and health. *Journal of the Institute of Brewing*, 122(4), pp. 553–68.

Menzi, H. 2002. Manure management in Europe: Results of a recent survey. In: G. Venglovský and J., Gréserová (Eds), *RAMIRAN, Recycling of Agricultural, Municipal and Industrial Residues in Agriculture*. FAO, European Cooperative Research Network, Strbské Pleso, High Tatras, Slovak Republic.

Monforti, F., Bódis, K., Scarlat, N. and Dallemand, J. F. 2013. The possible contribution of agricultural crop residues to renewable energy targets in Europe: A spatially explicit study. *Renewable and Sustainable Energy Reviews*, 19, pp. 666–77. Available at: http://www.sciencedirect.com/science/article/pii/S1364032112006740.

Monforti, F., Lugato, E., Motola, V., Bodis, K., Scarlat, N. and Dallemand, J. F. 2015. Optimal energy use of agricultural crop residues preserving soil organic carbon stocks in Europe. *Renewable and Sustainable Energy Reviews*, 44, pp. 519–29. Available at: https://www.sciencedirect.com/science/article/pii/S1364032114010855 (accessed 31 March 2018).

Morales, M., Quintero, J., Conejeros, R. and Aroca, G. 2015. Life cycle assessment of lignocellulosic bioethanol: Environmental impacts and energy balance. *Renewable and Sustainable Energy Reviews*, 42, pp. 1349–61. Available at: https://www.sciencedirect.com/science/article/pii/S1364032114009228 (accessed 25 May 2018).

Mussatto, S. I., Dragone, G. and Roberto, I. C. 2006. Brewers' spent grain: Generation, characteristics and potential applications. *Journal of Cereal Science*, 43(1), pp. 1–14. Available at: https://www.sciencedirect.com/science/article/pii/S0733521005000706 (accessed 18 April 2018).

Neeteson, J. J., Greenwood, D. J. and Draycott, A. 1987. A dynamic model to predict yield and optimum nitrogen fertilizer application rate for potatoes. *Proceedings of the Fertiliser Society*, No. 262. London, England, p. 31.

Nova. 2007. Rape cake: More than just a by-product – Bio-based News – The portal for bio-based economy and industrial biotechnology. Available at: http://news.bio-based.eu/rape-cake-more-than-just-a-by-product/ (accessed 21 May 2018).

Nova. 2018a. Bio-based economy – Services of the nova-Institut GmbH. Available at: http://bio-based.eu/ (accessed 7 July 2018).

Nova. 2018b. Home – Nova-Institute. Available at: http://nova-institute.eu/ (accessed 7 July 2018).

NVV. 2016. Mestnoteringen en Mestafzetprijzen. Available at: http://www.nvv.nl/mestafzetprijzen (accessed 18 June 2018).

Oil World. 2018. OIL WORLD ISTA Mielke GmbH: Independent global market analyses & forecasts since 1958. Available at: https://www.oilworld.biz/ (accessed 7 July 2018).

Panagiotopoulos, J. A., Bakker, R. R., de Vrije, T., Koukios, E. G. and Claassen, P. A. M. 2010. Prospects of utilization of sugar beet carbohydrates for biological hydrogen production in the EU. *Journal of Cleaner Production*, 18, pp. S9–S14. Available at: https://www.sciencedirect.com/science/article/pii/S0959652610000855 (accessed 10 April 2018).

Parajuli, R., Dalgaard, T., Jørgensen, U., Adamsen, A. P. S., Knudsen, M. T., Birkved, M., Gylling, M. and Schørring, J. K. 2015. Biorefining in the prevailing energy and materials crisis: A review of sustainable pathways for biorefinery value chains and sustainability assessment methodologies. *Renewable and Sustainable Energy Reviews*, 43, pp. 244–63. Available at: https://www.sciencedirect.com/science/article/pii/S1364032114009721 (accessed 23 May 2018).

Pawelzik, P., Carus, M., Hotchkiss, J., Narayan, R., Selke, S., Wellisch, M., Weiss, M., Wicke, B. and Patel, M. K. 2013. Critical aspects in the life cycle assessment (LCA) of bio-based materials – Reviewing methodologies and deriving recommendations. *Resources, Conservation and Recycling*, 73, pp. 211–28. Available at: http://www.sciencedirect.com/science/article/pii/S0921344913000359.

Pfister, S., Koehler, A. and Hellweg, S. 2009. Assessing the environmental impacts of freshwater consumption in LCA. *Environmental Science & Technology*, 43(11), pp. 4098–104. Available at: https://doi.org/10.1021/es802423e.

Piotrowski, S., Carus, M. and Essel, R. 2015. Global bioeconomy in the conflict between biomass supply and demand. Available at: http://online.liebertpub.com/doi/10.1089/ind.2015.29021.stp.

Plevin, R. J., Jones, A. D., Torn, M. S. and Gibbs, H. K. 2010. Greenhouse gas emissions from biofuels' indirect land use change are uncertain but may be much greater than previously estimated. *Environmental Science & Technology*, 44(21), pp. 8015–21. Available at: https://doi.org/10.1021/es101946t.

Plevin, R. J., Delucchi, M. A. and Creutzig, F. 2013. Using attributional life cycle assessment to estimate climate-change mitigation benefits misleads policy makers. *Journal of Industrial Ecology*, 18(1), pp. 73–83. Available at: https://doi.org/10.1111/jiec.12074.

Prapaspongsa, T., Christensen, P., Schmidt, J. H. and Thrane, M. 2010. LCA of comprehensive pig manure management incorporating integrated technology systems. *Journal of Cleaner Production*, 18(14), pp. 1413–22. Available at: http://dx.doi.org/10.1016/j.jclepro.2010.05.015.

Scarlat, N., Martinov, M. and Dallemand, J.-F. 2010. Assessment of the availability of agricultural crop residues in the European Union: Potential and limitations for bioenergy use. *Waste Management*, 30(10), pp. 1889–97. Available at: http://www.sciencedirect.com/science/article/pii/S0956053X10002436.

Singh, S., Kumar, A. and Ali, B. 2011. Integration of energy and water consumption factors for biomass conversion pathways. *Biofuels, Bioproducts and Biorefining*, 5, pp. 399–409.

Tamayo Tenorio, A., Gieteling, J., de Jong, G. A. H., Boom, R. M. and van der Goot, A. J. 2016. Recovery of protein from green leaves: Overview of crucial steps for utilisation. *Food Chemistry*, 203, pp. 402–8. Available at: http://www.sciencedirect.com/science/article/pii/S0308814616302643.

Ten Hoeve, M., Hutchings, N. J., Peters, G. M., Svanström, M., Jensen, L. S. and Bruun, S. 2014. Life cycle assessment of pig slurry treatment technologies for nutrient redistribution in Denmark. *Journal of Environmental Management*, 132, pp. 60–70. Available at: http://dx.doi.org/10.1016/j.jenvman.2013.10.023.

The Brewers of Europe. 2016. *Beer statistics 2017 edition* Euromonitor. Available at: http://portal.euromonitor.com/portal/statistics/tab.

Tijmensen, M. J. A., Mombarg, H., van den Broek, R. C. A. and Wasser, R. 2002. *Haalbaarheid van co-vergisting van oogstresten in de mestvergister in de Wieringermeer*, Utrecht, the Netherlands.

Tilman, D., Cassman, K. G., Matson, P. A., Naylor, R. and Polasky, S. 2002. Agricultural sustainability and intensive production practices. *Nature*, 418, p. 671. Available at: http://dx.doi.org/10.1038/nature01014.

Tonini, D., Hamelin, L. and Astrup, T. F. 2015. Environmental implications of the use of agro-industrial residues for biorefineries: Application of a deterministic model for indirect land-use changes. *GCB Bioenergy*, 8(4), pp. 690–706. Available at: https://doi.org/10.1111/gcbb.12290.

Tuck, C. O., Pérez, E., Horváth, I. T., Sheldon, R. A. and Poliakoff, M. 2012. Valorization of biomass: Deriving more value from waste. *Science (New York, N.Y.)*, 337(6095), pp. 695–9. Available at: http://www.ncbi.nlm.nih.gov/pubmed/22879509 (accessed 16 March 2018).

US Grains Council. 2012. A guide to distiller's dried grains with solubles (DDGS). Available at: http://www.grains.org/buyingselling/ddgs/ddgs-user-handbook.

Velghe, F. and Wierinck, I. 2013. Evaluatie van de vergisters in Nederland. Available at: https://www.rvo.nl/sites/default/files/2013/10/Evaluatie vergisting NL Fase 2 eindrapport_finaal_okt 2013.pdf.

Wiloso, E. I., Heijungs, R., Huppes, G. and Fang, K. 2016. Effect of biogenic carbon inventory on the life cycle assessment of bioenergy: Challenges to the neutrality assumption. *Journal of Cleaner Production*, 125, pp. 78–85. Available at: http://www.sciencedirect.com/science/article/pii/S0959652616301676.

WUR. 2018. Biobased products – WUR. Available at: https://www.wur.nl/en/Research-Results/Research-Institutes/food-biobased-research/about/Biobased-Products-2.htm (accessed 7 July 2018).

Index